POLYMERIZATION

Polymerization

Edited by:
Vikas Mishra

ARCLER
PRESS

www.arclerpress.com

Polymerization

Vikas Mishra

Arcler Press
2010 Winston Park Drive,
2nd Floor
Oakville, ON L6H 5R7
Canada
www.arclerpress.com
Tel: 001-289-291-7705
001-905-616-2116
Fax: 001-289-291-7601
Email: orders@arclereducation.com

ISBN: 978-1-77361-181-5 (Hardcover)

Declaration

About the Editor

Vikas obtained his Ph.D. from King George's Medical University, India in 2010. His interests span from bio-resources, fermentation technology to medical virology and neuroscience research. He is currently working as a research associate at the University of British Columbia.

Contents

List of Contributors

Vikas Mishra

Ozlem Ates
Department of Medical Services and Techniques, Nisantasi University, Istanbul, Turkey

Michele Greque de Morais
Laboratory of Microbiology and Biochemical, College of Chemistry and Food Engineering, Federal University of Rio Grande, Avenida Itália, Km 8, 96203-900 Rio Grande, RS, Brazil

Bruna da Silva Vaz
Laboratory of Microbiology and Biochemical, College of Chemistry and Food Engineering, Federal University of Rio Grande, Avenida Itália, Km 8, 96203-900 Rio Grande, RS, Brazil

Etiele Greque de Morais
Laboratory of Biochemical Engineering, College of Chemistry and Food Engineering, Federal University of Rio Grande, 96203-900 Rio Grande, RS, Brazil

Jorge Alberto Vieira Costa
Laboratory of Biochemical Engineering, College of Chemistry and Food Engineering, Federal University of Rio Grande, 96203-900 Rio Grande, RS, Brazil

Oladeji O. Ige
Department of Materials Science and Engineering, Obafemi Awolowo University, Ile-Ife 220282, Nigeria

Lasisi E. Umoru
Department of Materials Science and Engineering, Obafemi Awolowo University, Ile-Ife 220282, Nigeria

Sunday Aribo
Department of Metallurgical and Materials Engineering, Federal University of Technology, Akure 340252, Nigeria

Tizazu Mekonnen
Biorefining Conversions and Fermentations Laboratory, Department of Agricultural, Food and Nutritional Sciences, University of Alberta, Edmonton, AB T6G 2P5, Canada

Paolo Mussone
Biorefining Conversions and Fermentations Laboratory, Department of Agricultural, Food and Nutritional Sciences, University of Alberta, Edmonton, AB T6G 2P5, Canada

Hamdy Khalil
The Woodbridge Group, 8214 Kipling Avenue, Woodbridge, ON, Canada L4L 2A4

David Bressler
Biorefining Conversions and Fermentations Laboratory, Department of Agricultural, Food and Nutritional Sciences, University of Alberta, Edmonton, AB T6G 2P5, Canada

Jörg Schnauß
Institute for Experimental Physics I, Universität Leipzig, Linnéstraße 5, Leipzig 04103, Germany
Fraunhofer Institute for Cell Therapy and Immunology, Perlickstraße 1, Leipzig 04103, Germany

Tina Händler
Institute for Experimental Physics I, Universität Leipzig, Linnéstraße 5, Leipzig 04103, Germany
Fraunhofer Institute for Cell Therapy and Immunology, Perlickstraße 1, Leipzig 04103, Germany

Josef A. Käs
Institute for Experimental Physics I, Universität Leipzig, Linnéstraße 5, Leipzig 04103, Germany

Dietrich Braun
Deutsches Kunststoff-Institut, Schlossgartenstr. 6, 64289 Darmstadt, Germany

Eric Jean Fischer
Institute for Chemical and Bioengineering, ETH Zurich, Vladimir-Prelog-Weg 1, 8093 Zurich, Switzerland

Giuseppe Storti
Institute for Chemical and Bioengineering, ETH Zurich, Vladimir-Prelog-Weg 1, 8093 Zurich, Switzerland

Danilo Cuccato
Institute for Chemical and Bioengineering, ETH Zurich, Vladimir-Prelog-Weg 1, 8093 Zurich, Switzerland

Hidetaka Tobita
Department of Materials Science and Engineering, University of Fukui, Fukui 910-8507, Japan

Antonio Veloso
POLYMAT, University of the Basque Country UPV/EHU, Joxe Mari Korta R&D Ctr, Avda. Tolosa-72, Donostia-San Sebastián 20018, Spain

Wendy García
POLYMAT, University of the Basque Country UPV/EHU, Joxe Mari Korta R&D Ctr, Avda. Tolosa-72, Donostia-San Sebastián 20018, Spain

Amaia Agirre
POLYMAT, University of the Basque Country UPV/EHU, Joxe Mari Korta R&D Ctr, Avda. Tolosa-72, Donostia-San Sebastián 20018, Spain

Nicholas Ballard
POLYMAT, University of the Basque Country UPV/EHU, Joxe Mari Korta R&D Ctr, Avda. Tolosa-72, Donostia-San Sebastián 20018, Spain

Fernando Ruipérez
POLYMAT, University of the Basque Country UPV/EHU, Joxe Mari Korta R&D Ctr, Avda. Tolosa-72, Donostia-San Sebastián 20018, Spain

José C. de la Cal
POLYMAT, University of the Basque Country UPV/EHU, Joxe Mari Korta R&D Ctr, Avda. Tolosa-72, Donostia-San Sebastián 20018, Spain

José M. Asua
POLYMAT, University of the Basque Country UPV/EHU, Joxe Mari Korta R&D Ctr, Avda. Tolosa-72, Donostia-San Sebastián 20018, Spain

Bishnu P. Koiry
Rubber Technology Centre, Indian Institute of Technology Kharagpur, Kharagpur, India

Nikhil K. Singha
Rubber Technology Centre, Indian Institute of Technology Kharagpur, Kharagpur, India

Preface

As the human endeavors have advanced in the various fields of technology and science, so has advanced our need for new and improved material. From ancient natural materials like, rocks, stone, plant fibers, jute, hemp, hair, wool, silk and animal hide, mankind has moved to new materials like rubbers, resins, glass, synthetic fibers, plastics, and has now advanced to a variety of polymers. These polymeric material include foams, fibers, films, membranes, emulsions and coatings. These synthetic materials have wide range of applications that range from baby diaper, packaging, manufacturing cars, light weight aircrafts for aerospace industry, dental implants, medical devices and building construction material. Structurally, polymers are substances containing a large number of repeated small structural units, joined by a particular type of linkage. All polymers are not synthetic, like nylon, teflon, epoxy, polyethylene, synthetic rubber. Interestingly, there are many naturally occurring polymers, like DNA, proteins, wool, cellulose, silk, pectin, chitin etc. This book is an attempts to provide an overview of the current understanding of biopolymers and free radical polymerization process, as well as applications of polymers.

Editor

INTRODUCTION

Vikas Mishra

WHAT ARE POLYMERS?

The word polymer is a literal combination of two words, "poly", and stands for many and "mer" represents the units, so in simple words, polymer is a substance where a small, simple chemical unit is repeated many times to form a polymer molecule or macromolecule. The repeat unit may consist of a single atom or more commonly a small group of atoms joined by a specific linkage. The polymers can be natural and synthetic or man-made. The synthetic polymers, like, nylon, polyethylene, polyester, Teflon, and epoxy, are usually petroleum derivatives. On the other hand the natural polymers, like, DNA, protein, cellulose, chitin, pectin etc, are all water soluble.

Rubber is an interesting example, which is found in nature as a latex of more than 200 different trees, but *Hevea brasiliensis* or commonly called the rubber plant, is the most significant source of commercial natural rubber. Natural rubber, the white milky liquid, obtained from tree is essentially a polymer made from cis-1,4- isoprene units, with some impurities in it, such as resins, proteins, sugars and fatty acids, which function as weak antioxidants and accelerators in the natural rubber. On the other hand, the commonly known and widely used rubber is the synthetic version of the rubber. The synthetic rubber is made from the polymerization of a variety of monomers of petroleum origin, including isoprene. The most widely used rubber is styrene-butadien rubber, which is derived from polymerization of styrene and 1,3-butadiene.

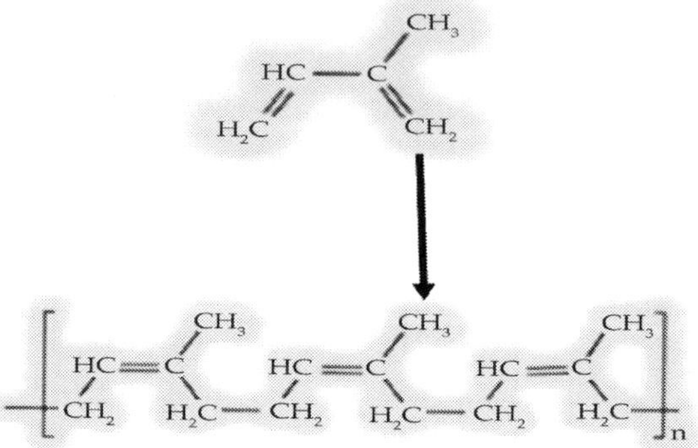

Figure: Natural rubber.

The natural rubber is inferior in quality to the synthetic rubber, as natural rubber is sticky and thermally unstable, making it particularly difficult to handle, nor does it have very good properties or durability as it can easily rot. So, natural rubber is usually vulcanized, a process by which the rubber is heated in the presence of sulfur, to improve its elasticity, resistance and stability. Even vulcanized rubber is sensitive to most of the organic liquids. Synthetic rubber is not having the inherent drawbacks of natural rubber and is preferable due to the control over the monomeric composition of the rubber and accordingly the physical, mechanical, and chemical properties can be controlled, making it particularly suitable for some applications. The physical and mechanical properties of rubber that are measured include, durometer or hardness range, tensile strength range, elongation range, abrasion resistance, adhesion to metal, adhesion to rigid materials, compression set, flex cracking resistance, impact resistance, resilience / rebound, tear resistance, vibration dampening etc. Thermal stability of rubber is also influenced by its composition, like range of highest temperature and lowest temperature tolerance (both in static and in constant use), brittle point, etc. In addition, depending upon the applications of rubber the chemical resistance to acids and alkalies (both concentrated and dilute), alcohols, aldehydes, amines, ketones, esters, ethers, oils (vegetable oil, diester oils), liquid petroleum gases and fuel oils, brake fluids, halogenated solvents, lacquer solvents, silicone oil, refrigerant (ammonia, halofluorocarbons and halofluorocarbons with oil), etc. All these properties define the performance of the polymer in the environment, where a lot of factors can be used to screen its applicability, like colorability, flame resistance, gas permeability, odor, ozone or oxidation resistance, radiation resistance, steam resistance, weather resistance (sunlight, freeze and thaw and water).

Some examples of synthetic rubber and their main properties with applications are as follows:

Rubber	Composition	Properties and common applications
Polyisoprene (IR)	Isoprene Cis or Trans	• Synthetic version of natural rubber • Natural rubber is 100% Cis, but in synthetic IR, the amount of cis units can be varied according to the catalyst used in the synthesis. • With an increase in cis units, crystallization is facilitated, the glass transition temperature decreases and strength properties improve • Abrasion resistant, high tensile strength, high resilience and resistant to inorganic chemicals • Used in tires, conveyor lines and transmission straps gaskets, tubes, paddings, footwear, sports equipment protective gloves, sealants and sealing materials.
Acrylic (ACM)	Copolymer of ethyl butyl acrylate	• heat and oil resistant • Used in automotive industry, industrial seals, hoses, and adhesive applications, wherever high temperature and oil resistance are needed in combination. • outstanding tensile strength over extended periods of use, with weather resistance • low oil swell • excellent vibration dampening

Polybutadiene (BR)	1,3-butadiene	• has a high resistance to wear, Excellent resistance to abrasion, compression and impact • Manufacture of tires, golf ball cores, as an additive to improve the toughness (impact modification) of plastics such as polystyrene and acrylonitrile butadiene styrene (ABS). Typically about 7% polybutadiene is added to the polymerization process to make these rubber-toughened resins.
Styrene Buta-diene Rubber (SBR) Neolite (Good-year)	Styrene, buta-diene	• Outstanding resistance to Abrasion, Impact, tear, tensile strength • Excellent adhesion to metal and other rigid surfaces, excellent low temperature flex-ibility and resilience • The styrene/butadiene ratio, polymerization conditions, and catalyst used, influences the properties of the polymer: with high styrene content, the rubbers are harder and less rubbery. • Applicable in car tires, footwear, conveyor belts, molded rubber goods, sponge and foamed products
Chlorosulfo-nated polyeth-ylene (CSM), Hyp-alon	reacting poly-ethylene with chlorine and sulfur dioxide	• resistance to chemicals, acids, alcohol, oil, abrasion, weather (especially moisture) and temperature extremes and ultraviolet light • Inclusion of chlorine in the polymer increas-es its resistance to fire (self-extinguishing), • Resistance to the microbial growth and exhibits good adhesion to various surfaces • used to make inflatable boats and folding kayaks. It is also used in roofing materials and as a surface coat material on radars ow-ing to its radar-transparent quality • Hypalon is also used in the construction of the decking of modern snowshoes

Polychloroprene (CR), Neoprene	Polymer of chloroprene	• Chloroprene is polymerized by emulsion free radical polymerization using potassium persulphate as free radical initiator. • Resistant to abrasion, ozone, oil, solvent, acids, alcohol, impact and tear • high tensile strength, good adhesion to metals and rigid material, low temperature stability, weather resistant • Used in construction of laptop sleeves, orthopedic braces (wrist, knee, etc.), electrical insulation of wires, conveyer belts, gaskets, wear suits, inflatables
Nitrile-Butadiene Rubber (NBR), Acrylonitrile Butadiene, Nitrile rubber	Copolymer of Butadiene and acrylonitrile	• Variation in the acrylonitrile content modifies the physical and chemical properties of the polymer • good oil, fuel and fat resistance • Increase in the acrylonitrile content improves oil resistance, hardness, abrasion resistance and heat resistance, but raises the glass transition temperature. • Used in automotive and aeronautical industry to utilize the oil and fuel resistance properties in making hoses, seals, grommets, and self-sealing fuel tanks • Molded rubber in footwear, adhesives, sealants and floor mats

Polydimethylsiloxane Rubber (VMQ, PMQ, or PVMQ), Silicone Rubbers	silicone and oxygen atoms – siloxane groups	• Inorganic polymers, as the main chain structure does not include carbon atoms, but some side groups, usually methyl groups are present on the main siloxane chain. • It is an elastomer, non-reactive, stable with alkalis', oils, ammonia, • Resistant to flame, ozone, oxidation, sunlight, water • Used in electrical equipment and technical products for high temperatures, sportswear, and footwear; electronics; medical devices and implants; sealants for aeronautical industry.

There are many more varieties of synthetic rubber. This gives the broad varieties of polymers available today and the multibillion dollar industry the polymers caters, that varies from agriculture, medicine, sports, aeronautics, construction etc. The range of applications of polymers has far exceeded the applicability of any other material ever used by mankind.

The realization of the long-chain molecular or the macromolecular concept of polymers in the 1920s proved to be a vital turning point in the synthetic polymer formation. Today polymerization is achieved by free radical reactions, anionic, cationic, photo polymerization etc. In this book the topics that are covered are, biopolymers and synthetic polymers produced by Free radical polymerization process.

SECTION I: BIOPOLYMERIZATION

CHAPTER 1

SYSTEMS BIOLOGY OF MICROBIAL EXOPOLYSACCHARIDES PRODUCTION

Ozlem Ates

Department of Medical Services and Techniques, Nisantasi University, Istanbul, Turkey

ABSTRACT

Exopolysaccharides (EPSs) produced by diverse group of microbial systems are rapidly emerging as new and industrially important biomaterials. Due to their unique and complex chemical structures and many interesting physicochemical and rheological properties with novel functionality, the microbial EPSs find wide range of commercial applications in various fields of the economy such as food, feed, packaging, chemical, textile, cosmetics and pharmaceutical industry, agriculture, and medicine. EPSs are mainly associated with high-value applications, and they have received considerable research attention over recent decades with their biocompatibility, biodegradability, and both environmental and human compatibility. However, only a few microbial EPSs have achieved to be used commercially due to their high production costs.

The emerging need to overcome economic hurdles and the increasing significance of microbial EPSs in industrial and medical biotechnology call for the elucidation of the interrelations between metabolic pathways and EPS biosynthesis mechanism in order to control and hence enhance its microbial productivity. Moreover, a better understanding of biosynthesis mechanism is a significant issue for improvement of product quality and properties and also for the design of novel strains. Therefore, a systems-based approach constitutes an important step toward understanding the interplay between metabolism and EPS biosynthesis and further enhances its metabolic performance for industrial application. In this review, primarily the microbial EPSs, their biosynthesis mechanism, and important factors for their production will be discussed. After this brief introduction, recent literature on the application of omics technologies and systems biology tools for the improvement of production yields will be critically evaluated. Special focus will be given to EPSs with high market value such as xanthan, levan, pullulan, and dextran.

Keywords: EPS, microbial production, exopolysaccharides, systems biology, xanthan, levan, pullulan, dextran

INTRODUCTION

Biopolymer (also called renewable polymers) is used as a term to describe polymers produced by biological systems and polymers that are not synthesized chemically but are derived from biological starting materials such as amino acids, sugars, and natural fats (Tang et al., 2012). Consequently, biopolymers can be classified as synthetic or natural polymers (Vroman and Tighzert, 2009). The biopolymers are superior to petrochemical-derived polymers in several aspects that include biocompatibility, biodegradability, and both environmental and human compatibility. Although petroleum-based polymers have negative effects to environment and humanity such as toxicity, defiance to biodegradation, and waste accumulation, they have been used in a variety of industrial applications. Therefore, in response to these problems, biopolymers are a suitable alternative that the researchers were looking (Keshavarz and Roy, 2010).

Today, several microorganisms are identified as microbial biopolymer producers and these polymers can be found as attached to the cell surface or extracted from the fermentation medium. Bacteria use these microbial biopolymers as storage materials in response to particular environmental stresses (Sanchez-Garcia et al., 2010). Due to their biological functions microbial polysaccharides can be generally classified as intracellular storage polysaccharides (glycogen), capsular polysaccharides (e.g., K30 O-Antigen), and extracellular bacterial polysaccharides (for example, levan, xanthan, sphingan, alginate, pullulan, cellulose, etc.), which are important for biofilm formation and pathogenicity (Schmid and Sieber, 2015).

Microbial polysaccharides that are produced by microorganisms and secreted out of the cell are defined as exopolysaccharides (EPSs). In nature, they have a significant role for protection of the cell, adhesion of bacteria tosolid surfaces, and participating in cell-to-cell interactions (Nicolaus et al., 2010). In recent years, there is a significant interest on microbial EPSs since they have different structural and functional properties (Morris and Harding, 2009). EPSs are important resources for hydrocolloids used in food, pharmaceutical, chemical, and many other industries (Ahmad et al., 2015). Due to their many interesting physicochemical and rheological properties with novel functionality, the microbial EPSs act as new biomaterials and find a wide range of applications in many industrial sectors such as textiles, detergents, adhesives, microbial enhanced oil recovery (MEOR), wastewater treatment, dredging, brewing, downstream processing, cosmetology, pharmacology, and food additives (Rühmann et al., 2015). Xanthan, dextran, and pullulan are examples of microbial polysaccharides with a considerable market due to their exceptional properties. However, plant and algal polysaccharides such as starch, galactomannans, pectin, carrageenan, and alginate still include a major part of the hydrocolloid market, which has a market value of 4 million US dollars in 2008, 3.9 billion US dollars in 2012, and this value is expected to reach 7 billion US dollars by 2019 (Williams et al., 2007; Patel and Prajapati, 2013). Since microbial EPSs enable fast and high yielding production processes under controlled conditions, they are economically competitive to the plant and algal origin polysaccharides, which are affected by climatological and geological environmental conditions (Kaur et al., 2014). Although there is an increased attraction for microbial EPSs in industrial and medical applications and they are related with high-value applications, only a few bacterial EPSs have achieved to be used commercially such as xanthan, gellan, and dextran due to high production costs (Freitas et al., 2011; Llamas et al., 2012).

Due to the exceptionally high production costs, microbial EPSs could never find their proper place in the polymer market and therefore, high-level EPSs producing microbial systems gain escalating industrial importance. Increasing significance of EPSs in industrial and medical biotechnology calls for the elucidation of the interrelations between metabolic pathways and biosynthesis mechanism in order to control and hence improve microbial productivity. Therefore, extensive interest has been dedicated to understand bacterial EPSs biosynthesis mechanism and pathways and enhance productivity within the past years. Omics technologies such as genome sequencing, functional genomics, protein structure analysis, and new bioinformatics tools have been used to identify new EPS biosynthesis pathways and understand the principles of EPS formation (Schmid and Sieber, 2015).

A modeling approach that linked omics data and the simulation of variable expression and enzyme activity will provide information about a cell's macromolecular machinery (Lerman et al., 2012). For this purpose, genome-based and genome-scale metabolic reconstructions can be used to understand and predict phenotypes of a microbial species (Hanemaaijer et al., 2015). Therefore, a systems biology approach constitutes an important step toward understanding the interplay between metabolism and microbial EPS biosynthesis and further enhances its metabolic performance for industrial application. Figure 1 demonstrated a brief summary of integration of omics studies with systems biology.

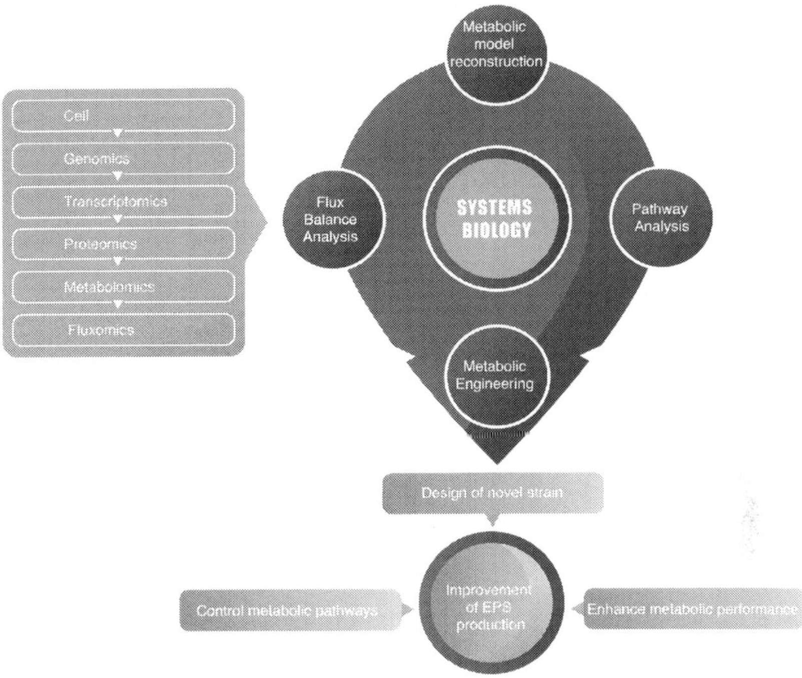

Figure 1: A schematic diagram of integration of omics studies with systems biology.

In this review after a brief description of the microbial EPSs, biosynthesis mechanism and important factors for their production, recent literature on the application of omics technologies and systems biology tools for the improvement of production yields will be critically evaluated. Microbial EPSs with high market value such as xanthan, dextran, scleroglucan, pullulan, and levan are specially focused.

MICROBIAL EXOPOLYSACCHARIDES

The biopolymers produced by microorganisms were categorized to four main groups: polyesters, polyamides, inorganic polyanhydrides, and polysaccharides (Crescenzi and Dentini, 1996; Rehm, 2009, 2010). Since the microbial biopolymers serve as reserve material or as part of a protective mechanism, the biopolymer producer microorganisms have significant advantages under certain environmental conditions (Rehm, 2010). The first bacterial polymer dextran was discovered by Pasteur

(1861) in the mid-nineteenth century as a microbial product in wine and the bacterium *Leuconostoc mesenteriodes* was identified by Van Tieghem (1878) as dextran producer strain.

The bacterial polysaccharides that are synthesized and secreted by various microorganisms into the extracellular environment either as soluble or insoluble polymers are defined as EPSs. Their compositions, functions, chemical, and physical properties that establish their primary conformation vary from one bacterial species to another. EPSs are composed of mainly of carbohydrates (a wide range of sugar residues) and some non-carbohydrate substituents (such as acetate, pyruvate, succinate, and phosphate) (Vu et al., 2009; Nicolaus et al., 2010; Llamas et al., 2012; Staudt et al., 2012).

Most of EPSs producer bacteria have been described to produce either homo or heteropolysaccharide (Kumar et al., 2007). On the other hand, bacteria (*Serratia marcescens*, *Aeromonas salominicida*, and *Pseudomonas* sp. strain NCIB 2021) that were able to produce two different polysaccharides have been reported (Kwon et al., 1994). Due to linkage bonds and nature of monomeric units, homopolysaccharides can be categorized as α-d-glucans, β-d-glucans, fructans, and polygalactan. d-glucose, d-galactose, l-rhamnose, and *N*-acetylglucosamine (GlcNAc), *N*-acetylgalactosamine (GalNAc), or glucuronic acid (GlcA) are the repeating units of heteropolysaccharides and occasionally non-carbohydrate substituent such as phosphate, acetyl, and glycerol. Homopolysaccharides and heteropolysaccharides are also dissimilar in synthetic enzymes and site of synthesis. Biosynthesis of homopolysaccharides requires specific substrates like sucrose, while the residues of heteropolysaccharide are produced intracellularly and precursors are located across the membrane by isoprenoid glycosyl carrier lipids for extracellular polymerization (Nwodo et al., 2012). EPSs have also been classified in seven categories based on their functionality by Flemming and Wingender (2010) as constructive or structural (serve in the matrix help water retention and cell protection), sorptive (composed of charged polymers), surface-active (including molecules with amphiphilic behavior), active, informative, redox-active, and nutritive.

EPS affords self-protection for cells from desiccation, predation, the effects of antibiotics, antimicrobial substances, antibodies, bacteriophages and adherence to other bacteria, animal, and plant tissues under different stress conditions such as biotic stress, competition, and abiotic stresses, including temperature, light intensity, or pH (Mata et al., 2006; Kumar et al., 2007; Kumar and Mody, 2009; Ordax et al., 2010; Donot et al., 2012; Staudt et al., 2012). Additionally, EPS supplies bacterial aggregation, surface attachment, and symbiosis of plant-microbe; hence, it is a crucial property for wastewater treatment and soil aggregation. Furthermore, pathogenicity of a microorganism is related with the production of capsular EPS and depends on the rate and amount of EPS synthesis (Kumar et al., 2007).

Microorganisms are often linked with a high cellular density biofilm and its stability is controlled by EPSs through interactions between the polysaccharide chains. Moreover, microbial diversity is biologically supported by EPS to constitute a substrate for the microbial growth. Biofilm formation plays a crucial role both in adhesion and in adaptation of bacteria to the physicochemical conditions of the environment (Donot et al., 2012).

Environmental factors and specific culture conditions such as pH, temperature, carbon-to-nitrogen (C/N) ratio, oxygenation rate, and carbon sources can impact EPS production. EPS composition can be altered by conditional changes (differing monosaccharides or by monosaccharide molar ratio). For instance, due to the carbon source, *Lactobacillus casei* has been shown to alter the chemical composition of its EPS (Staudt et al., 2012). Furthermore, production of microbial EPSs in bioreactors enables to optimize growth and production yields by studying physiology and genetic engineering (Delbarre-Ladrat et al., 2014).

Due to their unique and complex chemical structures that offers beneficial bioactive functions, biocompatibility, and biodegradability, microbial EPSs have find a wide range of application areas in chemical, food, pharmaceutical, cosmetics, packaging industries, agriculture, and medicine in which they can be used as adhesives, absorbents, lubricants, soil conditioners, cosmetic, drug delivery vehicles, textiles, high-strength materials, emulsifiers, viscosifiers, suspending, and chelating agents. In recent years, several novel bacterial EPSs have been isolated and identified; however, a few of them have achieved to have significant commercial value due to the high production costs (Mata et al., 2006; Kumar et al., 2007; Nicolaus et al., 2010; Freitas et al., 2011; Llamas et al., 2012; Delbarre-Ladrat et al., 2014). Bacterial EPSs such as xanthan, gellan, dextran, and curdlan with superior physical and chemical properties are used instead of plant (guar gum or pectin) or algae (e.g., carrageenan or alginate) polysaccharides in traditional applications (Kumar and Mody, 2009; Nicolaus et al., 2010; Freitas et al., 2011; Liang and Wang, 2015). Other bacterial EPSs such as levan, pullulan, and wellan with unique properties and biological activities have found new commercial opportunities (Freitas et al., 2011).

GalactoPol, which is synthesized by *Pseudomonas oleovorans* and composed mainly of galactose, and a fucose containing EPS FucoPol that is synthesized by *Enterobacter* A47 have been recently reported bacterial EPSs with great commercial potential. Almost 30 species of lactic acid bacteria (LAB) are also known as polysaccharide producers and one of the commercial EPS dextran producer *Leuconostoc mesenteroides* is a LAB; however, low production yields avoid LAB species to be exploited commercially. Besides, *lactobacilli* are GRAS (generally recognized as safe) bacteria and their EPS could be utilized in foods (Badel et al., 2011).

After the discovery of the various EPSs, the activities of enzymes related with EPS production were investigated and radioisotope-labeled precursors were used to elucidate the metabolic pathways for microbial biosynthesis. Moreover, understanding the molecular and regulatory mechanisms behind the biosynthesis of microbial polymers is an essential requirement for engineering bacteria leading to production of tailor-made biopolymers with high-value applications for industrial and medical applications with an economic cost (Rehm, 2009, 2010).

BACTERIAL SYNTHESIS OF EXOPOLYSACCHARIDES

Extensive progress has been made in elucidating the biosynthetic and genetic mechanisms of biosynthesis of bacterial polysaccharides in recent years. The mechanism of biosynthesis and the precursors required illustrate diversity for different classes of EPSs. EPSs are synthesized by bacteria extracellularly or intracellularly (Boels et al., 2001; Kumar et al., 2007; Badel et al., 2011; Freitas et al., 2011; Li and Wang, 2012). Genes required for EPS production are responsible for encoding regulation, chain-length determination, repeat-unit assembly, polymerization, and export. The mechanism regulating EPS biosynthesis is a challenged topic to be understood despite accumulating knowledge of EPS gene organization (Péant et al., 2005). Regulation of EPS biosynthesis is related with various physiological and metabolic parameters such as the availability of sugar precursors and the expression level of enzymes (Delbarre-Ladrat et al., 2014). Information on genetics of certain EPS like xanthan is abundant; however, genetic data for other EPS synthesis (i.e., pullulan) is still limited (Donot et al., 2012).

Bacterial EPSs are mostly generated intracellularly and exported to the extracellular environment with the exception of homopolysaccharides such as dextran, levan, and mutan that are synthesized outside the cells by the action of secreted enzymes that convert the substrate into the polymer. The enzymes involved in EPS synthesis are found at different regions of the cell and can be characterized into four categories. The first group is intracellular enzymes such as hexokinase, which phosphorylates glucose (Glc) to glucose-6-phosphate (Glc-6-P). They are also involved in other cellular metabolisms. The second group is required to catalyze conversion of sugar nucleotides. Uridine-5′-diphosphate (UDP)-glucose pyrophosphorylase that catalyzes the conversion of Glc-1-P to UDP-Glc, which is one of the key molecules in EPS synthesis can be given as an example for this class of enzymes. Another enzyme group is glycosyltransferases (GTFs) that are located in the cell periplasmic membrane. The sugar nucleotides are transferred by GTFs to a repeating unit attached to glycosyl carrier lipid. The enzymatic functions, the structures, and identification of the genes that encode GTFs has been investigated intense and due to amino acid sequence similarities more than 94 GTF families were reported in the Carbohydrate-Active EnZymes (CAZy) database (http://www.cazy.org) (Li and Wang, 2012). The last class is presumably involved in the polymerization of the macromolecules and situated outside the cell membrane and the cell wall (Kumar et al., 2007).

The general mechanisms for the production of bacterial polysaccharides are Wzx/Wzy-dependent pathway, the ATP-binding cassette (ABC) transporter-dependent pathway, the synthase-dependent pathway, and the extracellular synthesis by use of a single sucrase protein. Inside the cell, the precursor molecules are transformed by enzymes and produce activated sugars/sugar acids in the first three mechanisms.

Alternatively, in extracellular production pathway by direct addition of monosaccharides obtained by cleavage of di- or trisaccharides, the polymer strand is elongated (Schmid and Sieber, 2015).These general mechanisms of EPS biosynthesis were demonstrated in detail in Figure 2.

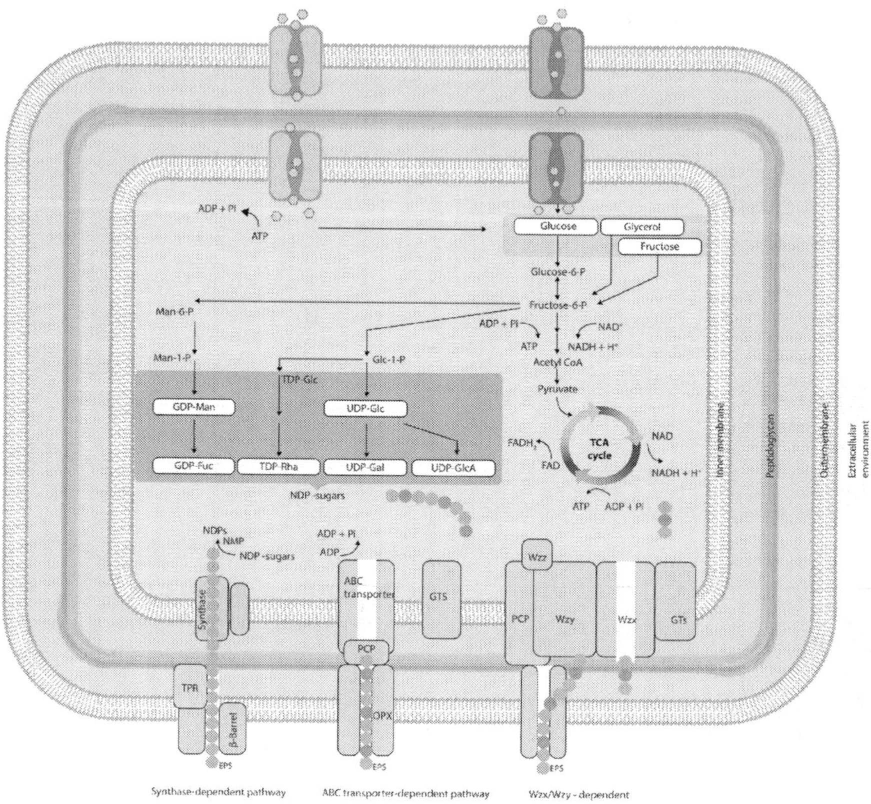

Figure 2: Simplified schematic diagram summarizing the biosynthetic pathways (Synthase dependent, ABC transporter, Wzx/Wzy dependent) involved in microbial EPS synthesis (OPX, outer-membrane polysaccharide; PCP, polysaccharide copolymerase; TPR, tetratricopeptide repeat proteins).

In the Wzx/Wzy-dependent process, activated sugars are linked in a specific sequence to a lipid carrier by GTFs until the repeating unit is formed involving a Wzy protein. In the Wzx/Wzy-independent (ABC transporter-dependent) pathway, polymerization occurs at the cytoplasmic side of the inner membrane. The genes, which are required for high-level polymerization and surface assembly, are described as wza (encoding an outer-membrane protein), wzb (encoding an acid phosphatase), and wzc (encoding an inner-membrane tyrosine autokinase). In most Gram-negative bacteria (i.e., *Erwinia* spp., *Methylobacillus* sp. strain 12S, *Rhizobium* spp., and *Xanthomonas campestris*), EPS biosynthesis and export have been reported to occur via the Wzx/Wzy-independent and Wzx/Wzy-dependent pathway (Arco et al., 2005; Cescutti et al., 2010; Freitas et al., 2011).

EPS secretion can occur in the presence or absence of a lipid acceptor molecule in the synthase-dependent pathway, which secretes complete polymer strands across the membranes and the cell wall, and is not dependent of a flippase for translocating repeat units. In this system, the polymerization and the translocation process are performed simultaneously by a single synthase protein, a membrane-embedded glycosyl transferase. These pathways are often utilized for the assembly of homopolymers requiring only one type of sugar precursor such as curdlan [β-(1-3)-linked glucose monomers] or bacterial cellulose [β-(1-4)-linked glucose units] (Rehm, 2010; Whitney and Howell, 2013; Schmid and Sieber, 2015). Regulation of polymerization is implemented by an inner-membrane receptor, in Gram-negative synthase-dependent secretion systems such as *P. aeruginosa* alginate and *Gluconacetobacter xylinus* cellulose (Whitney and Howell, 2013).

In extracellular synthesis, polymerization reaction occurs as transfer of a monosaccharide from a disaccharide to a growing polysaccharide chain in the extracellular environment. This type of production of EPSs is uncomplicated; independent of the central carbon metabolism besides there is a limited variation in structure. The extracellular EPS synthesis can occur for homopolysaccharides (dextran, levan, and mutan) by extracellular GTF (Boels et al., 2001; Finore et al., 2014).

The intracellular biosynthesis of homo- and heteropolysaccharides includes production of (ir)regular repeating units from sugar nucleotide precursors, which are also involved in the biosynthesis of several cell wall components and can therefore be considered essential for growth (Boels et al., 2001; Nicolaus et al., 2010; Li and Wang, 2012). Direct precursors for bacterial EPS biosynthesis are formed intracellulary from intermediates of the central carbon metabolism. The precursors and donor monomers for the biosynthesis of most repeating units are sugar nucleotides such as nucleoside diphosphate sugars (such as ADP-glucose), nucleoside diphosphate sugar acids (such as GDP-mannuronic acid), and nucleoside diphosphate sugar derivatives [such as UDP-glucose, UDP-*N*-acetyl glucosamine, UDP-galactose, and deoxythymidine diphosphate (dTDP)-rhamnose] (Barreto et al., 2005; Péant et al., 2005; Rehm, 2010). These sugars can be transported by basically three different ways: ATP hydrolysis coupled to sugar translocation via a sugar transport ATPase, the import with coupled to transport of ions and other solutes, and transport via the phosphoenolpyruvate (PEP) transport system (PTS) (Barreto et al., 2005; Péant et al., 2005).

GTFs (EC 2.4.x.y) catalyze heteropolysaccharide biosynthesis which has numerous intracellular steps and only the last step that polymerization of repeating units occurs is extracellular. Depending on substrate type, uptake of sugars is achieved through a passive or an active transport system by the cell in the first step. Subsequently, the substrate is catabolized in the cytoplasm through glycolysis and sugar nucleotides are formed. The biosynthesis of activated precursors [energy-rich monosaccharides, mainly nucleoside diphosphate sugars (NDP-sugars)], which are derived from phosphorylated sugars is occurred. Finally, EPS is secreted to extracellular environment therefore their secretion from cytoplasm through cell membrane without compromising the critical barrier properties is a challenging process (Badel et al., 2011; Freitas et al., 2011).

Conversely, homopolysaccharides are synthesized extracellularly by GTFs. This class is defined as glycansucrases class (E.C. 2.4.x.y) and dissimilar to classical Leloir-type GTF they utilize sucrose as donor substrate instead of nucleotide-sugars. The transfer of monosaccharides, generating a glycosidic bond, from activated molecules to an acceptor molecule is catalyzed by these enzymes. Energy released by degredation of sugars is used to catalyze transfer of a glycosyl residue on forming polysaccharide. Due to the product of biosynthesis, the enzymes can be differentiated between transglucosydases (EC 2.4.1.y) and transfructosydases (EC 2.4.1.y or 2.y). Transglucosidases class includes dextransucrase, mutansucrase, and reuteransucrase (EC 2.4.1.5), which are high molecular weight extracellular enzymes and catalyze hydrolysis of sucrose to glucose and fructose and glucosyl transfer on carbohydrate or non-carbohydrate compounds. EPS structures can be varied based on different enzymes intervention and the synthesis of each polysaccharide is catalyzed by a specific GTF; therefore, two products encoded by two genes of *gtf* result in two different EPS. In addition, the enzyme conformation affects branching degree of homopolysaccharides. Levansucrases (EC 2.4.1.10) and inulosucrases (EC 2.4.1.9) from transfructosidases class produce levan and inulin type fructans. *Ftf* genes are induced under stress conditions and sucrose hydrolysis and fructosyl transfer on fructan polymerized chain or syntheses of tri- or tetrasaccharides are catalyzed. In fructan, glucose is the non-terminal reducing residue (G-Fn) (Badel et al., 2011).

OMICS STUDIES AND SYSTEMS BIOLOGY AND OF MICROBIAL BIOPOLYMER AND MICROBIAL EPS PRODUCTION

Systems biology offers valuable application areas in molecular sciences, medicine, pharmacy, and engineering such as pathway-based biomarkers and diagnosis, systematic measurement and modeling of genetic interactions, systems biology of stem cells, identification of disease genes, drug design, strain development, bioprocess optimization (Medina-Cleghorn and Nomura, 2013).

Strain improvement using systems level analysis of metabolic, gene regulatory and signaling networks, and integration of omics data are the most focused subjects of systems biology. Biochemical and bioprocess engineering principles are applied for optimization of upstream-to-downstream bioprocesses at first stages of strain development (Barrett et al., 2006). Process development has been a supporter of the scientific achievements in systems biology, mostly in the areas of transcriptomics, proteomics, metabolomics, and fluxomics with availability of genome sequences for production organisms. The applications of systems biology in industry become a challenged subject (Otero and Nielsen, 2010).

In recent years, the enormous amounts of genome sequencing projects have resulted in accumulation of complete genome sequence information for a number of species. This information is valuable for understanding biological capabilities of organisms and biological processes such as signal transduction and cellular metabolism at the system levels. Therefore, an ever-increasing number of models for bacteria and more papers that describe new reconstruction tools and improvements have been published. Genome-scale constraint-based metabolic models have been reconstructed for several organisms and such constraint-based models can be quickly generated by software packages using an organism's genomic, biochemical, and physiological data. These metabolic models have been used to integrate high-throughput data understand cellular metabolism, to develop metabolic engineering strategies, to design media and processes, to consider theoretical capabilities, and to control the process online, which illustrates its usefulness for development and optimization of process. Metabolic models are used for generating new hypotheses and targeting promising areas in engineering field. Metabolic engineering studies have been performed to modify microbes to produce industrially relevant biochemicals and biofuels such as ethanol and higher alcohols, fatty acids, amino acids, shikimate precursors, terpenoids, polyketides, and polymer precursors (e.g., 1,4-butanediol) (Henry et al., 2010; Baart and Martens, 2012; Xu et al., 2013; Long et al., 2015; Simeonidis and Price, 2015).

Genome-scale metabolic networks have great achievement in development of metabolic engineering strategies for strain improvement mainly in five industrial fields: food and nutrients, biopharmaceuticals, biopolymer materials, microbial biofuels, and bioremediation. Metabolic models are built to improve the yield of fermentation by products and explore metabolic mechanisms and processes in food and nutrients industry. Several biopharmaceuticals and the productivity of useful biopolymers and their precursors have improved by genome-scale metabolic model-guided metabolic engineering strategy (Xu et al., 2013). Dupont's near-decade long optimization of *Escherichia coli* for bioproduction of 1,3-propanediol is an important genome-scale metabolic engineering application (Nakamura and Whited, 2003). The industrially optimized strain required up to 26 genomic changes including insertions, deletions, and regulatory modifications. Recent advances in constraint-based modeling have enabled *in silico* prediction of genomic targets for the enhancement of strain performance or product yield (Esvelt and Wang, 2013). The engineering strategies have been successfully implemented for the improvements in the yield or production process, alterations in the degree of polymerization, removal of side chains or non-sugar substituents, or heterologous expression of EPS biosynthesis gene clusters (Becker, 2015). Additionally, the gene clusters of significant EPSs were figured out in Figure 3.

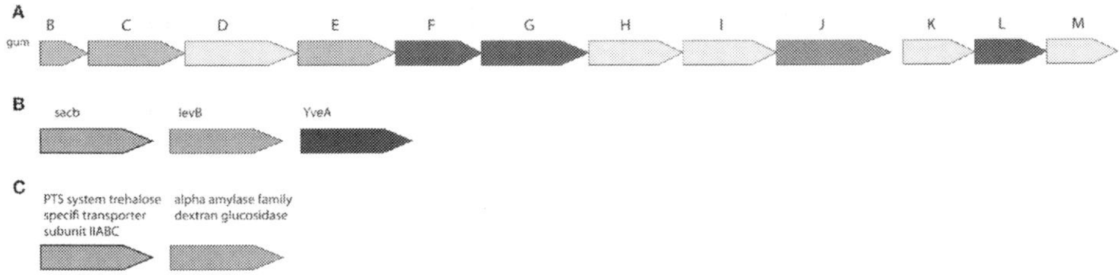

Figure 3: Gene clusters of significantly important microbial EPSs biosynthesis. **(A)**: Xanthan biosynthesis by *Xanthomonas campestris* pv. *campestris*ATCC 33913, **(B)**: Levan biosynthesis by *Bacillus subtilis* subsp. *subtilis* 6051-HGW (bsh), **(C)**: Dextran biosynthesis by *Streptococcus intermedius* JTH08 (sie). Conserved operan informations were obtained from ODB Operon Database (Okuda and Yoshizawa, 2011).

Due to their superior properties, wide application areas, there is a high demand to improve microbial EPS production with an economical cost. Therefore, the omics data and tools were utilized to perform systems biology approaches to understand and control EPS biosynthesis mechanism, design novel strains, and enhance productivity.

Natural or engineered microorganisms could synthesize many biopolymers and their monomers such as poly-3-hydroxyalkanates (PHAs), polylactic acid (PLA), polysaccaharides, carboxylic acids, and butanediols. Systems biology approach and genome-scale metabolic models-guided metabolic engineering strategies have been successfully employed to enhance the productivity of useful biopolymers and their precursors (Xu et al., 2013).

Jung et al. reported direct synthesis of PLA, which is a promising biomass-derived homopolymer and its copolymer, poly (3-hydroxybutyrateco- lactate), P(3HB-co-LA), by direct fermentation of metabolically engineered *E. coli*. In typical conditions, PLA production involves two steps fermentative production of lactic acid followed by chemical polymerization with low production yields. In this study, *in silico* genome-scale metabolic flux analysis was performed to determine metabolic engineering targets to improve *E. coli* strain. The engineering process was achieved by knocking out the ackA, ppc, and adhE genes and by replacing the promoters of the ldhA and acs genes with the trc promoter, and therefore, an 11 wt% enhancement of PLA production was obtained.

Polyhydroxyalkanoates (PHA) synthesizing capacity of *Pseudomonas putida* was investigated by genome-scale metabolic model of this microorganism and survival under anaerobic stress was achieved by introducing the *ackA* gene from *Pseudomonas aeruginosa* and *Escherichia coli*(Sohn et al., 2010).

Cai et al. (2011) reported the draft genome of the moderately halophilic bacterium *Halomonas*sp. TD01. In this study, several genes relevant to PHA and osmolytes biosynthesis were analyzed providing invaluable clues for understanding of the evolution and genes transfer, the strategic guidance of the genetic engineering of halophilic *Halomonas* sp. TD01 for co-production of PHA and ectoine.

The analysis of the genes required for the synthesis of the EPS mauran by *H. maura* strain S-30 was performed to identify gene cluster in this strain. Three conserved genes, *epsA*, *epsB*, and *eps*C, also a wzx homolog, *epsJ*, which indicates that mauran is formed by a Wzy-dependent system, were found. It was also reported that eps gene cluster reaches maximum activity during stationary phase, in the presence of high salt concentrations (5% w/v), which was investigated by transcriptional expression assays using a derivate of *H. maura* S-30, which carries an *epsA*: *lacZ* transcriptional fusion (Arco et al., 2005).

Generation of monomers including propanediols, butanediols, diamines, and terpenoids by microorganism through easier biosynthetic pathways was also reported (Lee et al., 2011, 2012; Curran and Alper, 2012). Additionally, production of monomers has been improved using systems biology approaches (Yim et al., 2011; Ng et al., 2012).

Genome-scale metabolic model-guided metabolic engineering approach has been performed and successfully used to improve production yields of various biopolymers and their precursors in synthetic material industry. The microbial production of monomers such as butanediols that are important raw materials in this industry are enhanced by genome-scale metabolic model strategies (Xu et al., 2013). For instance, Ng et al. (2012) have designed and constructed *S. cerevisiae* strains with improved production of 2,3-butanediol with gene deletion strategy, in which disruption of alcohol dehydrogenase (ADH) pathway is required, by performing *in silico*genome-scale metabolic analysis. Yim et al. (2011) have used biopathway prediction algorithm to elucidate possible pathways for 1,4-butanediol (BDO) biosynthesis. Strain development was performed by engineering the *E. coli* host to enhance anaerobic operation of the oxidative tricarboxylic acid cycle and drive the BDO pathway. The engineered strain was able to produce BDO from glucose, xylose, sucrose and biomass-derived mixed sugar streams. Furthermore, the productions of some important carboxylic acid monomers, used as raw materials in synthetic material industry, such as formic acid, malic acid, and succinic acid were improved in engineered *S. cerevisiae* or *E. coli* via genome-scale metabolic models-guided metabolic engineering strategies (Lee et al., 2005b; Wang et al., 2006; Moon et al., 2008; Kennedy et al., 2009). These successfully implemented studies will be helpful to improve microbial EPS production and to design industrial strategies.

Systems metabolic engineering of *E. coli* or *Corynebacterium glutamicum* as efficient cell factories has resulted in overproduction of 1,5-diaminopentane as building block for novel biopolymers (Kind and Wittmann, 2011). The importance of the Entner–Doudoroff pathway in PHB production was predicted by stoichiometric flux analysis of recombinant *E. coli* metabolic model and confirmed experimentally (Hong et al., 2003). The dynamics of PHA copolymer structure and properties were identified by mathematical models during its *in vivo*accumulation (Aldor and Keasling, 2003). Besides, an optimal carbon source switching strategy for the production of block copolymers was described by a population balance model in *Ralstonia eutropha* system (Mantzaris et al., 2001).

Previously genome-scale metabolic model reconstructions of biopolymer producer strain such as *Pseduomonas putida* (Nogales et al., 2008; Puchalka et al., 2008) and *Pseudomonas aeruginosa* (Oberhardt et al., 2008) have been published, as they could be used to elucidate biopolymer synthesis mechanism and improvement of production.

Thermophilic microorganism *Brevibacillus thermoruber* 423 is able to produce high levels of EPS (Yasar Yildiz et al., 2015). Recently, draft genome sequence and whole-genome analysis of this bacterium have been reported. Whole-genome analysis of this bacterium was performed by a systems-based approach to understand the biological mechanisms and whole-genome organization of thermophilic EPS producers. Therefore, strategies for the genetic and metabolic optimization of EPS production could be developed. Genome annotation was used to detect essential genes associated with EPS biosynthesis and a hypothetical mechanism for EPS biosynthesis was generated considering the experimental evidences. The genome sequence of *B. thermoruber* strain 423 is being used to reconstruct a genome-scale metabolic model to develop metabolic engineering strategies since the metabolic model will be used to optimize medium compositions, to modify the microorganism genetically, to improve production yields, and to modify EPS monomer composition (Yasar Yildiz et al., 2014).

Nadkarni et al. (2014) performed comparative genome analysis of *Lactobacillus rhamnosus*clinical isolates to identify EPS cluster. In this study, transcriptional orientation of the eps cluster genes, the presence of two genes homologous to priming glycosytransferases, the absence of *rmlACBD* genes involved in the dTDP-rhamnose biosynthetic pathway, and the presence of a family 2 GTF in the eps cluster of both clinical isolates of *L. rhamnosus*, is predicted to alter EPS composition and could influence pathogenicity.

The first complete genome sequence of Russia origin *Bifidobacterium longum* subsp. longum strain GT15, comparative genome analysis, identification, and characterization of regulatory genes, *in silico* analysis of all the most significant probiotic genes and considered genes have been reported. The genomic analysis for polysaccharides was also performed, and it was observed that most of the genes in the carbohydrate metabolism category were involved in the utilization of oligo-polysaccharides. The genome also contains genes predicted to encode proteins involved in the production of capsular EPS, which are most likely involved in bacteria–host interactions (Zakharevich et al., 2015).

Genome sequence of moderately halophilic and EPS-producing *Salipiger mucosus* DSM 16094T and the presence of a high number of genes associated with biosynthesis of EPSs have been reported. Genes associated with the synthesis of polyhydroxyalkanoates have been also found (Riedel et al., 2014).

Acidithiobacillus ferrooxidans was the first biomining microorganism whose genome was sequenced and the genes involved in the biosynthesis EPSs precursors have been studied (Valdés et al., 2008). The cluster of five genes proposed to be involved in the biosynthesis of EPSs precursors via the Leloir pathway have been also identified previously (Barreto et al., 2005).

A curdlan producer *Agrobacterium* sp. ATCC 31749's genome was sequenced and the curdlan biosynthesis operon (crdASC) was identified (Ruffing et al., 2011). Moreover, transcriptome analysis of this microorganism has been performed to understand the regulation of EPS biosynthesis (Ruffing and Chen, 2012). In this study, transcriptome profiling was used to identify genes that expressed during curdlan biosynthesis and carry out targeted gene knockouts to investigate their roles in the transcriptional regulation of curdlan production. The analysis showed that curdlan synthesis operon was upregulated by up to 100-fold upon nitrogen depletion. Moreover, novel regulation mechanisms including RpoN-independent NtrC regulation and intracellular pH regulation by acidocalcisomes were identified.

Engineering studies for LAB, particularly members of the genera *Lactococcus* and *Lactobacillus*, have been performed for the production of platform chemicals, such as lactate l- and d-stereoisomers, 1,3-propanediol, and 2,3-butanediol, food flavors and sweeteners, vitamins, and complex polysaccharides (Gaspar et al., 2013).

Exopolysaccharide biosynthetic pathways in LAB have been engineered to have challenges greater than manipulating specific steps in primary metabolism (Patel et al., 2011). Since EPSs enhance potential health benefits of fermented food products, many metabolic engineering approaches are employed to improve productivity and structure of EPS (Gaspar et al., 2013).

Omics studies can be used to improve the understanding of metabolism in food industry microorganisms. The metagenomic studies for fermented food were performed to analyze the metabolic potentials of LAB bacteria, which is very important in industrial fermentations. Jung et al. (2011) performed metagenomic studies changes in bacterial populations, metabolic potential, and the overall genetic features of the microbial community during a 29-day fermentation process of the traditional Korean food kimchi. The transcriptome response has been analyzed in yogurt fermentation (Sieuwerts et al., 2010) and in milk (Goh et al., 2011).

The first genome-scale model for *L. lactis* (Oliveira et al., 2005), and since whole-genome metabolic reconstructions for *Lb. plantarum* (Teusink et al., 2005) and *S. thermophilus*(Pastink et al., 2009) have been reported. Functional genomics and other studies have performed to investigate genomic diversity in LAB and the findings highlighted the variety of carbon substrates potentially used by LAB including simple sugars, complex carbohydrates such as xylan, starch, and fructans, α-galactosides (e.g., raffinose and stachyose), pentoses (D arabinose and D-xylose), and the cheap C3 carbon source glycerol (Teusink et al., 2009; Siezen and van Hylckama Vlieg, 2011).

Complete genomic sequence of *Lb. bulgaricus* 2038 has been reported and genomic analysis of EPS biosynthesis has been performed. Two neighboring *eps* clusters with significant differences were identified when compared with genome sequence of *Lb. bulgaricus* species by comparative genomic analysis (Hao et al., 2011).

Genomic studies microbial EPSs producer of deep-sea bacteria such as *Zunongwangia profunda* SM-A87, *Pseudoalteromonas* sp. SM9913, *Pseudoalteromonas haloplanktis* TAC125 (Qin et al., 2011), have been also performed and analyzed for EPS gene clusters (Finore et al., 2014). In addition, genome sequence of several deep-sea isolates such as *Idiomarina loihiensis*(Hou et al., 2004) and *Alteromonas macleodii* (Ivars-Martinez et al., 2008) demonstrated the EPS biosynthesis genes (Finore et al., 2014).

Such omic studies and works from systems biology perspective will play an important role both scientifically and economically, since there is a great need for developing efficient methodologies for enhanced EPS biosynthesis. More information on the genome of the microorganism will enable to develop strategies to successfully enhance production rate and also to engineer EPSs properties by modifying composition and chain length. Systems-based modeling approach constitutes an important step toward understanding the interplay between metabolism and EPS biosynthesis.

Systems biology approaches and metabolic reconstruction studies of microbial EPS have been also reported for two important EPSs: xanthan and levan. These studies and their findings were given in detail in the following sections. Besides, these two important EPSs researches on pullulan and dextran were also discussed and the general properties and applications for all these important EPSs were summarized in Table 1.

Table 1: General description of significantly important microbial EPSs

Microbial EPS	Monomeric units	Microorganism	Industrial applications	Omic studies	Metabolic model
Xanthan	Glucose, mannose and glucuronate	*Xanthamonas* sp.	Thickening, stabilizing agent, food additive, etc.	Genome sequence, Proteomics	Available (Schatschneider et al., 2013)
Levan	Fructose	*Halomonas smyrnensis* AAD6[T], *Zymomonas mobilis*, *Bacillus subtilis (natto)*	Emulsifier, stabilizer and thickener, encapsulating agent, food and feed additive, osmoregulator, and cryoprotector, etc.	Genome sequence	Available (Ates et al., 2013; Diken et al., 2015)
Pullulan	Glucose	*Aureobasidium pullulans*	Thickening, stabilizing, texturizing, gelling agents, etc.	Genome sequence, proteomics, genome shuffling	Not available
Dextran	Glucose	*Leuconostoc* spp. and *Streptococcus* spp.	Blood plasma extender and chromatography media	Genome sequence	Not available

XANTHAN

The commercially most important microbial EPS is known as Xanthan which is produced by the plant-pathogen-proteobacterium *Xanthomonas campestris* pv. campestris (Xcc) (Vorhölter et al., 2008; Frese et al., 2014). It is a heteropolysaccharide composed of repetitive pentasaccharide units consisting of monomeric units of two glucose, two mannose, and one GlcA residues with with a backbone chain consisting of (1-4)-β-d-glucan cellulose (Khan et al., 2007). Due to its superior properties and rheological characteristics, xanthan has found a wide range of applications as a thickening or stabilizing agent in food, cosmetics and oil drilling industries (Schatschneider et. al, 2013; Chivero et al., 2015). It has been described as a "benchmark" product based on its significance in food and non-food applications which include dairy products, drinks, confectionary, dressing, bakery products, syrups, and pet foods, as well as the oil, pharmaceutical, cosmetic, paper, paint, and textile industries (Patel and Prajapati, 2013; Cho and Yoo, 2015). Xanthan was also employed in non-food applications such stabilizing cattle feed supplements, calf milk substitutes, agricultural herbicides, fungicides, pesticides, and fertilizers, and to impart thixotropy into toothpaste preparations (Morris, 2006). Considering the commercial importance of this microbial EPS, omics studies and metabolic model reconstructions were performed to clarify xanthan biosynthesis mechanism.

The genomes of five *Xanthomonas* strains *X. campestris* pv. campestris strains ATCC 33913 (da Silva et al., 2002) and 8004 (Qian et al., 2005), *X. campestris* pv.vesicatoria strain 85-10 (Thieme et al., 2005), *X. oryzae* pv. oryzae strains KACC10331 (Lee et al., 2005a,b)

and MAFF 311018 (Ochiai et al., 2005), *X. axonopodis* pv. citri strain 306 (da Silva et al., 2002) have been sequenced. The complete genome sequence of the xanthan producer strain *Xanthomonas campestris* pv. campestris strain B100 and its use for mechanistic model for biosynthesis of xanthan have been also reported (Vorhölter et al., 2008). In this study, the gene products and metabolic pathways for xanthan polymerization were investigated in detail. The gene products of *gumJ, gumC, gumD*, and *gumE* were analyzed to establish detailed functions in a xanthan polymerization. Moreover, the mechanistic model for the biosynthesis of xanthan was established.

In vivo proteome analysis *X. campestris* pv. *campestris* has been performed to investigate protein expression of the microorganism during host–plant interaction. Peptide mass fingerprinting or *de novo* sequencing methods were utilized for identification of the functions of proteins. This approach will be used to determine the roles of proteins in pathogenicity mechanismsand also xanthan biosynthesis (Andrade et al., 2008).

Schatschneider et al. (2013) has reported the first large-scale metabolic model for *Xanthomonas campestris* (Xcc), which was reconstructed from genome data of, manually curated and further expanded in size. The impact of xanthan production was studied *in vivo* and *in silico* and compared with *gumD* mutant strain. This verified metabolic model is also the first model focusing on bacterial EPS synthesis and it can be used for detailed systems biology analyses and synthetic biology reengineering of Xcc. Moreover, draft genome of *X. campestris* B-1459, which was used in pioneering studies of xanthan biotechnology, and it can be used to analyze the genetic basis of xanthan biosynthesis has been reported recently (Wibberg et al., 2015).

LEVAN

Levan is a naturally occurring polymer that is composed of β-d-fructofuranose with β(2-6) linkages between fructose rings. It is synthesized by the action of a secreted levansucrase (EC 2.4.1.10) that directly converts sucrose into the polymer (Han and Clarke, 1990). As a homopolysaccharide with many distinguished properties such as high solubility in oil and water, strong adhesivity, good biocompatibility, and film-forming ability, it has great potential as a novel functional biopolymer in foods, feeds, cosmetics, pharmaceutical, and chemical industries (Kang et al., 2009; Kazak Sarilmiser et al., 2015). In fact, a recent literature analysis on microbial EPSs attributed levan together with xanthan, curdlan, and pullulan as the most promising polysaccharides for various industrial sectors (Donot et al., 2012).

Due to its exceptionally high production costs, levan could never find its proper place in the polymer market, and therefore, high-level levan producing microbial systems gain escalating industrial importance. Levan is produced as an EPS from sucrose-based substrates by a variety of microorganisms, including the halophilic bacterium *Halomonas smyrnensis* AAD6^T, which has been reported as the first levan producer extremophile (Poli et al., 2009).

The gram-negative halophilic bacterium *H. smyrnensis* AAD6^T, which was isolated from Çamaltı Saltern Area in Turkey (Poli et al., 2009, 2013), was found to excrete high levels of levan (Poli et al., 2009). With this microbial system, productivity levels were further improved by use of cheap sucrose substitutes such as molasses (Kucukasik, 2010; Kucukasik et al., 2011) as well as other cheap biomass resources (Toksoy Oner, 2013) as fermentation substrate. Further research on the potential uses of levan produced by *H. smyrnensis* AAD6^T as a bioflocculating agent (Sam et al., 2011), its nanostructured thin films (Sima et al., 2011, 2012), its suitability for peptide-based drug nanocarrier systems (Sezer et al., 2011), and its adhesive mulitilayer films (Costa et al., 2013) have been reported.

Increasing significance of levan in industrial and medical biotechnology calls for the elucidation of the interrelations between metabolic pathways and levan biosynthesis mechanism in order to control and hence enhance its microbial productivity. However, there is very limited information about the mechanisms involved in the biosynthesis of levan from extremophiles (Nicolaus et al., 2010) and no report about a systematic approach to analyze levan production by *H. smyrnensis* AAD6^T. Considering this fact, systems-based approaches were applied to improve the levan production capacity of *H. smyrnensis* AAD6^T cultures (Ates et al., 2013).

The genome sequence forms the basis for metabolic model reconstruction; however, there was a lack for genome information of *H.*

smyrnensis AAD6^T. Only recently, its draft genome sequence has been announced. *De novo* assembly of the whole sequencing reads were carried out in this study. Consequently, several genes related to EPS biosynthesis, including the genes for levansucrase and *ExoD* were revealed by genome analysis (Sogutcu et al., 2012). Due to the absence of genomic information, first, comprehensive metabolic model of a taxonomically close halophilic bacterium, namely, *C. salexigens* DSM3043 have been reconstructed (Ates et al., 2011). Then, in order to investigate levan biosynthesis by a metabolic systems analysis approach, the genome-scale metabolic network of *C. salexigens* was recruited and adopted to *H. smyrnensis* AAD6^T via integration of the available biochemical, physiological, and phenotypic features of *H. smyrnensis* AAD6^T. The *in silico* metabolic model was verified with dynamic experimental data on different medium compositions and was then systematically analyzed to identify critical network elements (i.e., enzymes, biochemical transformations, and metabolites) related to levan biosynthesis mechanism. The findings manifested mannitol as a significant metabolite for levan biosynthesis, which was further verified experimentally. In the previous study, 1.844 g/L levan yield from the stationary phase bioreactor cultures using a defined media containing sucrose as sole carbon source with almost fourfold increase of levan production was reported (Ates et al., 2013).

Recently, Diken et al. (2015) performed the whole-genome analysis of *H. smyrnensis* AAD6^T to investigate biological mechanisms, and furthermore, the genome-scale metabolic model *i*KYA1142, which included 980 metabolites and 1142 metabolic reactions, was reconstructed. The genomic analysis figured out the biotechnological potential of this microorganism as a result of its capacity to produce levan, Pel exopolysaccharide, polyhydroxyalkanoates (PHA), and osmoprotectants. Genes related to EPS biosynthesis and intracellular PHA biosynthesis were detected. *Hs_SacB* gene encoding the extracellular levansucrase enzyme (EC 2.4.1.10), Pel polysaccharide gene cluster (*PelA, PelB, PelC, PelD, PelE, PelF*, and *PelG*), Alginate lyase precursor (EC 4.2.2.3), and "Alginate biosynthesis protein Alg8" genes were predicted. The genome information and metabolic model will have a significant role on levan biosynthesis since they will be utilized to improve levan production by metabolic engineering strategies and medium optimization.

PULLULAN

A fungal glucan "pullulan" that is a linear homopolysaccharide composed of maltotriose reduplicative units connected by α-1,4-linkages is produced by *Aureobasidium pullulans* (Singh et al., 2008; Sheng et al., 2014; Özcan et al., 2014). Although pullulan production studies mostly focused on *A. pullulans*, other producer strains such as *Remella mesenterica*, *Cryphonectria parasitica*, and *Teloschistes flavicans* were also reported (Cheng et al., 2011). Biosynthesis of pullulan was occurred intracellularly at the cell wall or cell membrane and microorganisms secreted this EPS out to the cell surface to form a loose and slimy layer (Ma et al., 2015). Its solubility in water is excellent as a result of the linkage pattern. Moreoever, it has outstanding chemical and physical properties, such as low viscosity, non-toxicity, slow digestibility, high plasticity, and excellent film-forming (Cheng et al., 2011; Sheng et al., 2014).

The major market for pullulan is food industry and also its potential applications in pharmaceutical, agricultural, chemical, cosmetic, biomedical, and environmental remediation areas have been reported (Özcan et al., 2014; Ma et al., 2015). Due to its superior properties, pullulan can be used as non-polluting wrapping material for food supplements, oxygen impermeable, edible, and biodegradable and highly water soluble films, denture adhesives, emulsifiers and stabilizers for various food products, binder, lubricant, gelling agent, oral care products, and blood plasma substitute. Since it is non-immunogenic, non-toxic, non-carcinogenic, and non-mutagenic, pullulan can find novel application areas such as gene therapy, targeting drugs, and gene delivery (Cheng et al., 2011; Singh et al., 2015). Pullulan has been commercialized in various countries, used as a safe food ingredient and pharmaceutical bulking agent and in Japan.

The molecular biosynthesis of pullulan is a complex metabolic process therefore the molecular basis of this process is not clearly identified. The three significant enzymes for pullulan biosynthesis are α-phosphoglucose mutase (PGM, EC 5.4.2.2), UDP-glucose pyrophosphorylase (UGP, EC 2.7.7.9), and glucosyltransferase (FKS, EC 2.4.1.34). Nitrogen is the major component for the cultivation of *A. pullulans* in pullulan fermentation (Wang et al., 2015). Wang et al. (2015) analyzed gene expression of the key enzymes (PGM, UGP, FKS) to improve pullulan production. They have reported the improved pullulan production correlated to the high activities of PGM and FKS, increased the activities of α-phosphoglucose mutase and glucosyltransferase, and upregulated the transcriptional levels of pgm1 and fks genes by nitrogen limitation.

Kang et al. (2011) studied genome shuffling *A. pullulans* N3.387 by ethyl methane sulfonaten (EMS) and ultraviolet (UV) mutagenesis to improve pullulan biosynthesis and developed a mutant that could produce more pullulan than wild type strain.

Sheng et al. (2014) performed proteomic studies of pullulan production in *A. pullulans* to understand the effect of different concentrations of $(NH_4)_2SO_4$ which would be useful to optimize industrial pullulan production. The proteomic studies demonstrated the expression of antioxidant related enzymes and energy-generating enzymes and the depression of the enzymes concerning amino acid biosynthesis, glycogen biosynthesis, glycolysis, protein transport, and transcriptional regulation under nitrogen limitation, resulted in conversion of metabolic flux from the glycolysis pathway to the pullulan biosynthesis pathway.

Draft genome of *Aureobasidium pullulans* AY4 was determined and genome analysis revealed the presences of genes coding for commercially important enzymes such as pullulanases, dextranases, amylases, and cellulases (Chan et al., 2012). Gostin et al. (2014) performed *de novo*genome sequencing and genome analysis of the four varieties of *A. pullulans* to investigate the genomic basis of pullulan biosynthesis potential. Single-copy genes for phosphoglucose mutase and uridine diphosphoglucose pyrophosphorylase, which are the key enzymes for converting glucose units into pullulan, were present in four varieties of *A. pullulans*. Genomic analysis also revealed that these microorganisms included all of the putative enzymes that were proposed to be involved in pullulan biosynthesis (Gostin et al., 2014). The genome sequences of *A. pullulans*species will be facilitated to clarify biosynthesis mechanism of pullulan and improve its production.

DEXTRAN

Dextran, which is a homopolysaccharides composed of α-1,6 glycosidic linkages, is produced by *L. mesenteroides* hydrolases in the precense of sucrose. Various strains of bacteria such as *Streptococcus* and *Acetobacter* have been also found to produce dextran bacteria (e.g., *Leuconostoc* and *Streptococcus*), through the use of specific enzymes like glucansucrases (Patel et al., 2011; Casettari et al., 2015). Since dextran has a flexible structure as a result of free rotation of glycosidic bond and it is highly soluble in water, biocompatible, and biodegradable, it becomes a functional hydrocolloid (Ahmad et al., 2015). It has commercial applications in food, pharmaceutical and chemical industries as adjuvant, emulsifier, thickener, carrier, and stabilizer. Moreoever, dextran is used as therapeutic agent to restore blood volume, for the matrix preparation of chromatography columns, for synthesizing dextran sulfate for blood coagulation prevention and blood flow facilitation, as osmotic agents, lubricant in eye drops and for increase of blood sugar levels, iron carrier, anticoagulant for pharmacy (Patel et al., 2011; Han et al., 2014).

Dextrans have also found use in different areas such as veterinary medicines, biosensor material, food syrup stabilizers, dough improvers, metal-plating processes, enhanced oil recovery, stabilizing coating for protecting metal nanoparticles against oxidation, and coating on biomaterials to prevent undesirable protein absorption. Dextran and its derivatives such as cyclodextran are utilized in the pharmaceutical industry like cariostatic, anti-HIV, and anti-ulcer agent (Ahmad et al., 2015; Casettari et al., 2015).

Various genome sequence and genome analysis studies were carried out with dextran producer strains. For instance, the genome of LAB *Leuconostoc gasicomitatum* was sequenced and genome analysis was performed to understand the growth and spoilage potentials. The genomic analysis revealed genes for two dextransucrases catalyzing the formation of dextran from sucrose: *epsA* (LEGAS_0699) is part of a large EPS cluster, while *dsrA* (LEGAS_1012) is located as a single gene in the chromosome (Johansson et al., 2011). Saulnier et al. (2011) performed the metabolic pathway reconstruction, genome profiling, genomic, and transcriptomic comparisons of the *Lactobacillus reuteri* strains to define functional probiotic features. Genome-wide comparison resulted in dextranase gene that predicted to encode the synthesis of EPS. The first genome sequence of *Weissella* species has been announced for *Weissella confusa* (formerly *Lactobacillus confusus*) LBAE C39-2, which was also found promising for *in situ* production of dextran in sourdoughs (Amari et al., 2012). A draft genome sequence of dextran producer *Leuconostoc lactis* EFEL005 has been announced and genomic analysis was performed to understand its probiotic properties as a starter for fermented foods (Moon et al., 2015). The increased number of genome sequences will accelerate systems-based works for dextran biosynthesis mechanism.

CONCLUSION AND RECOMMENDATIONS

In microbial EPS production, a better understanding of biosynthesis mechanism is a significant issue for optimization of production yields, improvement of product quality and properties, and also for the design of novel strains. As most of the novel bacterial EPS with unique properties have expensive production costs and economic hurdles need to be overcome, this valuable information about biosynthesis is also be important to lower these charges.

More information on the genome of the EPS producer microorganisms will enable to develop additional strategies to successfully enhance EPSs production rate and also to engineer their properties by modifying composition and chain length. Since genome-scale reconstruction includes every reaction of target organism through integrating genome annotation and biochemical information, a systems-based metabolic modeling approach constitutes an important step toward understanding the interplay between metabolism and EPSs biosynthesis. Since microbial biopolymer biosynthesis is a result of a complex system of many metabolic processes, systems-based approaches are needed to control and optimize production in order to improve the formerly reported yields.

Furthermore, the genome-scale metabolic model based on genome sequence will have the capacity to consider gene expression, metabolomics, and proteomics data to get accurate prediction at different environmental conditions.

CONFLICT OF INTEREST STATEMENT

The author declares that the research was conducted in the absence of any commercial or financial relationships that could be construed as a potential conflict of interest.

ACKNOWLEDGMENTS

The financial support provided by TUBITAK through project 114M239 is gratefully acknowledged.

REFERENCES

1. Ahmad, N. H., Mustafa, S., and Che Man, Y. B. (2015). Microbial polysaccharides and their modification approaches: a review. *Int. J. Food Prop.* 18, 332–347. doi: 10.1080/10942912.2012.693561
2. Aldor, I. S., and Keasling, J. D. (2003). Process design for microbial plastic factories: metabolic engineering of polyhydroxyalkanoates. *Curr. Opin. Biotechnol.* 14, 475–483. doi:10.1016/j.copbio.2003.09.002
3. Amari, M., Laguerre, S., Vuillemin, M., Robert, H., Loux, V., Klopp, C., et al. (2012). Genome sequence of *Weissella confusa*LBAE C39-2, isolated from a wheat sourdough. *J. Bacteriol.* 194, 1608–1609. doi:10.1128/JB.06788-11
4. Andrade, A. E., Silva, L. P., Pereira, J. L., Noronha, E. F., Reis, F. B. Jr., Bloch, C. Jr., et al. (2008). In vivo proteome analysis of *Xanthomonas campestris* pv. campestris in the interaction with the host plant *Brassica oleracea*. *FEMS Microbiol. Lett.* 281, 167–174. doi:10.1111/j.1574-6968.2008.01090.x
5. Arco, Y., Llamas, I., Martinez-Checa, F., Argandona, M., Quesada, E., and del Moral, A. (2005). epsABCJ genes are involved in the biosynthesis of the exopolysaccharide mauran produced by *Halomonas maura*. *Microbiology* 151, 2841–2851. doi:10.1099/mic.0.27981-0
6. Ates, O., Arga, K. Y., and Toksoy Öner, E. (2013). The stimulatory effect of mannitol on levan biosynthesis: lessons from metabolic systems analysis of *Halomonas smyrnensis* AAD6^T. *Biotechnol. Prog.* 29, 1386–1397. doi:10.1002/btpr.1823
7. Ates, O., Toksoy Oner, E., and Arga, K. Y. (2011). Genome-scale reconstruction of metabolic network for a halophilic extremophile, *Chromohalobacter salexigens* DSM 3043. *BMC Syst. Biol.* 5:12. doi:10.1186/1752-0509-5-12
8. Baart, G. J., and Martens, D. E. (2012). "Genome-scale metabolic

models: reconstruction and analysis," in Christodoulides M., editor. *Neisseria Meningitidis* (London: Humana Press), 107–126.

9. Badel, S., Bernardi, T., and Michaud, P. (2011). New perspectives for *Lactobacilli* exopolysaccharides. *Biotechnol. Adv.* 29, 54–66. doi:10.1016/j.biotechadv.2010.08.011
10. Barreto, M., Jedlicki, E., and Holmes, D. S. (2005). Identification of a gene cluster for the formation of extracellular polysaccharide precursors in the chemolithoautotroph *Acidithiobacillus ferrooxidans*. *Appl. Environ. Microbiol.* 71, 2902–2909. doi:10.1128/AEM.71.6.2902-2909.2005
11. Barrett, C. L., Kim, T. Y., Kim, H. U., Palsson, B., and Lee, S. Y. (2006). Systems biology as a foundation for genome-scale synthetic biology. *Curr. Opin. Biotechnol.* 17, 488–492. doi:10.1016/j.copbio.2006.08.001
12. Becker, A. (2015). Challenges and perspectives in combinatorial assembly of novel exopolysaccharide biosynthesis pathways. *Front. Microbiol.* 6:687. doi:10.3389/fmicb.2015.00687
13. Boels, I. C., Kranenburg, R. V., Hugenholtz, J., Kleerebezem, M., and de Vos, W. M. (2001). Sugar catabolism and its impact on the biosynthesis and engineering of exopolysaccharide production in lactic acid bacteria. *Int. Dairy J.* 11, 723–732. doi:10.1016/S0958-6946(01)00116-9
14. Cai, L., Tan, D., Aibaidula, G., Dong, X., Chen, J. C., Tian, W. D., et al. (2011). Comparative genomics study of polyhydroxyalkanoates (PHA) and ectoine relevant genes from Halomonas sp. TD01 revealed extensive horizontal gene transfer events and co-evolutionary relationships. *Microb Cell Fact.* 10, 88.
15. Casettari, L., Bonacucina, G., Morris, G. A., Perinelli, D. R., Lucaioli, P., Cespi, M., et al. (2015). Dextran and its potential use as tablet excipient. *Powder Technol.* 273, 125–132. doi:10.1016/j.powtec.2014.12.030
16. Cescutti, P., Foschiatti, M., Furlanis, L., Lagatolla, C., and Rizzo, R. (2010). Isolation and characterisation of the biological repeating unit of cepacian, the exopolysaccharide produced by bacteria of the *Burkholderia cepacia* complex. *Carbohydr. Res.* 345, 1455–1460. doi:10.1016/j.carres.2010.03.029

17. Chan, G. F., Bamadhaj, H. M., Gan, H. M., and Rashid, N. A. A. (2012). Genome sequence of *Aureobasidium pullulans* AY4, an emerging opportunistic fungal pathogen with diverse biotechnological potential. *Eukaryotic Cell* 11, 1419–1420. doi:10.1128/EC.00245-12

18. Cheng, K. C., Demirci, A., and Catchmark, J. M. (2011). Pullulan: biosynthesis, production, and applications. *Appl. Microbiol. Biotechnol.* 92, 29–44. doi:10.1007/s00253-011-3477-y

19. Chivero, P., Gohtani, S., Yoshii, H., and Nakamura, A. (2015). Effect of xanthan and guar gums on the formation and stability of soy soluble polysaccharide oil-in-water emulsions. *Food Res. Intern.* 70, 7–14. doi:10.1016/j.foodres.2015.01.025

20. Cho, H. M., and Yoo, B. (2015). Rheological characteristics of cold thickened beverages containing xanthan gum-based food thickeners used for dysphagia diets. *J. Acad. Nutr. Diet.* 115, 106–111. doi:10.1016/j.jand.2014.08.028

21. Costa, R. R., Neto, A. I., Calgeris, I., Correia, C. R., Pinho, A. C., Fonseca, J., et al. (2013). Adhesive nanostructured multilayer films using a bacterial exopolysaccharide for biomedical applications. *J. Mater. Chem. B.* 1(18), 2367–2374.

22. Crescenzi, V., and Dentini, M. (1996). "Microbes in polymer chemistry," in *Hydrogels and Biodegradable Polymers for Bioapplications, ACS Symposium Series*, Vol. 627, ed. Ottenbrite, R. M., Huang S. J., and Park K. (Washington, DC: American Chemical Society), 221–232.

23. Curran, K. A., and Alper, H. S. (2012). Expanding the chemical palate of cells by combining systems biology and metabolic engineering. *Metab. Eng.* 14, 289–297. doi:10.1016/j.ymben.2012.04.006

24. da Silva, A. R., Ferro, J. A., Reinach, F. C., Farah, C. S., Furlan, L. R., Quaggio, R. B., et al. (2002). Comparison of the genomes of two *Xanthomonas* pathogens with differing host specificities. *Nature* 417, 459–463. doi:10.1038/417459a

25. Delbarre-Ladrat, C., Sinquin, C., Lebellenger, L., Zykwinska, A., and Colliec-Jouault, S. (2014). Exopolysaccharides produced by marine bacteria and their applications as glycosaminoglycan-like

molecules. *Front. Chem.* 2:85. doi:10.3389/fchem.2014.00085

26. Diken, E., Ozer, T., Arikan, M., Emrence, Z., Oner, E. T., Ustek, D., et al. (2015). Genomic analysis reveals the biotechnological and industrial potential of levan producing halophilic extremophile, *Halomonas smyrnensis* AAD6^T. *Springerplus* 4, 393. doi:10.1186/s40064-015-1184-3
27. Donot, F., Fontana, A., Baccou, J. C., and Schorr-Galindo, S. (2012). Microbial exopolysaccharides: main examples of synthesis, excretion, genetics and extraction. *Carbohydr. Polym.* 87, 951–962. doi:10.1016/j.carbpol.2011.08.083
28. Esvelt, K. M., and Wang, H. H. (2013). Genome-scale engineering for systems and synthetic biology. *Mol. Syst. Biol.* 9, 641. doi:10.1038/msb.2012.66
29. Finore, I., Di Donato, P., Mastascusa, V., Nicolaus, B., and Poli, A. (2014). Fermentation technologies for the optimization of marine microbial exopolysaccharide production. *Mar. Drugs* 12, 3005–3024. doi:10.3390/md12053005
30. Flemming, H. C., and Wingender, J. (2010). The biofilm matrix. *Nat. Rev. Microbiol.* 8, 623–633. doi:10.1038/nrmicro2415
31. Freitas, F., Alves, V. D., and Reis, M. A. (2011). Advances in bacterial exopolysaccharides: from production to biotechnological applications. *Trends Biotechnol.* 29, 388–398. doi:10.1016/j.tibtech.2011.03.008
32. Frese, M., Schatschneider, S., Voss, J., Vorhölter, F. J., and Niehaus, K. (2014). Characterization of the pyrophosphate-dependent 6-phosphofructokinase from *Xanthomonas campestris* pv. campestris. *Arch. Biochem. Biophys.* 546, 53–63. doi:10.1016/j.abb.2014.01.023
33. Gaspar, P., Carvalho, A. L., Vinga, S., Santos, H., and Neves, A. R. (2013). From physiology to systems metabolic engineering for the production of biochemicals by lactic acid bacteria. *Biotechnol. Adv.* 31, 764–788. doi:10.1016/j.biotechadv.2013.03.011
34. Goh, Y. J., Goin, C., O'Flaherty, S., Altermann, E., and Hutkins, R. (2011). Specialized adaptation of a lactic acid bacterium to the milk environment: the comparative genomics of *Streptococcus thermophilus* LMD-9. *Microb. Cell Fact.* 10(Suppl. 1), S22.

doi:10.1186/1475-2859-10-S1-S22

35. Gostin, C., Ohm, R. A., Kogej, T., Sonjak, S., Turk, M., Zajc, J., et al. (2014). Genome sequencing of four *Aureobasidium pullulans* varieties: biotechnological potential, stress tolerance, and description of new species. *BMC Genomics* 15:549. doi:10.1186/1471-2164-15-549
36. Han, J., Hang, F., Guo, B., Liu, Z., You, C., and Wu, Z. (2014). Dextran synthesized by *Leuconostoc mesenteroides* BD1710 in tomato juice supplemented with sucrose. *Carbohydr. Polym.* 112, 556–562. doi:10.1016/j.carbpol.2014.06.035
37. Han, Y. W., and Clarke, M. A. (1990). Production and characterization of microbial levan. *J. Agric. Food Chem.* 38, 393–396. doi:10.1021/jf00092a011
38. Hanemaaijer, M., Röling, W. F., Olivier, B. G., Khandelwal, R. A., Teusink, B., and Bruggeman, F. J. (2015). Systems modeling approaches for microbial community studies: from metagenomics to inference of the community structure. *Front. Microbiol.* 6:213. doi:10.3389/fmicb.2015.00213
39. Hao, P., Zheng, H., Yu, Y., Ding, G., Gu, W., Chen, S., et al. (2011). Complete sequencing and pan-genomic analysis of *Lactobacillus delbrueckii* subsp. *Bulgaricus* reveal its genetic basis for industrial yogurt production. *PLoS ONE* 6:e15964. doi:10.1371/journal.pone.0015964
40. Henry, C. S., DeJongh, M., Best, A. A., Frybarger, P. M., Linsay, B., and Stevens, R. L. (2010). High-throughput generation, optimization and analysis of genome-scale metabolic models. *Nat. Biotechnol.* 28, 977–982. doi:10.1038/nbt.1672
41. Hong, S. H., Park, S. J., Moon, S. Y., Park, J. P., and Lee, S. Y. (2003). In silico prediction and validation of the importance of the Entner-Doudoroff pathway in poly (3-hydroxybutyrate) production by metabolically engineered *Escherichia coli*. *Biotechnol. Bioeng.* 83, 854–863. doi:10.1002/bit.10733
42. Hou, S., Saw, J. H., Lee, K. S., Freitas, T. A., Belisle, C., Kawarabayasi, Y. M., et al. (2004). Genome sequence of the deep-sea γ-proteobacterium *Idiomarina loihiensis* reveals amino acid fermentation as a source of carbon and energy. *Proc. Natl. Acad.*

Sci. U.S.A. 101, 18036–18041. doi:10.1073/pnas.0407638102

43. Ivars-Martinez, E., Martin-Cuadrado, A. B., D'Auria, G., Mira, A., Ferriera, S., Johnson, J., et al. (2008). Comparative genomics of two ecotypes of the marine planktonic copiotroph *Alteromonas macleodii* suggests alternative lifestyles associated with different kinds of particulate organic matter. *ISME J.* 2, 1194–1212. doi:10.1038/ismej.2008.74
44. Johansson, P., Paulin, L., Säde, E., Salovuori, N., Alatalo, E. R., Björkroth, K. J., et al. (2011). Genome sequence of a food spoilage lactic acid bacterium, *Leuconostoc gasicomitatum* LMG 18811T, in association with specific spoilage reactions. *Appl. Environ. Microbiol.* 77, 4344–4351. doi:10.1128/AEM.00102-11
45. Jung, J. Y., Lee, S. H., Kim, J. M., Park, M. S., Bae, J. W., and Hahn, Y. (2011). Metagenomic analysis of kimchi, a traditional Korean fermented food. *Appl. Environ. Microbiol.* 77, 2264–2274. doi:10.1128/AEM.02157-10
46. Kang, J. X., Chen, X. J., Chen, W. R., Li, M. S., Fang, Y., Li, D. S., et al. (2011). Enhanced production of pullulan in *Aureobasidium pullulans* by a new process of genome shuffling. *Process Biochem.* 46, 792–795. doi:10.1016/j.procbio.2010.11.004
47. Kang, S. A., Jang, K. H., Seo, J. W., Kim, K. H., Kim, Y. H., Rairakhwada, D., et al. (2009). "Levan: applications and perspectives," in *Microbial Production of Biopolymers and Polymer Precursors: Applications and Perspectives*, ed. Rehm B. (Norfolk: Caister Academic Press, Inc.), 145–161.
48. Kaur, V., Bera, M. B., Panesar, P. S., Kumar, H., and Kennedy, J. F. (2014). Welan gum: microbial production, characterization, and applications. *Int. J. Biol. Macromol.* 65, 454–461. doi:10.1016/j.ijbiomac.2014.01.061
49. Kazak Sarilmiser, H., Ates, Ö, Özdemir, G., Arga, K. Y., and Toksoy Öner, E. (2015). Effective stimulating factors for microbial levan production by *Halomonas Smyrnensis* AAD6^{T}. *J. Biosci. Bioeng.* 119, 455–463. doi:10.1016/j.jbiosc.2014.09.019
50. Kennedy, C. J., Boyle, P. M., Waks, Z., and Silver, P. A. (2009). Systems-level engineering of nonfermentative metabolism in yeast. *Genetics* 183, 385–397. doi:10.1534/genetics.109.105254

51. Keshavarz, T., and Roy, I. (2010). Polyhydroxyalkanoates: bioplastics with a green agenda. *Curr. Opin. Microbiol.* 13, 321–326. doi:10.1016/j.mib.2010.02.006

52. Khan, T., Park, J. K., and Kwon, J. H. (2007). Functional biopolymers produced by biochemical technology considering applications in food engineering. *Korean J. Chem. Eng.* 24, 816–826. doi:10.1007/s11814-007-0047-1

53. Kind, S., and Wittmann, C. (2011). Bio-based production of the platform chemical 1, 5-diaminopentane. *Appl. Microbiol. Biotechnol.* 91, 1287–1296. doi:10.1007/s00253-011-3457-2

54. Kucukasik, F. (2010). Evaluation of Different Waste Streams as Fermentation Substrates for Microbial Biopolymer Production. Ph.D. thesis, Institute of Environmental Sciences, Istanbul, Turkey.

55. Kucukasik, F., Kazak, H., Guney, D., Finore, I., Poli, A., Yenigun, O., et al. (2011). Molasses as fermentation substrate for Levan production by Halomonas sp. *Appl. Microbiol. Biotechnol.* 89(6), 1729–1740.

56. Kumar, A. S., Mody, K., and Jha, B. (2007). Bacterial exopolysaccharides-a perception. *J. Basic Microbiol.* 47(2), 103–117.

57. Kumar, A. S., and Mody, K. (2009). "Microbial exopolysaccharides: variety and potential applications," in *Microbial Production of Biopolymers and Polymer Precursors: Applications and Perspectives*, ed. B. Rehm (Norfolk: Caister Academic Press, Inc.), 229–253.

58. Kwon, K. J., Park, K. J., Kim, J. D., Kong, J. Y., and Kong, I. S. (1994). Isolation of two different polysaccharides from halophilic *Zoogloea* sp. *Biotechnol. Lett.* 16, 783–788. doi:10.1007/BF00133954

59. Lee, B. M., Park, Y. J., Park, D. S., Kang, H. W., Kim, J. G., Song, E. S., et al. (2005a). The genome sequence of *Xanthomonas oryzae pathovar oryzae* KACC10331, the bacterial blight pathogen of rice. *Nucleic Acids Res.* 33, 577–586. doi:10.1093/nar/gki206

60. Lee, S. J., Lee, D. Y., Kim, T. Y., Kim, B. H., Lee, J., and Lee, S. Y. (2005b). Metabolic engineering of *Escherichia coli* for enhanced production of succinic acid, based on genome comparison and *in*

silico gene knockout simulation. *Appl. Environ. Microbiol.* 71, 7880–7887. doi:10.1128/AEM.71.12.7880-7887.2005

61. Lee, J. W., Kim, H. U., Choi, S., Yi, J., and Lee, S. Y. (2011). Microbial production of building block chemicals and polymers. *Curr. Opin. Biotechnol.* 22, 758–767. doi:10.1016/j.copbio.2011.02.011
62. Lee, J. W., Na, D., Park, J. M., Lee, J., Choi, S., and Lee, S. Y. (2012). Systems metabolic engineering of microorganisms for natural and non-natural chemicals. *Nat. Chem. Biol.* 8, 536–546. doi:10.1038/nchembio.970
63. Lerman, J. A., Hyduke, D. R., Latif, H., Portnoy, V. A., Lewis, N. E., Orth, J. D., et al. (2012). *In silico* method for modelling metabolism and gene product expression at genome scale. *Nat. Commun.* 3, 929. doi:10.1038/ncomms1928
64. Li, J., and Wang, N. (2012). The *gpsX* gene encoding a glycosyltransferase is important for polysaccharide production and required for full virulence in *Xanthomonas citri* subsp. Citri. *BMC Microbiol.* 12:31. doi:10.1186/1471-2180-12-31
65. Liang, T. W., and Wang, S. L. (2015). Recent advances in exopolysaccharides from *Paenibacillus* spp.: production, isolation, structure, and bioactivities. *Mar. Drugs* 13, 1847–1863. doi:10.3390/md13041847
66. Llamas, I., Amjres, H., Mata, J. A., Quesada, E., and Béjar, V. (2012). The potential biotechnological applications of the exopolysaccharide produced by the halophilic bacterium *Halomonas almeriensis*. *Molecules* 17, 7103–7120. doi:10.3390/molecules17067103
67. Long, M. R., Ong, W. K., and Reed, J. L. (2015). Computational methods in metabolic engineering for strain design. *Curr. Opin. Biotechnol.* 34, 135–141. doi:10.1016/j.copbio.2014.12.019
68. Ma, Z. C., Liu, N. N., Chi, Z., Liu, G. L., and Chi, Z. M. (2015). Genetic modification of the marine-isolated yeast *Aureobasidium melanogenum* P16 for efficient pullulan production from inulin. *Mar. Biotechnol.* 17, 511–522. doi:10.1007/s10126-015-9638-8
69. Mantzaris, N. V., Kelley, A. S., Srienc, F., and Daoutidis, P. (2001). Optimal carbon source switching strategy for the production of PHA copolymers. *Am. Inst. Chem. Eng. J.* 47, 727–743. doi:10.1002/

aic.690470319

70. Mata, J. A., Béjar, V., Llamas, I., Arias, S., Bressollier, P., Tallon, R., et al. (2006). Exopolysaccharides produced by the recently described halophilic bacteria *Halomonas ventosae* and *Halomonas anticariensis*. *Res. Microbiol.* 157, 827–835. doi:10.1016/j.resmic.2006.06.004
71. Medina-Cleghorn, D., and Nomura, D. K. (2013). Chemical approaches to study metabolic networks. *Pflugers Arch.* 465, 427–440. doi:10.1007/s00424-012-1201-0
72. Moon, J. S., Choi, H. S., Shin, S. Y., Noh, S. J., Jeon, C. O., and Han, N. S. (2015). Genome sequence analysis of potential probiotic strain *Leuconostoc lactis* EFEL005 isolated from kimchi. *J. Microbiol.* 53, 337–342. doi:10.1007/s12275-015-5090-8
73. Moon, S. Y., Hong, S. H., Kim, T. Y., and Lee, S. Y. (2008). Metabolic engineering of *Escherichia coli* for the production of malic acid. *Biochem. Eng. J.* 40, 312–320. doi:10.1016/j.bej.2008.01.001
74. Morris, G., and Harding, S. E. (2009). "Polysaccharides, microbial," in *Encyclopedia of Microbiology*, ed. Schaechter M. (Amsterdam: Elsevier), 482–494.
75. Morris, V. J. (2006). "Bacterial polysaccharides," in *Food Polysaccharides and Their Applications*, eds Stephen, A. M. Philips G. O., and Williams P. A. (New York: Taylor Francis), 413–454.
76. Nadkarni, M. A., Chen, Z., Wilkins, M. R., and Hunter, N. (2014). Comparative genome analysis of *Lactobacillus rhamnosus*clinical isolates from initial stages of dental pulp infection: identification of a new exopolysaccharide cluster. *PLoS ONE*9:e90643. doi:10.1371/journal.pone.0090643
77. Nakamura, C. E., and Whited, G. M. (2003). Metabolic engineering for the microbial production of 1, 3-propanediol. *Curr. Opin. Biotechnol.* 14, 454–459. doi:10.1016/j.copbio.2003.08.005
78. Ng, C. Y., Jung, M. Y., Lee, J., and Oh, M. K. (2012). Production of 2, 3-butanediol in *Saccharomyces cerevisiae* by *in silic*o aided metabolic engineering. *Microb. Cell Fact.* 11, 68. doi:10.1186/1475-2859-11-68
79. Nicolaus, B., Kambourova, M., and Toksoy Oner, E. (2010). Exopolysaccharides from extremophiles: from fundamentals to

biotechnology. *Environ Technol.* 31(10), 1145–1158.

80. Nogales, J., Palsson, B. O., and Thiele, I. (2008). A genome-scale metabolic reconstruction of *Pseudomonas putida*KT2440: iJN746 as a cell factory. *BMC Syst. Biol.* 2:79. doi:10.1186/1752-0509-2-79
81. Nwodo, U. U., Green, E., and Okoh, A. I. (2012). Bacterial exopolysaccharides: functionality and prospects. *Int. J. Mol. Sci.*13, 14002–14015. doi:10.3390/ijms131114002
82. Oberhardt, M. A., Puchalka, J., Fryer, K. E., Martins dos Santos, V. A. P., and Papin, J. A. (2008). Genome-scale metabolic network analysis of the opportunistic pathogen *Pseudomonas aeruginosa* PAO1. *J. Bacteriol.* 190, 2790–2803. doi:10.1128/JB.01583-07
83. Ochiai, H., Inoue, Y., Takeya, M., Sasaki, A., and Kaku, H. (2005). Genome sequence of *Xanthomonas oryzae* pv. *Oryzae*suggests contribution of large numbers of effector genes and insertion sequences to its race diversity. *Jpn. Agric. Res. Q.* 39, 275–287. doi:10.6090/jarq.39.275
84. Okuda, S., and Yoshizawa, A. C. (2011). ODB: a database for operon organizations, 2011 update. *Nucleic Acids Res.*39(Suppl. 1), D552–D555. doi:10.1093/nar/gkq1090
85. Oliveira,A.P.,Nielsen,J.,andForster,J.(2005).Modeling*Lactococcus lactis* using a genome-scale flux model. *BMC Microbiol.* 5:39. doi:10.1186/1471-2180-5-39
86. Ordax, M., Marco-Noales, E., López, M. M., and Biosca, E. G. (2010). Exopolysaccharides favor the survival of *Erwinia amylovora* under copper stress through different strategies. *Res. Microbiol.* 161, 549–555. doi:10.1016/j.resmic.2010.05.003
87. Otero, J. M., and Nielsen, J. (2010). Industrial systems biology. *Biotechnol. Bioeng.* 105, 439–460. doi:10.1002/bit.22592
88. Özcan, E., Sargin, S., and Goksungur, Y. (2014). Comparison of pullulan production performances of air-lift and bubble column bioreactors and optimization of process parameters in air-lift bioreactor. *Biochem. Eng. J.* 92, 9–15. doi:10.1016/j.bej.2014.05.017
89. Pasteur, L. (1861). On the viscous fermentation and the butyrous fermentation. *Bulletin de la Société Chimique de France*11, 30–31; (in French).

90. Pastink, M. I., Teusink, B., Hols, P., Visser, S., de Vos, W. M., and Hugenholtz, J. (2009). Genome-scale model of *Streptococcus thermophilus* LMG18311 for metabolic comparison of lactic acid bacteria. *Appl. Environ. Microbiol.* 75, 3627–3633. doi:10.1128/AEM.00138-09

91. Patel, A., and Prajapati, J. B. (2013). Food and health applications of exopolysaccharides produced by lactic acid bacteria. *Adv. Dairy Res.* 1, 2. doi:10.4172/2329-888X.1000107

92. Patel, S., Majumder, A., and Goyal, A. (2011). Potentials of exopolysaccharides from lactic acid bacteria. *Indian J. Microbiol.*52, 3–12. doi:10.1007/s12088-011-0148-8

93. Péant, B., LaPointe, G., Gilbert, C., Atlan, D., Ward, P., and Roy, D. (2005). Comparative analysis of the exopolysaccharide biosynthesis gene clusters from four strains of *Lactobacillus rhamnosus*. *Microbiology* 151, 1839–1851. doi:10.1099/mic.0.27852-0

94. Poli, A., Kazak, H., Gürleyendag, B., Tommonaro, G., Pieretti, G., Toksoy Öner, E., et al. (2009). High level synthesis of levan by a novel *Halomonas* species growing on defined media. *Carbohydr. Polym.* 78, 651–657. doi:10.1016/j.carbpol.2009.05.031

95. Poli, A., Nicolaus, B., Denizci, A. A., Yavuzturk, B., and Kazan, D. (2013). *Halomonas smyrnensis* sp. nov., a moderately halophilic, exopolysaccharide-producing bacterium from camalti saltern area, Turkey. *Int. J. Syst. Evol. Microbiol.* 63, 10–18. doi:10.1099/ijs.0.037036-0

96. Puchalka, J., Oberhardt, M. A., Godinho, M., Bielecka, A., Regenhardt, D., Timmis, K. N., et al. (2008). Genome-scale reconstruction and analysis of the *Pseudomonas putida* KT2440 metabolic network facilitates applications in biotechnology. *PLoS Comput. Biol.* 4:e1000210. doi:10.1371/journal.pcbi.1000210

97. Qian, W., Jia, Y., Ren, S. X., He, Y. Q., Feng, J. X., Lu, L. F., et al. (2005). Comparative and functional genomic analyses of the pathogenicity of phytopathogen *Xanthomonas campestris* pv. Campestris. *Genome Res.* 15, 757–767. doi:10.1101/gr.3378705

98. Qin, Q. L., Li, Y., Zhang, Y. J., Zhou, Z. M., Zhang, W. X., Chen, X. L. Z., et al. (2011). Comparative genomics reveals a deep-sea sediment-

adapted life style of *Pseudoalteromonas* sp. SM9913. *ISME J.* 5, 274–284. doi:10.1038/ismej.2010.103

99. Rehm, B. (2009). *Microbial Production of Biopolymers and Polymer Precursors: Applications and Perspectives.* Norfolk: Caister Academic.
100. Rehm, B. H. (2010). Bacterial polymers: biosynthesis, modifications and applications. *Nat. Rev. Microbiol.* 8, 578–592. doi:10.1038/nrmicro2354
101. Riedel, T., Spring, S., Fiebig, A., Petersen, J., Kyrpides, N. C., Göker, M., et al. (2014). Genome sequence of the exopolysaccharide-producing *Salipiger mucosus* type strain (DSM 16094T), a moderately halophilic member of the *Roseobacter clade. Stand. Genomic Sci.* 9, 1333. doi:10.4056/sigs.4909790
102. Ruffing, A. M., Castro-Melchor, M., Hu, W. S., and Chen, R. R. (2011). Genome sequence of the curdlan-producing Agrobacterium sp. strain ATCC 31749. *J. Bacteriol.* 193(16), 4294–4295.
103. Ruffing, A. M., and Chen, R. R. (2012). Transcriptome profiling of a curdlan-producing *Agrobacterium* reveals conserved regulatory mechanisms of exopolysaccharide biosynthesis. *Microb. Cell Fact.* 11, 1–13. doi:10.1186/1475-2859-11-17
104. Rühmann, B., Schmid, J., and Sieber, V. (2015). High throughput exopolysaccharide screening platform: from strain cultivation to monosaccharide composition and carbohydrate fingerprinting in one day. *Carbohydr. Polym.* 122, 212–220. doi:10.1016/j.carbpol.2014.12.021
105. Sam, S., Kucukasik, F., Yenigun, O., Nicolaus, B., Toksoy Oner, E., and Yukselen, M. A. (2011). Flocculating performances of exopolysaccharides produced by a halophilic bacterial strain cultivated on agro-industrial waste. *Bioresour. Technol.* 102, 1788–1794. doi:10.1016/j.biortech.2010.09.020
106. Sanchez-Garcia, M. D., Lagaron, J. M., and Hoa, S. V. (2010). Effect of addition of carbon nanofibers and carbon nanotubes on properties of thermoplastic biopolymers. *Compos. Sci. Technol.* 70, 1095–1105. doi:10.1016/j.compscitech.2010.02.015
107. Saulnier, D. M., Santos, F., Roos, S., Mistretta, T. A., Spinler, J. K., Molenaar, D., et al. (2011). Exploring metabolic

pathway reconstruction and genome-wide expression profiling in *Lactobacillus reuteri* to define functional probiotic features. *PLoS ONE* 6:e18783. doi:10.1371/journal.pone.0018783

108. Schatschneider, S., Persicke, M., Watt, S. A., Hublik, G., Pühler, A., Niehaus, K., et al. (2013). Establishment *in silico*analysis, and experimental verification of a large-scale metabolic network of the xanthan producing *Xanthomonas campestris* pv. Campestris strain B100. *J. Biotechnol.* 167, 123–134. doi:10.1016/j.jbiotec.2013.01.023

109. Schmid, J., and Sieber, V. (2015). Enzymatic transformations involved in the biosynthesis of microbial exopolysaccharides based on the assembly of repeat units. *Chembiochem* 16, 1141–1147. doi:10.1002/cbic.201500035

110. Sezer, A. D., Kazak, H., Toksoy Oner, E., and Akbuğa, J. (2011). Levan-based nanocarrier system for peptide and protein drug delivery: optimization and influence of experimental parameters on the nanoparticle characteristic. *Carbohydr. Polym.* 84, 358–363. doi:10.1016/j.carbpol.2010.11.046

111. Sheng, L., Zhu, G., and Tong, Q. (2014). Comparative proteomic analysis of *Aureobasidium pullulans* in the presence of high and low levels of nitrogen source. *J. Agric. Food Chem.* 62, 10529–10534. doi:10.1021/jf503390f

112. Sieuwerts, S., Molenaar, D., van Hijum, S. A., Beerthuyzen, M., Stevens, M. J., and Janssen, P. W. (2010). Mixed-culture transcriptome analysis reveals the molecular basis of mixed-culture growth in *Streptococcus thermophilus* and *Lactobacillus bulgaricus*. *Appl. Environ. Microbiol.* 76, 7775–7784. doi:10.1128/AEM.01122-10

113. Siezen, R. J., and van Hylckama Vlieg, J. E. (2011). Genomic diversity and versatility of *Lactobacillus plantarum*, a natural metabolic engineer. *Microb. Cell Fact.* 10(Suppl. 1), S3. doi:10.1186/1475-2859-10-S1-S3

114. Sima, F., Axente, E., Sima, L. E., Tuyel, U., Eroglu, M. S., Serban, N., et al. (2012). Combinatorial matrix-assisted pulsed laser evaporation: single-step synthesis of biopolymer compositional gradient thin film assemblies. *Appl. Phys. Lett.* 101, 233705. doi:10.1063/1.4769987

115. Sima, F., Mutlu Cansever, E., Eroglu, M. S., Sima Livia, E., Serban, N., Ristoscu, C., et al. (2011). Levan nanostructured thin films by MAPLE assembling. *Biomacromolecules* 12, 2251–2256. doi:10.1021/bm200340b

116. Simeonidis, E., and Price, N. D. (2015). Genome-scale modeling for metabolic engineering. *J. Ind. Microbiol. Biotechnol.*42, 327–338. doi:10.1007/s10295-014-1576-3

117. Singh, R. S., Kaur, N., and Kennedy, J. F. (2015). Pullulan and pullulan derivatives as promising biomolecules for drug and gene targeting. *Carbohydr. Polym.* 123, 190–207. doi:10.1016/j.carbpol.2015.01.032

118. Singh, R. S., Saini, G. K., and Kennedy, J. F. (2008). Pullulan: microbial sources, production and applications. *Carbohydr. Polym.* 73, 515–531. doi:10.1016/j.carbpol.2008.01.003

119. Sogutcu, E., Emrence, Z., Arikan, M., Cakiris, A., Abaci, N., Öner, E. T., et al. (2012). Draft genome sequence of Halomonas smyrnensis AAD6T. *J. Bacteriol.*, 194(20), 5690–5691.

120. Sohn, S. B., Kim, T. Y., Park, J. M., and Lee, S. Y. (2010). *In silico* genome-scale metabolic analysis of *Pseudomonas putida*KT2440 for polyhydroxyalkanoate synthesis, degradation of aromatics and anaerobic survival. *Biotechnol. J.* 5, 739–750. doi:10.1002/biot.201000124

121. Staudt, A. K., Wolfe, L. G., and Shrout, J. D. (2012). Variations in exopolysaccharide production by *Rhizobium tropici*. *Arch. Microbiol.* 194, 197–206. doi:10.1007/s00203-011-0742-5

122. Tang, X. Z., Kumar, P., Alavi, S., and Sandeep, K. P. (2012). Recent advances in biopolymers and biopolymer-based nanocomposites for food packaging materials. *Crit. Rev. Food Sci. Nutr.* 52, 426–442. doi:10.1080/10408398.2010.500508

123. Teusink, B., van Enckevort, F. H., Francke, C., Wiersma, A., Wegkamp, A., Smid, E. J., et al. (2005). *In silico* reconstruction of the metabolic pathways of *Lactobacillus plantarum*: comparing predictions of nutrient requirements with those from growth experiments. *Appl. Environ. Microbiol.* 71, 7253–7262. doi:10.1128/JB.187.21.7254-7266.2005

124. Teusink, B., Wiersma, A., Jacobs, L., Notebaart, R. A., and Smid, E. J. (2009). Understanding the adaptive growth strategy of *Lactobacillus plantarum* by *in silico* optimisation. *PLoS Comput. Biol.* 5:e1000410. doi:10.1371/journal.pcbi.1000410

125. Thieme, F., Koebnik, R., Bekel, T., Berger, C., Boch, J., Büttner, D., et al. (2005). Insights into genome plasticity and pathogenicity of the plant pathogenic bacterium *Xanthomonas campestris* pv. Vesicatoria revealed by the complete genome sequence. *J. Bacteriol.* 187, 7254–7266. doi:10.1128/JB.187.21.7254-7266.2005

126. Toksoy Oner, E. (2013). "Microbial production of extracellular polysaccharides from biomass," in *Pretreatment Techniques for Biofuels and Biorefineries*, ed. Fang Z. (Berlin: Springer), 35–56.

127. Valdés, J., Pedroso, I., Quatrini, R., Dodson, R. J., Tettelin, H., Blake, R., et al. (2008). *Acidithiobacillus ferrooxidans*metabolism: from genome sequence to industrial applications. *BMC Genomics* 9:597. doi:10.1186/1471-2164-9-597

128. Van Tieghem, P. (1878). On sugar mill gum. *Ann. Sci. Nat. Bot. Biol. Veg.* 7, 180–203.

129. Vorhölter, F. J., Schneiker, S., Goesmann, A., Krause, L., Bekel, T., Kaiser, O., et al. (2008). The genome of *Xanthomonas campestr*is pv. Campestris B100 and its use for the reconstruction of metabolic pathways involved in xanthan biosynthesis. *J. Biotechnol.* 134, 33–45. doi:10.1016/j.jbiotec.2007.12.013

130. Vroman, I., and Tighzert, L. (2009). Biodegradable polymers. *Materials* 2, 307–344. doi:10.3390/ma2020307

131. Vu, B., Chen, M., Crawford, R. J., and Ivanova, E. P. (2009). Bacterial extracellular polysaccharides involved in biofilm formation. *Molecules* 14, 2535–2554. doi:10.3390/molecules14072535

132. Wang, D., Chen, F., Wei, G., Jiang, M., and Dong, M. (2015). The mechanism of improved pullulan production by nitrogen limitation in batch culture of *Aureobasidium pullulans*. *Carbohydr. Polym.* 127, 325–331. doi:10.1016/j.carbpol.2015.03.079

133. Wang, Q., Chen, X., Yang, Y., and Zhao, X. (2006). Genome-scale *in silic*o aided metabolic analysis and flux comparisons

of *Escherichia coli* to improve succinate production. *Appl. Microbiol. Biotechnol.* 73, 887–894. doi:10.1007/s00253-006-0535-y

134. Whitney, J. C., and Howell, P. L. (2013). Synthase-dependent exopolysaccharide secretion in Gram-negative bacteria. *Trends Microbiol.* 21, 63–72. doi:10.1016/j.tim.2012.10.001

135. Wibberg, D., Alkhateeb, R. S., Winkler, A., Albersmeier, A., Schatschneider, S., Albaum, S., et al. (2015). Draft genome of the xanthan producer *Xanthomonas campestris* NRRL B-1459 (ATCC 13951). *J. Biotechnol.* 204, 45–46. doi:10.1016/j.jbiotec.2015.03.026

136. Williams, P. A., Phillips, G. O., and Seisun, D. (2007). *Overview of the Hydrocolloid Market*. Cambridge UK: RSC Publishing.

137. Xu, C., Liu, L., Zhang, Z., Jin, D., Qiu, J., and Chen, M. (2013). Genome-scale metabolic model in guiding metabolic engineering of microbial improvement. *Appl. Microbiol. Biotechnol.* 97, 519–539. doi:10.1007/s00253-012-4543-9

138. Yasar Yildiz, S., Anzelmo, G., Ozer, T., Radchenkova, N., Genc, S., Di Donato, P., et al. (2014). *Brevibacillus themoruber*: a promising microbial cell factory for exopolysaccharide production. *J. Appl. Microbiol.* 116, 314–324. doi:10.1111/jam.12362

139. Yasar Yildiz, S. Y., Radchenkova, N., Arga, K. Y., Kambourova, M., and Oner, E. T. (2015). Genomic analysis of *Brevibacillus thermoruber* 423 reveals its biotechnological and industrial potential. *Appl. Microbiol. Biotechnol.* 99, 2277–2289. doi:10.1007/s00253-015-6388-5

140. Yim, H., Haselbeck, R., Niu, W., Pujol-Baxley, C., Burgard, A., Boldt, J., et al. (2011). Metabolic engineering of *Escherichia coli* for direct production of 1, 4-butanediol. *Nat. Chem. Biol.* 7, 445–452. doi:10.1038/nchembio.580

141. Zakharevich, N. V., Averina, O. V., Klimina, K. M., Kudryavtseva, A. V., Kasianov, A. S., Makeev, V. J., et al. (2015). Complete genome sequence of *Bifidobacterium longum* GT15: identification and characterization of unique and global regulatory genes. *Microb. Ecol.* 70, 819–834. doi:10.1007/s00248-015-0603-x

CHAPTER

2

BIOLOGICAL EFFECTS OF SPIRULINA (ARTHROSPIRA) BIOPOLYMERS AND BIOMASS IN THE DEVELOPMENT OF NANOSTRUCTURED SCAFFOLDS

Michele Greque de Morais[1] Bruna da Silva Vaz[1] Etiele Greque de Morais[2] and Jorge Alberto Vieira Costa[2]

[1]Laboratory of Microbiology and Biochemical, College of Chemistry and Food Engineering, Federal University of Rio Grande, Avenida Itália, Km 8, 96203-900 Rio Grande, RS, Brazil

[2]Laboratory of Biochemical Engineering, College of Chemistry and Food Engineering, Federal University of Rio Grande, 96203-900 Rio Grande, RS, Brazil

ABSTRACT

Spirulina is produced from pure cultures of the photosynthetic prokaryotic cyanobacteria Arthrospira. For many years research centers throughout the world have studied its application in various scientific fields, especially in foods and medicine. The biomass produced from Spirulina cultivation contains a variety of biocompounds, including biopeptides, biopolymers, carbohydrates, essential fatty acids, minerals, oligoelements, and sterols. Some of these compounds are bioactive and have anti-inflammatory, antibacterial, antioxidant, and antifungal properties. These compounds can be used in tissue engineering, the interdisciplinary field that combines techniques from cell science, engineering, and materials science and which has grown in importance over the past few decades. Spirulina biomass can be used to produce polyhydroxyalkanoates (PHAs), biopolymers that can substitute synthetic polymers in the construction of engineered extracellular matrices (scaffolds) for use in tissue cultures or bioactive molecule construction. This review describes the development of nanostructured scaffolds based on biopolymers extracted from microalgae and biomass from Spirulina production. These scaffolds have the potential to encourage cell growth while reducing the risk of organ or tissue rejection.

INTRODUCTION

Tissue engineering, the interdisciplinary field that combines techniques from cell science, engineering, and materials science, has the potential to reconstitute damaged tissues and organs using cells that are supported on scaffolds where the components of the extracellular matrix can segregate during tissue and organ formation. It is important to choose a scaffold material that stimulates cells to produce structures [1]; therefore, much research has been carried out on natural organic materials. The natural compounds that are incorporated into the scaffolds can act as a substratum for cellular growth by stimulating cell growth and anchoring with lower risk of tissue rejection when compared with synthetic sources.

Spirulina is a prokaryotic microalga, order Cyanophyceae, division Cyanophyta (Cyanobacteria). It has a distinctive arrangement of multicellular cylindrical trichomes in an open helix throughout its length. The helical shape of the trichomes is characteristic of the genus, but the length and size of the helix vary with species [2]. In 1981, the Food and Drug Administration (FDA) declared “Spirulina is source of protein and contains several vitamins and minerals. It can be legally marketed as a food or a food supplement if it is precisely defined and free from contaminants and adulterants" and is categorized by the FDA as "Generally Recognized as Safe" (GRAS) [3].

Spirulina biomass stimulates important biological processes and exhibits antiallergenic, antibacterial, antifungal, anti-inflammatory, antioxidant, and immunomodulating properties [4]. Thus, Spirulina LEB 18 biomass incorporated into scaffolds stimulates cell growth and tissue regeneration [5–7].

Nanofiber scaffolds have the potential to be used in tissue engineering because they can reproduce the structure and function of the native extracellular matrix [8]. Electrospun scaffolds have attracted attention because of their characteristics: they have a high surface area in relation to fiber diameter, a high porosity that stimulates cell growth and connections between cells, and good nutrient diffusion and they encourage angiogenesis/vascularization during tissue regeneration [9].

The synthetic polymers normally used to produce nanofiber scaffolds can be replaced by Spirulina biopolymers, which are biodegradable and biocompatible with cells and tissues [5]. Spirulina biomass can be added to the polymer solutions used in nanofiber production to produce scaffolds that incorporate Spirulina's properties; this is possible because electrospinning does not involve extreme temperatures or pH that would reduce the biological activity of the biomass or its nutrients. Depending on the solvent used to prepare the polymer, the internal components (proteins, fatty acids, and biopolymers) of the biomass can be made available within the scaffolds to stimulate cells or tissues [4].

Polyhydroxyalkanoates (PHAs), a family of biopolymers that includes polyhydroxybutyrate (PHB), can be extracted from various microorganisms, including Spirulina, and used to provide atoxic biocompatible scaffolding for human tissue and organ culture. Low molecular weight PHB has been detected bound to human serum albumin and low-density lipoproteins. It degrades into (R)-β-hydroxybutyric acid, a naturally occurring mammalian metabolite present at serum concentrations of 3 mg dL^{-1} to 10 mg dL^{-1} in adult humans and presents no health risks. The fact that PHB degrades into such atoxic compounds may explain its biocompatibility with cultured cells and tissues. Not only do Spirulina biopolymer nanofiber scaffolds have a lower risk of rejection in human tissue culture but they also contain advantageous bioactive compounds that are present in the Spirulina biomass [10, 11].

This review describes the progress made in tissue engineering when Spirulina biomass and biopolymers are used in the production of nanostructured scaffolds that promote cell growth while decreasing the risk of tissue and organ rejection.

TISSUE ENGINEERING

Since its beginnings, this field has focused on the development of biological substitutes for the recuperation, regeneration, or substitution of defective tissues [12]. Permanent implants often result in chronic inflammation, which can lead to severe clinical complications. Implants developed using biomaterials could be a viable alternative to reestablish the normal functions of damaged tissues and organs [13].

The process of using tissue engineering to restore or substitute tissues or organs damaged by accidents, congenital defects, or diseases involves the in vitro propagation of viable cells attached to biological or synthetic supports, known as scaffolds [6]. After cellular cultivation, the scaffold is implanted into the patient and degrades when the new organ or tissue is formed [13–15].

Scaffolds

Scaffolds are three-dimensional structures that guide tissue development in situ at the site of interest depending only on the growth of the surrounding tissue [16]. The scaffold should be selected according to the type of repair and the tissues or organs to be reconstituted, and the scaffold surface is selected according to the desired interactions between the cells and the scaffold [13, 17].

For effective tissue reconstruction, scaffolds must conform to specific requirements. High porosity and pore interconnectivity are fundamental characteristics for increasing the available specific surface area, which is important not only for cell anchorage and the internal growth of tissues but also for facilitating the distribution and transportation of oxygen, nutrients, and cellular residues [18].

The degradability is a parameter closely related to the solubility of the molds. If the solubility is too high, the scaffolds will be reabsorbed by the body fluids too quickly without accompanying tissue regeneration. However, if the solubility is too low, it will remain for too long in the body and impede regeneration. Therefore, the degradability is associated with the stability of the biomaterial in vivo and an appropriate time is extremely important for proper regeneration [6].

The nature of the scaffold's surface can also affect cellular responses that influence the speed of formation and quality of new tissue [17]. The most appropriate scaffold material should be biocompatible and biodegradable, so that it is nonimmunogenic to avoid further surgical intervention when tissue regeneration is complete [19]. Scaffolds are designed to have a cell structure that is similar to the natural ECM and therefore have characteristics that are suitable for cell culture.

EXTRACELLULAR MATRIX

The extracellular matrix (ECM) is a component of the connective tissue. It is produced by cells and supports the morphological organization and physiological functions that occur during tissue formation [20, 21]. ECM produces the biochemical and biomechanical signals necessary for tissue morphogenesis, differentiation, and homeostasis [22].

Due to their versatile properties, decellularized extracellular matrices have been widely used as a source of biological scaffolds in tissue engineering and regenerative medicine [23–29]. The most important limitations in regenerative medicine are the shortage of autologous tissue and organ donors and the negative immunological responses and pathogen transfer whenever allogeneic or xenogeneic tissues or organs are used [30, 31]. Another advantage of scaffolds obtained from decellularized tissues and organs is the retention of the structure of the original tissue and organs.

Scaffolds made using the electrospinning process mimic the natural extracellular matrix's mechanical and architectural characteristics, enabling the anchoring and migration of cells. Growth factors, drugs, viruses, and proteins can be incorporated in the matrix. The microalga Spirulina is a good choice for incorporation in the production of scaffolds [6]. It is important to choose a scaffold material that stimulates cells to produce structures of the correct format and size. Such scaffolds are generally developed from synthetic polymers, which can be incompatible with human cells [1]. Therefore, much research has been carried out on natural organic materials.

Extracellular Matrices Made of Natural or Synthetic Polymers

Polymers are the raw materials for scaffold production in tissue engineering, and several types of biodegradable polymers are utilized in the development of artificial skin, surgical sutures, vascular grafts, bone joining devices, and controlled-release pharmaceuticals [32].

These materials can be classified as natural polymers (including polysaccharides such as alginate, chitin and chitosan, and starch and hyaluronic acid derivatives); proteins (such as collagen, fibrin gel, and soy and silk proteins); synthetic polymers (such as poly(lactic acid) (PLA), poly(glycolic acid) (PGA), and polycaprolactone (PCL)) [1, 33]; and microbial polymers (biopolymers) (such as polyhydroxyalkanoates (PHAs)).

Synthetic polymers are available in potentially unlimited amounts. Their physicochemical properties can be controlled, with their degradation rates and mechanical properties subject to chemical modification [34]. However, many synthetic scaffolds have hydrophobic surfaces that hinder cell recognition by native cells. Natural and microbial polymers are biologically recognized, which makes it easier to reproduce the properties of the tissues to be regenerated, such as their mechanical and cellular anchoring properties [18].

NANOTECHNOLOGY AND SCAFFOLD DEVELOPMENT

Cells interact with their environment via thousands of nanometric interactions. In tissues and organs, cells are located in three-dimensional microenvironments surrounded by other cells and the extracellular matrix. The ECM contains collagen and elastin, which are organized in nanostructures with specific bioactive functions that regulate cellular homeostasis. An essential stage of scaffold development is the creation of synthetic microenvironments that facilitate the formation of a three-dimensional structure to control cell behavior and promote specific cell interactions [35].

Nanotechnology has been used in several biomedical applications, including pharmaceutical transport, biological detection, disease diagnosis, clinical images resolution, and scaffold development [36]. Nanometric tissue engineering can produce biomaterials that regulate the interactions between cells and their microenvironments by the emission of molecular signals [37, 38]. The biometric and physicochemical properties of nanomaterials enable them to stimulate cell growth and regenerate injured tissue [39].

Developing nanofiber scaffolds using the electrospinning process enables the reproduction of the principal extracellular architecture and makes it easy for the cells to unite to tissue because such scaffolds have similar mechanical properties to natural structures [5]. Several authors have studied the application of nanofiber scaffolds in various processes, including the rebuilding of nerves [40] and brain tissue [41], the transport of pharmaceuticals through oral mucosa [42], and the cultivation of stem cells [6].

Production of Nanostructured Scaffolds via Electrospinning

The greatest challenge in the area of tissue engineering is the development of scaffolds that reproduce nanometric tissue architecture. The electrospinning process is the most widely adopted technique for the formation of polymer nanofibers [39], due to the repeatability of this method and the simplicity of scaling it up.

Nanostructured scaffolds obtained via electrospinning have been attracting attention due to their high porosity. They contain interlinked voids that can increase both cellular development and the connections between cells as well as nutrient diffusion, angiogenesis, and vascularization during tissue regeneration [9]. Electrospinning produces nanofibers with a diameter of between 3 nm and 1000 nm and it can be used to process several types of polymers [15, 43].

Electrospinning is carried out by applying a high voltage to a polymer solution in a process that results in nanofiber formation and lengthening due to electrostatic repulsion. The polymer solution is fed at a constant flow rate through a capillary charged with a high voltage (10 kV to 30 kV). When the electric field attains enough energy to overcome surface tension at the tip of the capillary, a "Taylor Cone" forms and the nanofibers are deposited in a stationary or rotating collector where the solvent evaporates and the nanofibers collect [44, 45].

Many parameters influence this process, including the properties of the polymers, the solvent, and the environment. Some of these parameters are viscosity, elasticity, conductivity, solution flow rate, surface tension, capillary diameter, distance between the capillary tip and the collector, polymer concentration, temperature, humidity, and air flow rate [33]. In addition, the manner in which the nanofibers are collected can influence their orientation, with the fibers being deposited either randomly or in alignment [8]. Electrospinning can be accomplished under laboratory conditions to produce a sterile product and can easily be scaled up [46]. Another advantage is the possibility of incorporating growth factors, drugs, viruses, proteins, and other properties into the nanofibers. Thus, Spirulina is a good candidate for incorporation into scaffolds [6].

THE SPIRULINA (ARTHROSPIRA) MICROALGA

Phytoplankton are aquatic photosynthetic photoautotrophs characterized by the presence of various colored pigments [47]. Photosynthetic phytoplankton include the eukaryotic algae and the prokaryotic cyanobacteria. Biotechnological processes based on phytoplankton have come to the fore due to their potential to produce a wide range of byproducts, including carbohydrates, lipids, minerals, pigments, proteins, and vitamins [48], many of which are natural products, have a high nutritional value, and are commercially important [47].

Spirulina is the common name for the product produced from pure cultures of the photosynthetic prokaryotic cyanobacteria Arthrospira, the natural habitat of which is alkaline lakes. Spirulina was originally given this name due to the spiral nature of its filaments (Figure 1(a)) and was thought to be eukaryotic algae. However, it was later found to be a prokaryotic cyanobacterium belonging to the genus Arthrospira. In South America, it has been cultivated as a food supplement since the time of the Aztecs, about 400 years ago [49].

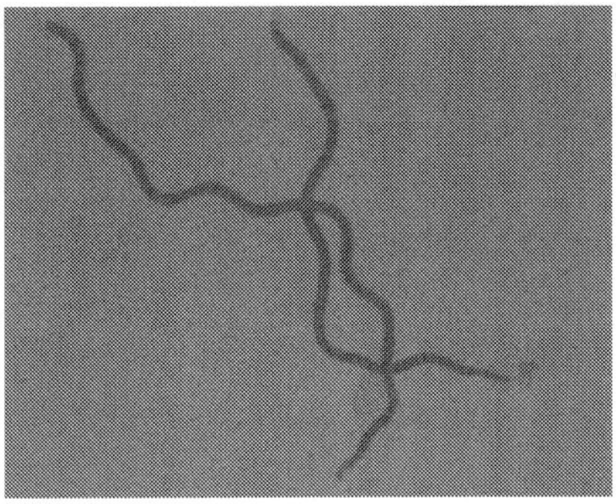

(a)

(b)

Figure 1: Spirulina LEB 18 (a) and culture at the Laboratory of Biochemical Engineering (b).

Medical and nutritional studies about Spirulina have proliferated since the 1970s, because it is a good source of high quality protein (the concentration of which can reach 70% of the biomass), vitamins (B12 and provitamin A), minerals (especially iron), phenolics, and essential fatty acids [50]. For many years, Spirulina has been investigated for application in several fields, especially for use in foods and medicine. It is thus frequently used as a nutritional supplement and is generally regarded as safe when cultivated under conditions of appropriate hygiene. Toxicological studies have demonstrated that Spirulina is safe for human consumption.

Spirulina has been produced on a large scale in several countries for application as a food supplement and a pharmaceutical product [5]. In Germany (BlueBiotech International GmbH) and the United States

(Cyanotech, Eathrise Nutritionals, and Phycobiologics) Spirulina is cultivated on a commercial scale for use as dietary supplement [2]. Clinical studies have demonstrated that Spirulina biomass has therapeutic properties and that it may be used to treat allergies [51], cancer [52], and HIV [53] and for the reduction of LDL cholesterol [54, 55]. It has also been reported to stimulate the immune system and intestinal lactobacilli, reduce hyperlipidemia and obesity, and counteract the effects of radiation, drugs, and heavy metals [48, 56]. Noninsulin dependent diabetics have shown a reduction in hypoglycemia when Spirulina was added to their diet [55].

Since 1996, the Laboratory of Biochemical Engineering (LEB) of the Federal University of Rio Grande (FURG), Brazil, has been running a research program on the cultivation of microalgae and other phytoplankton (Figure 1(b)).

Spirulina LEB 18 Biopolymers Used in Nanostructured Scaffolds

The first step in tissue reconstruction consists of the selection of the support material for the cells. During this phase, consideration must be made regarding the type of lesion being repaired along with its location in the body and the extension of the lesion. Permanent implants can cause inflammation, which, although a normal response to a foreign body, can result in more severe clinical complications such as tissue contraction [13].

Collagen-based scaffolds are currently substituted by supports produced from biodegradable polymers [57]. Biodegradable microbial polymers have potential application in the formation of nanostructured scaffolds. Such polymers include the polyhydroxyalkanoates (PHAs), a group of about 150 polymers that has been attracting medical interest. This group includes polyhydroxybutyrate (PHB), poly(3-hydroxybutyrate) (P3HB) and its copolymers, poly(4-hydroxybutyrate) (P4HB), 3-hydroxyvalerate (PHBV), and 3-hydroxyhexanoate (PHBHHx) [58]. The most studied polymer of the group is PHB, which is biodegradable, thermoplastic, and easily processed, making it a good candidate for the development of biodegradable scaffolds [59].

Human biocompatibility is one of the advantages of biopolymers compared with synthetic biodegradable polymers, and they can be used to produce scaffolds that facilitate the anchoring of implanted cells to the tissue that is being regenerated [60]. The United States Food and Drug Administration (FDA) has approved PHB for food packaging. Furthermore, because PHB is biocompatible with cells and tissues and is easily absorbed by the human body it can be used in the medical-pharmaceutical field for sutures, bone prostheses, cardiovascular grafts, orthopedic pins, and implants, as well as in tissue regeneration and repair [61]. Formulations of PHB biopolymer can be used as a matrix for the development of controlled-release medications such as hormones and other pharmaceuticals. The sodium salt of PHB can also be used as an anesthetic [62]. Neural stem cells have been produced in PHB scaffolds and have the potential to repair central nervous system lesions. PHB can also be used to regenerate bones, cartilage [58], and nervous and cardiovascular tissues [63].

PHAs are biodegradable and therefore they do not produce any toxic substances during their metabolism. In the environment, bacteria and fungi secrete an extracellular depolymerase that readily degrades polymeric PHAs into their monomers. In mammalian tissues, degradation products are absorbed through the cellular wall and metabolized [64]. The degradation rate of PHAs depends on many factors. Some factors, such as temperature, humidity, pH, and nutrient supply, are related to the environment, while other factors, such as additives composition, crystallinity, and surface area, are intrinsic to the biopolymer [65].

During PHB synthesis, two acetyl-CoA molecules are joined in a condensation reaction catalyzed by the enzyme 3-(β-ketothiolase) to form acetoacetyl-CoA. This enzyme competes for acetyl-CoA with several other metabolic pathways, such as acetate and citrate formation and fatty acid synthesis [66].

The product is reduced to 3-hydroxybutyryl-CoA in a reaction catalyzed by acetoacetyl NADPH-dependent reductase. High concentrations of NADPH and NADH inhibit the enzyme citrate synthase, which is responsible for feeding acetyl-CoA into the tricarboxylic (TCA) cycle, making acetyl-CoA available to 3-β-ketothiolase and enabling PHB to be synthesized by the polymerisation of 3-hydroxybutyryl-CoA units of by PHA synthase. The biosynthesis of PHB-HV proceeds with precursors such as acetic, itaconic, propionic, oleic, or valeric acid. Propionic acid is the valerate precursor that is most commonly employed in PHB-HV biosynthesis [66].

Polyhydroxybutyrate can be produced by prokaryotic microorganisms such as Spirulina, where it functions as carbon and energy reserve [67].

Since 2007, our team has studied PHB from microalgae. These studies investigated different genera and species of microalgae that produce this biopolymer, as well as physicochemical characterization (scanning electron microscopy, gas chromatography, thermal analysis, differential scanning calorimetry, color, and opacity), optimization of the extraction/purification process, and applications of the biopolymer in the development of biofilms, nanofibers, and nanocapsules (Figure 2).

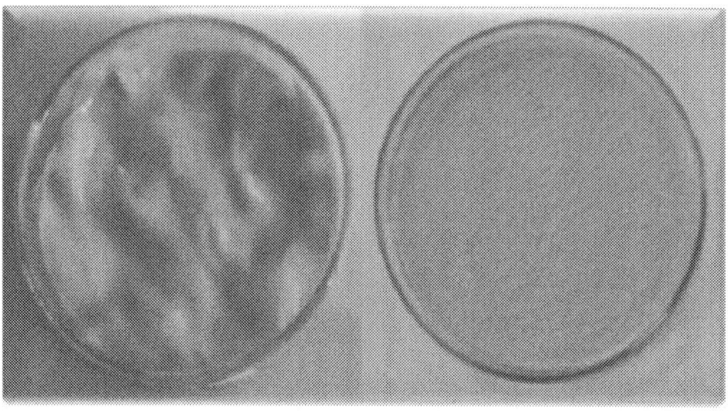

(a)

(b)

Figure 2: Polyhydroxybutyrate biopolymer produced from the biomass of Spirulina strain LEB 18. Biopolymer before and after drying (a) and scanning electron microscopy of the surface of PHB with 2,000x magnification (b).

Physicochemical and Biological Properties and Stem Cell Cultivation of the Scaffolds Made with the Incorporation of Spirulina (Arthrospira) Biomass or Biopolymers Obtained from the Microalgal Biomass

Since 2007, our team has studied the development of nanofibers produced from PLA, polyethylene oxide (PEO), and PHB extracted from LEB 18 and the incorporation of LEB 18 biomass or some of its metabolites (such as C-phycocyanin).

The formation of nanofibers via electrospinning is dependent upon the properties of the solution used and the electrical set-up. Morais et al. [5] observed that the addition of Spirulina LEB 18 biomass to nanofibers results in a strong increase in conductivity. Nanofibers with Spirulina LEB 18 biomass free of beads were produced with diameters of 107 nm (Figure 3(a)) [5].

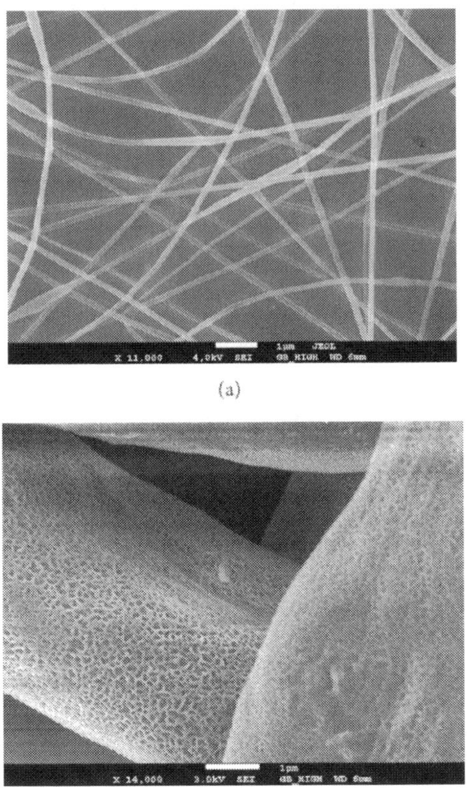

(a)

(b)

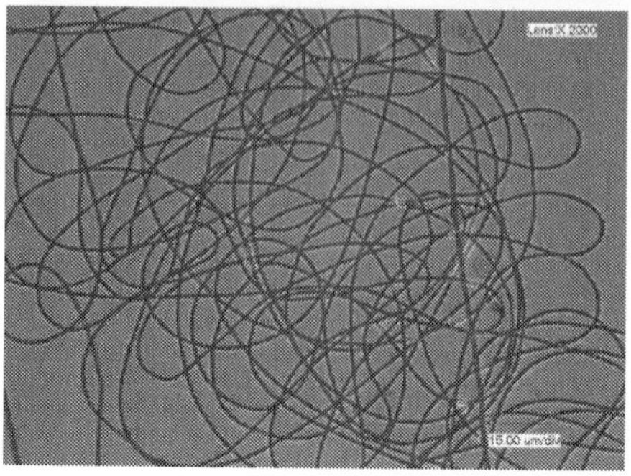

(c)

Figure 3: Nanofibers produced with 65% polyethylene oxide and 35% Spirulina LEB 18 biomass (a), nanofibers incorporating 25% LEB 18 polyhydroxybutyrate and 5% LEB 18 biomass (b), and optical image of PHB nanofibers incorporating Spirulina LEB 18 biomass. 2,000x magnification (c).

The elasticity, tensile strength, and breaking elongation of Spirulina LEB 18 PHB nanofibers were higher than those of commercial PHB samples. The general finding is that nanofibers composed of Spirulina LEB 18 PHB have surprisingly enhanced mechanical properties when compared with nanofibers composed of commercial PHB. These nanofibers had high porosity, and the cells filled the matrix structure, thus enabling the arrival of nutrients and growth factors and removal of metabolic products (data not published) (Figure 3(b)) [68].

The PHB extracted from Spirulina and commercial PHB-HV5 and PHB-HV12 were electrospun with and without the addition of sodium chloride or LEB 18 biomass. Electrospinning of 22% w/w Spirulina PHB without the addition of sodium chloride or LEB 18 biomass produced uniform nanofibers with a diameter of about 750 nm, while the addition of sodium chloride reduced the diameter to about 480 nm, and the addition of 5% w/w LEB 18 biomass reduced it to about 310 nm. It is important to note that if biomass is added to the spinning solution, PHB nanofibers can be spun with PHB concentrations as low as 7% w/w.

This produces nanofibers with markedly reduced fiber diameters, which could be of importance for membranes produced from such nanofibers. One reason for the reduced nanofiber diameter in the presence of LEB 18 biomass may be that the biomass also contained some PHB, although this cannot be the main reason because the amount of biomass was small. This phenomenon needs further research. Spinning nanofibers using lower concentrations of PHB would reduce production costs.

In another study carried out by our team, all of the electrospinning conditions used for the development of nanofibers with PHB extracted from Spirulina sp. LEB 18 were tested for commercial PHB. The commercial biopolymer did not form fibers under any of the conditions, forming only drops, while the PHB extracted from LEB 18 produced nanofibers with a diameter of 470.1 nm. The conditions that formed the smallest diameters were PHB polymer solution extracted from LEB 18 with a concentration of 20% (w/v), flow rate of 150 $\mu L \cdot h^{-1}$, capillary diameter of 0.45 mm, and voltage of 24.1 kV.

The addition of LEB 18 biomass can provide additional optical functionalization of the nanofibers and affect the transmission of light because it produces a nanofiber with a strong green color (Figure 3(c)).

In recent studies where nanofibers incorporating phycocyanin were developed, resistance to the thermal degradation of this biopigment increased when compared with the phycocyanin alone. This showed that the nanofibers produced via electrospinning may protect the added bioactive compounds (data not published) [69].

Scaffolds of poly-D,L-lactic acid (PDLLA) associated (or not) with Spirulina LEB 18 biomass (PDLLA/Sp) were developed with the aim of closely mimicking the natural ECM [6]. This resulted in nanofibers ranging from 163 to 581 nm in the PDLLA matrices and from 91 to 576 nm in the PDLLA/Sp scaffolds [6]. The physicochemical and biological properties of the nanofibers produced with PDLLA/Sp showed that these scaffolds had a high porosity and a large number of interconnected pores. They also had a greater and faster wettability when compared with the PDLLA matrix, and the cells had greater adhesion to PDLLA/Sp scaffolds than to PDLLA alone. The results of the cytotoxic assay showed there was not an increase in cell death. The degradability test showed that the PDLLA/Sp scaffolds had a rapid degradation rate (50% degraded within 60 days). Steffens et al. [6] observed that Spirulina LEB 18 biomass was released from the nanofiber while the fiber was being degraded. The authors showed that the PDLLA/Sp was capable of increasing the number of viable cells when compared with scaffolds made of PDLLA alone [6].

In another study, Steffens et al. [70] promoted the cultivation of stem cells with the PDLLA/Sp scaffold produced for testing in an animal model of skin injury. The PDLLA/Sp scaffolds were more moldable and had better adherence to the wound when compared with the PDLLA. The authors observed that PDLLA/Sp was adequate for use in animals because it supported the suture and the mechanical stress, all of the animals survived, and there were no complications related to the procedure [70].

Scaffolds made from 0–25 μg/mL of Spirulina nanofibers have been used to produce artificial tissue andSpirulina enabled the proliferation of mouse fibroblasts. No cytotoxic effects were encountered [71].

Antibacterial and anti-inflammatory effects are critical when scaffolds are used in humans. This is especially true in patients with serious burns where the external protective skin barrier has been completely lost [6]. The application of 0.1% Spirulina extract reduces the levels of the bacteria Escherichia coli and Staphylococcus aureusto insignificant levels within 30 minutes. A methanolic extract of Arthrospira platensis had higher antimicrobial activity than dichloromethane, petroleum ether or ethyl acetate extracts, and volatile antibacterial compounds [7].

Spirulina stimulates lymphocytes and other cells involved in the immune response. Phycocyanin, a blue pigment associated with the chlorophyll of this organism, exhibits antioxidant and anti-inflammatory properties (due to the inhibition of the release of histamine) [72].

Biomass containing phycocyanin increased the immunity of mice and stimulated haematopoiesis by affecting the glycoprotein hormone erythropoietin and increasing the production of white blood cells. Spirulina C-phycocyanin can eliminate free radicals because it is a cyclooxygenase-2 inhibitor that induces apoptosis in the macrophages of mouse via the activation of lipopolysaccharide- (LPS-) induced macrophage [73].

CONCLUSION

The development of nanostructured scaffolds using polyhydroxybutyrate biopolymer and the incorporation ofSpirulina biomass is a significant advance in the field of tissue engineering. This progress is exemplified by the nanofiber architecture, which reproduces the extracellular matrix while reducing tissue and organ rejection during restructuring because of the biocompatible nature of the matrix. This matrix also stimulates cell growth, better nutrient diffusion, and specific cellular interactions due to the properties of LEB 18 biomass. The use of this technology may result in the development of scaffolds that do not require tissue or organ donors.

ACKNOWLEDGMENT

The authors would like to thank the National Council of Technological and Scientific Development (CNPq) for their Scholarship.

REFERENCES

1. E. Sachlos and J. T. Czernuszka, "Making tissue engineering scaffolds work. Review: the application of solid freeform fabrication technology to the production of tissue engineering scaffolds," European Cells & Materials, vol. 30, no. 5, pp. 29–39, 2003. ·
2. J. A. C. Costa and M. G. Morais, "Microalgae for food production," in Fermentation Process Engineering in the Food Industry, C. R. Soccol, A. Pandey, and C. Larroche, Eds., p. 486, Taylor & Francis, 2013.
3. R. Fox, Spirulina Production & Potencial, Edisud, Aix-en-Provence, France, 1996.
4. Z. Khan, P. Bhadouria, and P. S. Bisen, "Nutritional and therapeutic potential of Spirulina," Current Pharmaceutical Biotechnology, vol. 6, no. 5, pp. 373–379, 2005.
5. M.G.deMorais,C.Stillings,R.Derschetal.,"Preparationofnanofibers containing the microalga Spirulina (Arthrospira)," Bioresource Technology, vol. 101, no. 8, pp. 2872–2876, 2010. ··
6. D. Steffens, M. Lersch, A. Rosa et al., "A new biomaterial of nanofibers with the microalga spirulina as scaffolds to cultivate with stem cells for use in tissue engineering," Journal of Biomedical Nanotechnology, vol. 9, no. 4, pp. 710–718, 2013.
7. M. A. Qureshi, N. T. Kidd, and R. A. Ali, "Spirulina platensis extract enhances chicken macrophage functions after in vitro exposure," Journal of Nutritional Immunology, vol. 3, no. 4, pp. 35–45, 1995. ·

8. S. A. Sell, P. S. Wolfe, K. Garg, J. M. McCool, I. A. Rodriguez, and G. L. Bowlin, "The use of natural polymers in tissue engineering: a focus on electrospun extracellular matrix analogues," Polymers, vol. 2, no. 4, pp. 522–553, 2010.
9. J. Ramier, T. Bouderlique, O. Stoilova et al., "Biocomposite scaffolds based on electrospun poly(3-hydroxybutyrate) nanofibers and electrosprayed hydroxyapatite nanoparticles for bone tissue engineering applications," Materials Science and Engineering: C, vol. 38, pp. 161–169, 2014. ·
10. K. Sudesh, H. Abe, and Y. Doi, "Synthesis, structure and properties of polyhydroxyalkanoates: biological polyesters," Progress in Polymer Science, vol. 25, no. 10, pp. 1503–1555, 2000.
11. M. Jau, S. Yew, P. S. Y. Toh et al., "Biosynthesis and mobilization of poly(3-hydroxybutyrate) [P(3HB)] by Spirulina platensis," International Journal of Biological Macromolecules, vol. 36, no. 3, pp. 144–151, 2005.
12. B. P. Chan and K. W. Leong, "Scaffolding in tissue engineering: general approaches and tissue-specific considerations," European Spine Journal, vol. 17, no. 4, pp. S467–S479, 2008.
13. S. H. Barbanti, C. A. C. Zavaglia, and E. A. R. Duek, "Polímeros Bioreabsorvíveis na Engenharia de Tecidos," Polímeros: Ciência e Tecnologia, vol. 15, no. 1, pp. 13–21, 2005.
14. H.-Y. Cheung, K.-T. Lau, T.-P. Lu, and D. Hui, "A critical review on polymer-based bio-engineered materials for scaffold development," Composites B, vol. 38, no. 3, pp. 291–300, 2007. ··
15. R. Vasita and D. S. Katti, "Nanofibers and their applications in tissue engineering," International Journal of Nanomedicine, vol. 1, no. 1, pp. 15–30, 2006.
16. N. Pallua and C. V. Suscheck, Tissue Engineering-From Lab to Clinic, Springer, 2010.
17. I. O. Smith, X. H. Liu, L. A. Smith, and P. X. Ma, "Nanostructured polymer scaffolds for tissue engineering and regenerative medicine," Wiley Interdisciplinary Reviews: Nanomedicine and Nanobiotechnology, vol. 1, no. 2, pp. 226–236, 2009.
18. I. Armentano, M. Dottori, E. Fortunati, S. Mattioli, and J. M. Kenny, "Biodegradable polymer matrix nanocomposites for tissue

engineering: a review," Polymer Degradation and Stability, vol. 95, no. 11, pp. 2126–2146, 2010.

19. J. M. Holzwarth and P. X. Ma, "Biomimetic nanofibrous scaffolds for bone tissue engineering," Biomaterials, vol. 32, no. 36, pp. 9622–9629, 2011.
20. Z. Xiao, J. Blonder, M. Zhou, and T. D. Veenstra, "Proteomic analysis of extracellular matrix and vesicles," Journal of Proteomics, vol. 72, no. 1, pp. 34–45, 2009.
21. C. Frantz, K. M. Stewart, and V. M. Weaver, "The extracellular matrix at a glance," Journal of Cell Science, vol. 123, no. 24, pp. 4195–4200, 2010.
22. H. Järveläinen, A. Sainio, M. Koulu, T. N. Wight, and R. Penttinen, "Extracellular matrix molecules: potential targets in pharmacotherapy," Pharmacological Reviews, vol. 61, no. 2, pp. 198–223, 2009.
23. F. Bolland, S. Korossis, S. Wilshaw et al., "Development and characterisation of a full-thickness acellular porcine bladder matrix for tissue engineering," Biomaterials, vol. 28, no. 6, pp. 1061–1070, 2007.
24. R. Chen, H. Ho, Y. Tsai, and M. Sheu, "Process development of an acellular dermal matrix (ADM) for biomedical applications," Biomaterials, vol. 25, no. 13, pp. 2679–2686, 2004.
25. T. Hoshiba, H. Lu, N. Kawazoe, and G. Chen, "Decellularized matrices for tissue engineering," Expert Opinion on Biological Therapy, vol. 10, no. 12, pp. 1717–1728, 2010.
26. H. C. Ott, T. S. Matthiesen, S. K. Goh et al., "Perfusion-decellularized matrix: using nature's platform to engineer a bioartificial heart," Nature Medicine, vol. 14, no. 2, pp. 213–221, 2008.
27. T. H. Petersen, E. A. Calle, L. Zhao et al., "Tissue-engineered lungs for in vivo implantation," Science, vol. 329, no. 5991, pp. 538–541, 2010.
28. B. E. Uygun, A. Soto-Gutierrez, H. Yagi et al., "Organ reengineering through development of a transplantable recellularized liver graft using decellularized liver matrix," Nature Medicine, vol. 16, no. 7, pp. 814–820, 2010.

29. P. W. Whitlock, T. L. Smith, G. G. Poehling, J. S. Shilt, and M. van Dyke, "A naturally derived, cytocompatible, and architecturally optimized scaffold for tendon and ligament regeneration," Biomaterials, vol. 28, no. 29, pp. 4321–4329, 2007.
30. H. Lu, T. Hoshiba, N. Kawazoe, and G. Chen, "Autologous extracellular matrix scaffolds for tissue engineering," Biomaterials, vol. 32, no. 10, pp. 2489–2499, 2011.
31. H. Lu, T. Hoshiba, N. Kawazoe, I. Koda, M. Song, and G. Chen, "Cultured cell-derived extracellular matrix scaffolds for tissue engineering," Biomaterials, vol. 32, no. 36, pp. 9658–9666, 2011.
32. R. L. Órefice, M. M. Pereira, and H. S. Mansur, Biomateriais: fundamentos e aplicações, Cultura Médica, Guanabara Koogan, Rio de Janeiro, Brazil, 1st edition, 2012.
33. S. Bose, M. Roy, and A. Bandyopadhyay, "Recent advances in bone tissue engineering scaffolds," Trends in Biotechnology, vol. 30, no. 10, pp. 546–554, 2012.
34. J. J. Song and H. C. Ott, "Organ engineering based on decellularized matrix scaffolds," Trends in Molecular Medicine, vol. 17, no. 8, pp. 424–432, 2011.
35. C. Cunha, S. Panseri, and S. Antonini, "Emerging nanotechnology approaches in tissue engineering for peripheral nerve regeneration," Nanomedicine: Nanotechnology, Biology, and Medicine, vol. 7, no. 1, pp. 50–59, 2011.
36. N. Barkalina, C. Charalambous, C. Jones, and K. Coward, "Nanotechnology in reproductive medicine: emerging applications of nanomaterials," Nanomedicine: Nanotechnology, Biology and Medicine, vol. 10, no. 5, pp. 921–938, 2014.
37. I. Wheeldon, A. Farhadi, A. G. Bick, E. Jabbari, and A. Khademhosseini, "Nanoscale tissue engineering: spatial control over cell-materials interactions," Nanotechnology, vol. 22, no. 21, Article ID 212001, 2011.
38. B. D. Fahlman, Materials Chemistry, Springer, Dordrecht, The Netherlands, 2007.
39. M. Zhang and J. Gunn, "Polyblend nanofibers for biomedical applications: perspectives and challenges," Trends in Biotechnology, vol. 28, no. 4, pp. 189–197, 2010.

40. X. Zhan, M. Gao, Y. Jiang et al., "Nanofiber scaffolds facilitate functional regeneration of peripheral nerve injury," Nanomedicine: Nanotechnology, Biology and Medicine, vol. 9, pp. 305–315, 2013.

41. G. K. K. Leung, Y. C. Wang, and W. Wu, "Peptide nanofiber scaffold for brain tissue reconstruction," Methods in Enzymology, vol. 508, pp. 177–190, 2012.

42. D. C. Aduba Jr., J. A. Hammer, Q. Yuan, W. Andrew Yeudall, G. L. Bowlin, and H. Yang, "Semi-interpenetrating network (sIPN) gelatin nanofiber scaffolds for oral mucosal drug delivery," Acta Biomaterialia, vol. 9, no. 5, pp. 6576–6584, 2013.

43. S. Ramakristna, W. Teo, T. Lim, and Z. Ma, An Introduction to Electrospinning and Nanofibers, World Scientific, River Edge, NJ, USA, 2005.

44. T. J. Sill and H. A. von Recum, "Electrospinning: applications in drug delivery and tissue engineering," Biomaterials, vol. 29, no. 13, pp. 1989–2006, 2008.

45. B. M. Baker, R. P. Shah, A. M. Silverstein, J. L. Esterhai, J. A. Burdick, and R. L. Mauck, "Sacrificial nanofibrous composites provide instruction without impediment and enable functional tissue formation," Proceedings of the National Academy of Sciences of the United States of America, vol. 109, no. 35, pp. 14176–14181, 2012.

46. C. P. Barnes, S. A. Sell, E. D. Boland, D. G. Simpson, and G. L. Bowlin, "Nanofiber technology: designing the next generation of tissue engineering scaffolds," Advanced Drug Delivery Reviews, vol. 59, no. 14, pp. 1413–1433, 2007.

47. R. B. Derner, S. Ohse, M. Villela, S. M. de Carvalho, and R. Fett, "Microalgae, products and applications," Ciencia Rural, vol. 36, no. 6, pp. 1959–1967, 2006.

48. J. A. Vieira Costa, L. M. Colla, and P. F. Duarte Filho, "Improving Spirulina platensis biomass yield using a fed-batch process," Bioresource Technology, vol. 92, no. 3, pp. 237–241, 2004.

49. A. Vonshak, Spirulina platensis (Athrospira) Physiology, Cell-Biology and Biotechnoloby, Taylor & Francis, New York, NY, USA, 1997.

50. J. A. V. Costa, M. G. Morais, and M. R. Andrade, "Microalgal engineering: the metabolic products and the bioprocess," in Microalgae Biotechnology, Microbiology and Energy—Marine Biology, M. N. Johansen, Ed., chapter 4, p. 475, Nova Science, 2012.
51. T. K. Mao, J. van de Water, and M. E. Gershwin, "Effects of a Spirulina-based dietary supplement on cytokine production from allergic rhinitis patients," Journal of Medicinal Food, vol. 8, no. 1, pp. 27–30, 2005.
52. B. Mathew, R. Sankaranarayanan, P. P. Nair et al., "Evaluation of chemoprevention of oral cancer with Spirulina fusiformis," Nutrition and Cancer, vol. 24, no. 2, pp. 197–202, 1995.
53. T. Hirahashi, M. Matsumoto, K. Hazeki, Y. Saeki, M. Ui, and T. Seya, "Activation of the human innate immune system by Spirulina: augmentation of interferon production and NK cytotoxicity by oral administration of hot water extract of Spirulina platensis," International Immunopharmacology, vol. 2, no. 4, pp. 423–434, 2002.
54. N. Nakaya, Y. Homa, and Y. Goto, "Cholesterol lowering effect of Spirulina," Atherosclerosis, vol. 37, no. 6, pp. 1329–1337, 1988.
55. U. V. Mani, S. Desai, and U. Iyer, "Studies on the long-term effect of spirulina supplementation on serum lipid profile and glycated proteins in NIDDM patients," Journal of Nutraceuticals, Functional and Medical Foods, vol. 2, no. 3, pp. 25–32, 2000.
56. C. Jiménez, B. R. Cossío, and F. X. Niell, "The feasibility of industrial production of Spirulina(Arthrospira) in southern Spain," Aquaculture, vol. 221, pp. 331–345, 2003.
57. M. Okamoto and B. John, "Synthetic biopolymer nanocomposites for tissue engineering scaffolds," Progress in Polymer Science, vol. 38, pp. 1487–1503, 2013.
58. D. Puppi, F. Chiellini, A. M. Piras, and E. Chiellini, "Polymeric materials for bone and cartilage repair," Progress in Polymer Science, vol. 35, no. 4, pp. 403–440, 2010.
59. T. Freier, C. Kunze, C. Nischan et al., "In vitro and in vivo degradation studies for development of a biodegradable patch based on poly(3-

hydroxybutyrate)," Biomaterials, vol. 23, no. 13, pp. 2649–2657, 2002.

60. J. M. B. T. Cavalheiro, R. S. Raposo, M. C. M. D. de Almeida et al., "Effect of cultivation parameters on the production of poly(3-hydroxybutyrate-co-4-hydroxybutyrate) and poly(3-hydroxybutyrate-4-hydroxybutyrate-3-hydroxyvalerate) by Cupriavidus necator using waste glycerol," Bioresource Technology, vol. 111, pp. 391–397, 2012.
61. A. Shrivastav, S. K. Mishra, and S. Mishra, "Polyhydroxyalkanoate (PHA) synthesis by Spirulina subsalsafrom Gujarat coast of India," International Journal of Biological Macromolecules, vol. 46, no. 2, pp. 255–260, 2010.
62. A. Althuri, J. Mathew, R. Sindhu, R. Banerjee, A. Pandey, and P. Binod, "Microbial synthesis of poly-3-hydroxybutyrate and its application as targeted drug delivery vehicle," Bioresource Technology, vol. 145, pp. 290–296, 2013.
63. G. Adamus, W. Sikorska, H. Janeczek, M. Kwiecień, M. Sobota, and M. Kowalczuk, "Novel block copolymers of atactic PHB with natural PHA for cardiovascular engineering: synthesis and characterization," European Polymer Journal, vol. 48, no. 3, pp. 621–631, 2012.
64. E. Rudnik, "Biodegradability testing of compostable polymer materials," in Handbook of Biopolymers and Biodegradable Plastics: Properties, Processing and Applications, pp. 213–263, 2013.
65. S. K. Sharma and A. A. Mudhoo, Handbook of Applied Biopolymer Technology: Synthesis, Degradation and Applications, 2011.
66. S. Khanna and A. K. Srivastava, "Recent advances in microbial polyhydroxyalkanoates," Process Biochemistry, vol. 40, no. 2, pp. 607–619, 2005.
67. L. Sharma, A. Kumar Singh, B. Panda, and N. Mallick, "Process optimization for poly-β-hydroxybutyrate production in a nitrogen fixing cyanobacterium, Nostoc muscorum using response surface methodology," Bioresource Technology, vol. 98, no. 5, pp. 987–993, 2007.
68. M. G. Morais, "Bioengenharia microalgal na utilização de gás de combustão e extração de biopolímeros para desenvolvimento de

nanofibras," Programa de Pós Graduação em Engenharia e Ciência de Alimentos, Rio Grande, 2009.

69. F. S. Figueira, Purificação de C-ficocianina e sua incorporação em nanofibras [Tese de doutorado], Programa de Pós Graduação em Engenharia e Ciência de Alimentos, Rio Grande, Brazil, 2014.
70. D. Steffens, D. Leonardi, P. R. L. Soster et al., "Development of a new nanofiber scaffold for use with stem cells in a third degree burn animal model," BURNS, 2014. ·
71. S. Jung, D. S. Kim, J. H. Ju, and H. S. Shin, "Assessment of Spirulina-PCL nanofiber for the regeneration of dermal fibroblast layers," In Vitro Cellular and Developmental Biology, vol. 49, no. 1, pp. 27–33, 2013.
72. V. B. Bhat and K. M. Madyastha, "Scavenging of peroxynitrite by phycocyanin and phycocyanobilin from Spirulina platensis: protection against oxidative damage to DNA," Biochemical and Biophysical Research Communications, vol. 285, no. 2, pp. 262–266, 2001.
73. M. C. Reddy, J. Subhashini, S. V. K. Mahipal et al., "C-Phycocyanin, a selective cyclooxygenase-2 inhibitor, induces apoptosis in lipopolysaccharide-stimulated RAW 264.7 macrophages," Biochemical and Biophysical Research Communications, vol. 304, no. 2, pp. 385–392, 2003.

CHAPTER 3

NATURAL PRODUCTS: A MINEFIELD OF BIOMATERIALS

Oladeji O. Ige[1] Lasisi E. Umoru[1] and Sunday Aribo[2]

[1]Department of Materials Science and Engineering, Obafemi Awolowo University, Ile-Ife 220282, Nigeria

[2]Department of Metallurgical and Materials Engineering, Federal University of Technology, Akure 340252, Nigeria

ABSTRACT

The development of natural biomaterials is not regarded as a new area of science, but has existed for centuries. The use of natural products as a biomaterial is currently undergoing a renaissance in the biomedical field. The major limitations of natural biomaterials are due to the immunogenic response that can occur following implantation and the lot-to-lot variability in molecular structure associated with animal sourcing.

The chemical stability and biocompatibility of natural products in the body greatly accounts for their utilization in recent times. The paper succinctly defines biomaterials in terms of natural products and also that natural products as materials in biomedical fields are considerably versatile and promising. The various types of natural products and forms of biomaterials are highlighted. Three main areas of applications of natural products as materials in medicine are described, namely, wound management products, drug delivery systems, and tissue engineering. This paper presents a brief history of natural products as biomaterials, various types of natural biomaterials, properties, demand and economic importance, and the area of application of natural biomaterials in recent times.

INTRODUCTION

A biomaterial is regarded as any nondrug material that can be used to treat, enhance or replace any tissue, organ, or function in an organism [1]. While the definition of biomaterial was reframed as a nondrug substance suitable for inclusion in systems which augment or replace the function of bodily tissues or organs [2]. This definition explicitly described biomaterial in relation to drugs and as such, there is a need to clarify the impression that natural products are synonymous with drugs. The definition implies that natural products can be applied as biomaterials by eliminating the ambiguity always associated with natural products as drugs. It must be emphasized that this definition is not regarded as one of the most popular and is not often cited as this one which defines biomaterial as a nonviable material that intends to interact with physiological environment [3]. However, in this study, the following definition will be adopted: biomaterial can be defined as any substance (other than a drug) or combination of substances synthetic or natural in origin, which can be used any time, as a whole or as a part of a system which treats, augments, or replaces any tissue, organ or function of the body [4]. It must be noted that in this study the substances are natural in origin.

Economy

The field of biomaterials working under biological constraints is rapidly expanding. It is considered that this domain represents 2-3% of the overall health expenses in developed countries [5]. There are over 13 million medical device implants annually in USA. However, all these devices are prone to incomplete or nonspecific cellular healing which may lead to the ultimate failure of the device [6].

History

The new interest in natural biomaterials could really be considered as a renaissance. Historians have traced the use of sutures made from animal sinew to ancient Egypt, while some say they were used even earlier [7]. The ancient Egyptians were regarded as the first to use biomedical materials, employing coconut shells to repair injured skulls; wood and ivory as false teeth and these dated as far back as 3000 BC [8]. Some of the earliest biomaterial applications were as far back as ancient Phoenicia where loose teeth were bound together with gold wires, tying artificial teeth to neighbouring teeth [2].

As early as the first century AD in both Greece and India, physicians were using natural biomaterials while performing plastic surgery to repair mutilations from battle and punishment. There are even accounts of some of these physicians treating disemboweled soldiers to good effect [7]. The earliest recorded hip replacement surgery took place in Germany in 1891 AD, and in this instance ivory was used. This material is favoured for its relatively inexpensive cost although it also turned out to have useful biomechanical bonding qualities, which makes it well suited to work with human body tissue [8]. From as early as a century ago artificial, materials and devices have been developed to a point where they can replace various components of the human body. These materials are capable of being in contact with body fluids and tissues for prolonged periods of time, whilst eliciting little if any adverse reactions occur [9].

In the early 1900's, bone plates were successfully implemented to stabilize bone fractures and to accelerate their healing. While by the time 1950's and 60's, blood vessel replacement was in clinical trials artificial heart valves and hip joint were in development [2]. It is believed that by knowing the history of the development of biomaterials we can have a better perspective of how the state of the art is progressing. The big question with biomaterials is How will this product perform better than those available today? It is all too easy when performing research in today's compartmentalized and specialised society to forge the bigger picture. In the past, the research approach to biomaterials was spurred on by innovative and creative doctors who used whatever they thought might work to fix a problem sometimes with wonderful results but with extremely high risk. Today's researchers are much more focused into improving medicine methodically and with as little risk to patients as possible. The road has been long and treacherous; nature has given us many amazing remedies put together by millions of years of evolution [10].

Theoretically, any material natural or man made can be a biomaterial as long as it serves the stated medical and surgical purposes. The development of biomaterials is not a new area. It encompasses elements of medicine, biology, chemistry, tissue engineering, and materials science. Nevertheless, the demand for biocompatible, biodegradable, and bioresorbable materials has increased dramatically during the last decade. An ideal biomaterial is one that is nonimmunogenic, biocompatible, and biodegradable, which can be functionalized with bioactive proteins and chemicals. In particular, biodegradability is one of the essential properties of the biomaterials [11]. It must be emphasized that the key factors in biomaterial usage are its biocompatibility, biofunctionality, and availability to a lesser extent.

Biomaterials are ultimately intended for implantation in or on the human body. They can be designed with a range of properties that are capable of either promoting or inhibiting specific host cell and tissue responses [12]. Biomaterials also refer to biologically-derived materials used for their structural rather than biological properties, for example, collagen (a protein found in the skin, connective tissues, and bone) as a cosmetics ingredient. Also carbohydrates (biotechnologically modified) are being used as lubricants for biomedical applications and as bulking agents in the food industry [1]. Not until recent times have naturally derived biomaterials been explored as facilitators and promoters of healing and regeneration. Today, biomaterials of all types are being used for everything from wound dressing to tendon and ligament repair. Extensive experimentation has been undertaken to identify the composition, mechanical properties, and in vivo response of naturally derived biomaterials. The choice of biomaterials depends on the type of procedure being performed, the severity of the patient's condition, and the surgeon's preference. To be successful, the implant should effectively repair the defect it covers without eliciting an adverse tissue reaction while maintaining mechanical and biological integrity for a desired amount of time from a few weeks to several years. The prime reason biomaterials have come about is to provide a remedy for surgical problems [10].

Biomaterials research is one of the most important fields of modern medicine. Biomaterials are used in organ implants, wound healings, drug delivery, and so forth. It was suggested that for many reasons natural biomaterials are the most preferred ones; they are biodegradable, biocompatible, and nontoxic [13]. Biomaterials can be divided into four major classes of materials, namely; polymers, metals, ceramics (including carbons, glass-ceramics, and glasses), and natural materials (including those from both plants and animals). Composite materials are the ones which are comprised of two or more different classes of materials and they are regarded as the fifth class of biomaterials [14]. It is observed that metals rarely occur in nature as a single entity; hence, they are synthesized from their compound ores such as oxides, sulphides, and carbonates with the exception of rare metals like gold, and platinum. Natural metals have found application as dental materials in ancient times. Recently, the applications of metals as natural biomaterials are not documented. There is an international resurgence of interest in natural products as a source of novel bioactive substances for the development of novel drugs and therapies, the same trend is observed in the application of natural products as materials in medicine. By 2020, Ireland will have a leading capacity in the utilization of natural biomaterials and nutraceuticals [15].

There are several materials derived from the animal or plant world being considered for use as biomaterials and they are called natural biomaterials. One of the advantages of using natural materials for implants is that they are similar to materials, and are familiar to the body systems. In this regard, the field of biomimetics (or mimicking nature) is growing. However, natural materials can be subjected to problems of immunogenicity. Another problem faced by these materials, essentially natural polymers, is their tendency to denature or decompose at temperatures below their melting point. This severely limits their fabrication into implants of different sizes and shapes. Natural materials do not usually offer the problems of toxicity often faced by synthetic materials. Also, they may carry specific protein binding sites and other biochemical signals that may assist in tissue healing or integration [16].

This paper intends to highlight the different types of natural products that are currently being employed as biomaterials, state their advantages and disadvantages, and also to summarize the areas of application in the medical industry.

NATURAL PRODUCTS

Biological structures have always been a source of inspiration for solving technical challenges in architecture, mechanical engineering, or materials science. Nature has developed—with comparatively few base substances, mainly polymers and minerals—a range of materials with remarkable functional properties [17]. Natural was defined as something that is present in or produced by nature and not artificial or man made, and most often the definition is assumed to mean something good or pure [18].

The term natural products is quite commonly understood to refer to herbs, herbal concoctions, dietary supplements, traditional medicine, or alternative medicine [19]. It must be stated that while the stories of herbs and drugs are very much intertwined, it needs to be fully appreciated that the use of herbs as natural product therapy is different from their use on the platform of drug discovery and further development [20]. Natural products are regarded as chemicals, but they are not just accidents or products of convenience of nature. More than likely, they are a natural expression of the increase in complexity of organisms [21]. There are several definitions of natural products, and the common trend is that a natural product is a chemical compound or substance produced by a living organism. The living organism is regarded to be found in nature, and that they have pharmacological or biological activity for use in pharmaceutical drug discovery and drug design. A natural product can be considered as such, even if it can be prepared by total synthesis. It must be recognized that not all natural products can be fully synthesized, and many natural products have very complex structures that are too difficult and expensive to synthesize on an industrial scale. Such compounds can only be harvested from natural sources—a process which can be tedious, time-consuming, and expensive as well as being wasteful of the natural resource [22, 23]. It will be realised that natural products are regarded as only beneficial to the pharmaceutical industry at the expense of others like agriculture, food and even the chemical industry.

Hence, there is a need to redefine natural products, and according to the biology dictionary definition, natural products depend on the industry; medicine and pharmaceuticals; healthcare and nutritional supplement; agrochemical; food and flavouring. In this study, natural products are regarded as chemical substances that are produced by living organisms with the intent of their application in/as biomedical materials.

It must be noted that interest in natural sources to provide treatments for pain, palliatives, or curatives for a variety of maladies or recreational use dates back to the earliest parts of history. Also, several sources of information on natural products such as pharmaceutical products are well recorded and documented [20]. There are numerous studies on the applications of natural products as biomaterials, but a comprehensive summary has not been reported. Hence this paper intends to give a comprehensive review of natural product usage as biomedical materials.

Natural products can come from anywhere; generally, they are either of prebiotic origin or originate from microbes, plants/animal sources [24]. Suffice it to say that natural products can come from any point or level on the phylogenetic tree. When searching for natural products, one should never feel that a form of life is too low, simple or grotesque to provide a compound of interest. However, before one goes marching out into the woods, sailing out into the sea, climbing the highest mountains, or descending into the deepest caves, it is appropriate to perform a little bit of research, and hence a visit to the library becomes the first step in any search for a natural product. Ideally, it is important to know the history, folklore, origin of use, source, chemical structure, availability, and method of preparation, pharmacology, toxicology, and therapeutics of any natural products [20]. It is observed that it can be challenging to obtain information from practitioners of traditional medicine—which is always well versed in natural products—unless a genuine long-term relationship is made. Natural products may be extracted from tissues of terrestrial plants, marine organisms, or microorganism fermentation broths. A crude extract from any one of these sources typically contains novel, structurally diverse chemical compounds. The sources of natural products are as outlined after which a brief introduction of biomaterials will come up in the next section.

It is noted that several opportunities abound in nature that can provide natural products, and these opportunities can present themselves from almost any niche of nature and most likely some that have not even yet been discovered.

Microorganisms have proven to be an excellent source of natural products including polyketide and peptide antibiotics as well as classes of other biological active compounds [25]. It is noteworthy that some of these compounds when originally discovered failed in their development and their original uses as either antibiotics or as agricultural fungicides [20]. Microbes can be of any sources be it aerial, terrestrial, or even marine. It has been well reported in a wide range of applications in drug synthesis and design. Also, it is used in the biomedical industry as materials for wound management and drug delivery system among others. The diversity of microorganisms is of a staggering quantity, and only an extremely small proportion of bacteria and fungi have been examined for the production of potentially useful secondary metabolites. Bacteria, smuts, nests, yeasts, moulds, fungi, and many other forms of what we consider to be primitive life can be very useful [20].

People most commonly think of plants first when talking about natural products, but trees and shrubs can also provide excellent sources of biomaterials. Plants produce a variety of different types of compounds including biologically active proteins. Some of these types of compounds are even shared with other organisms, and they include such chemical families as lectins, defensins, cyclotides, and ribosome-inactivating proteins [25]. Ribosome-inactivating proteins are a group of proteins exhibiting a wide spectrum of biological activities, including a ribonucleolytic activity for which the group is named. Plant antimicrobial peptides comprise another large group of biologically active compounds. This group of compounds can be further subdivided into thionins, defensins, cyclotides, and lectins. Thionins are small proteins that selectively form disulfides bridges with other proteins or form ion channels in membranes. This ability to make membranes more permeable suggests the potential for antimicrobial activity. Defensins are cysteine-rich peptides that also permeabilize membranes but appear to be very specific in their activity. Finally, lectins are proteins that have a noncatalytic domain that binds reversibly to specific carbohydrates; this activity encompasses potentially a wide spectrum of biological activities. Plants have been, over time, an extremely popular source of natural products [26–29]. Compounds isolated and identified from this source will undoubtedly continue to make strong contributions. Plants as biomedical materials have found applications in wound management, drug delivery systems, and medical fibres and textiles.

Various polypeptides of interest have been isolated from the venom of arachnids and anthropods that prey on insects [25]. Insect peptides have been the subject of research into the immune defense of insects but have not yet been investigated for effects and potential benefit in humans. Compounds from this peptide group include such sources as the termite (Pseudacanthotermes Spiniger), the mosquito (Anopheles gambiae), the moth (Heliothis virescens) and the beetle (Oryctes rhinoceros) [20]. Insect derived natural products offer another strong potential avenue for the development of future biomaterials. However, silk produced by silkworm, Bombyx mori, has excellent properties such as biocompatibility, biodegradation, nontoxicity, and adsorption properties. It has been commercially used as a biomaterial suture for decade and its application also includes wound management, enzyme immobilization matrices, vascular prostheses, structural implants among others [11, 30–35].

Animals, whether highly developed or poorly developed, whether they live on land, in the sea, or in the air can be excellent sources of natural products [20]. Research into a variety of antimicrobial peptides, such as megamins, defensins, cathelicidins, and protegrins generated by vertebrates has, over recent time, become popular. Cathelicidins-type peptides are a broad range of antimicrobial proteins that have been isolated from rabbits, mice, sheep, and humans. Some compounds have been the subject of substantial research, and they offer potential opportunities in the areas of cardiovascular function, immune and central nervous system functions. A typical application is in the area of tissue engineering and these include the use of tissue replacement from animals, dead corpses, and the man himself. These methods are known as xenograft, autograft, and allograft respectively.

The first discovery of a marine-based biologically active compound of interest was really quite by accident approximately 10 years after the end of World War II [36]. The marine environment, arguably the original source of all life, is a rich source of bioactive compounds [37–43]. More than 70% of our planet's surface is covered by oceans, and some experts feel that the potentially available biodiversity on the deep sea floor or coral reefs are greater than those existing in the rainforests [44]. The search for new biomedicals from marine organisms resulted in the isolation of more or less 10,000 metabolites with a broad spectrum of biological activities [45, 46]. There are wide applications of natural products obtained from the marine world such as biomedical materials. A typical example is the successful use of natural coral as a bone graft substitute in tissue engineering.

BIOCERAMICS

It has been accepted that no foreign material placed within a living body is completely compatible. The only substances that conform are those manufactured by the body itself (autogeneous) and any other substances are recognized as foreign, they initiate some type of reaction (host tissue response). Bioceramics are classified according to their bioactivity, and they are, namely, bioinert (alumina dental implant), bioactive [hydroxyapatite, $Ca(PO_4)_2$], surface active (bioglass), and bioresorbable [tricalcium phosphate implant, $Ca_3(PO_4)_2$] [2].

In the early 70’s, bioceramics were employed to perform singular biologically inert roles, such as to provide parts for bone replacement. The demands of bioceramics have changed from maintaining an essentially physical function without eliciting a host response to providing a more integrated interaction with the host. Bioceramics potentially can be used as body interactive materials, helping the body to heal, or promoting regeneration of tissues, thus restoring physiological functions. Ultimately, the field of bioceramics is fundamental to advances in the performance and function of medical devices and is a critical part of medicine and surgery. The correlations between material properties and biological performance will be useful in the design of improved bioceramics, particularly to overcome the problems of implant rejection and related infection [2]. Bioceramics have been proposed for biomedical applications such as dental restorations, middle ear reconstruction, rebuilding of facial and cranial bones, and filling of bony defects, to name a few [2, 47]. It was reported that commercially available porous bioceramics originate from two sources, namely; hydroxyapatite (e.g., Pro osteon) or bone (e.g., Endobon) [48].

In this paper, bioceramics that are sourced from natural products are to be considered and there are potentials in shells, coral, bone and soil/ mineral.

Bone

Bone consists of a compact tissue type (cortical bone) and a spongy porous material (trabecullar bone). In both tissue types, the basic building block is a bone lamella, typically about 5 μm thick. In cortical bone, lamellae form laminated cylindrical composite structures built around blood vessels, which are denoted as secondary osteons [49]. The mechanical performance of bone, often coined bone quality, does not only depend on the shape and the amount of the bone (as estimated by the bone mineral density, BMD), but also on its architecture and on the quality of the bone material [50, 51].

Bone can be utilized as a biomedical material when it is used to substitute a damaged part. The grafting involving the use of the patient's own bone in replacing the fractured part is known as autografting, whileallografting is the use of another human being's bone and often it involved the utilization of cadaver. Lastly, when animal bones such as rabbit, pig, dog among others are used as replacement this is known as xenografting [14].

Bone typically consists by weight of 25% water, 15% organic materials and 60% mineral phases. The mineral phase consists primarily of calcium and phosphates ions, with traces of magnesium, carbonate, hydroxyl, chloride, fluoride, and citrate ions. Hence, calcium phosphates occur naturally in the body, but they occur also within nature as mineral rocks, and certain compounds can be synthesized in the laboratory. Also, certain compounds are useful for implantation in the body since both their solubility and speed of hydrolysis increase with decreasing calcium-to-phosphorus ratio [48]. Table 1 summarizes the mineral name, chemical name, and compositions of various phases of calcium phosphates. Driessens stated that those compounds with a Ca/P ratio of less than 1 : 1 are not suitable for biological application [52].

Table 1: Chemical composition of some calcium phosphates minerals

Ca:P	Mineral name	Formula	Chemical use
1.0	Monetite	$CaHPO_4$	Dicalcium phosphate (DCP)
1.0	Brushite	$CaHPO_4 \cdot 2H_2O$	Dicalcium phosphate dihydrate (DCPD)
1.33	—	$Ca_8(HPO_4)_2(PO_4)_4 \cdot 5H_2O$	Octacalcium phosphate (OCP)
1.43	Whitlockisite	$Ca_{10}(HPO_4)(PO_4)_6$	—
1.5	—	$Ca_3(PO_4)_2$	Tricalcium phosphate (TCP)
1.67	Hydroxyapatite	$Ca_{10}(PO_4)_6(OH)_2$	—
2.0	—	$Ca_4P_2O_9$	Tetracalcium phosphate

The main crystalline component of the mineral phase of bone is a calcium-deficient carbonate hydroxyapatite.

Hydroxyapatite

This has been a well-known biomaterial since the 1960's, and it is universally accepted for the excellent biocompatibility, nondegradability, good haemocompatibility, noncarcinogenic, and nonimmunogenic reaction. It also releases calcium and phosphate ions, and it leads to osteoinduction. Hydroxyapatites are inorganic materials and based on their properties that are widely used in tissue engineering and bone replacement, it is also used as a coating for fixing prostheses [53, 54]. Hydroxyapatite has been used clinically in a wide range of forms and applications. However, one of the major contributions of clinical application of hydroxyapatite is in the form of filler in a polymer matrix [48]. The natural sources of hydroxyapatite are regarded to be safer due to their cross-reaction and other immunological reaction as compared to synthetic hydroxyapatite [54].

Coral

Natural coral (Porites) consists of a mineral phase principally calcium carbonate in the structural form of aragonite with impurities such as strontium, magnesium and fluoride ions, and an organic matrix. Commercially available coral (Biocoral) is used as bone graft material and has been reported to be biocompatible and resorbable [55, 56]. Natural coral graft substitutes are derived from the exoskeleton of marine madreporic corals. Researchers first started evaluating coral as a potential bone graft substitute in the early 1970s in animals and in 1979 in humans. The structure of the commonly used coral Porite, is similar to that of cancellous bone, and its initial mechanical properties resemble those of bone. The exoskeleton of these high content calcium carbonate scaffolds has since been shown to be biocompatible, osteoconductive, and biodegradable at variable rates depending on the exoskeleton porosity, the implantation site and the species. Although not osteoinductive or osteogenic, coral grafts act as an adequate carrier for growth factors and allow cell attachment, growth spreading, and differentiation. When applied appropriately and when selected to match the resorption rate with the bone formation rate of the implantation site, natural coral exoskeletons have been found to be impressive bone graft substitutes [57]. There are several natural sources of coral and coral-derived materials, and they are resorbed slowly and substituted by host bone. Only two of the naturally occurring coral-based materials will be discussed.

Coralline apatites can be derived from sea coral and they are a naturally occurring structure with optimal strength and structural characteristics. It is known that the pores structure of coralline calcium phosphate produced by certain species is similar to human cancellous bone, making it suitable material for bone graft applications. Coral and converted coralline hydroxyapatite have been used as bone grafts and orbital implants since the 1980s, as the porous nature of the structure allows ingrowth of blood vessels to supply blood to the bone, which eventually infiltrates the implant. Several methods have been developed in producing hydroxyapatite directly from corals but two of such processes are widely utilized and reported. These techniques are hydrothermal and microwave processes, although others such as sol-gel coating are also in use among others [53].

Coral sands have the potential for a range of biomedical applications where calcium phosphates are used. There is difficulty associated with manufacturing of spherical porous powders which has increased the interest in discovering possible alternatives in nature. It was reported that under controlled hydrothermal exchange, a mix product of calcite and β-tricalcium phosphate can be derived from coral sand grains, and they were able to retain their porous structures. This makes them suitable for potential biomedical application as biodegradable material [58].

Corals, including those of Indian and Australian origin, have been converted to coralline hydroxyapatite successfully in the past [59, 60]. Considerable efforts are needed in order to consider the possibilities of converting African corals to coralline hydroxyapatite especially in the west coast of the continent where these substances are present in abundance. Coral has architectural properties such as interconnected macroporosity, which can be adapted to the biological requirements of the receiver bone. Natural coral implanted into bony tissue is gradually resorbed and replaced by newly formed bone [61].

Coral has been used clinically with good results in spinal fusion or to fill periodontal defects [62–64]. The exoskeleton of coral contained a high proportion of calcium carbonate, and this has been shown to be biocompatible, bioactive, osteoconductive and biodegradable at variable rates depending on the exoskeleton porosity, the implantation site, and the species [55, 65]. Natural coral, submitted to rigorous protocols of preparation and purification can be used as a replacement for bone grafts in both orthopaedic surgery and maxilla-cranial-facial surgery.

Shell

There are numerous shells in nature, and some of them have been evaluated for possible biomedical applications. Barnacle shells are very complex and strong composite bioceramic composed of different structural units which consist of calcite microcrystals of very uniform size. These are organized in a massive microstructure toward the external shell and in mineral layers separated by organic sheets toward the internal shell. It was observed that in contrast to nacre, barnacle shell contains calcite microcrystals (instead of aragonite microcrystals and has considerable porosity when compared with nacre which is a very dense material. The different composition and microstructural characteristics should affect dissolution behaviour in the body fluids (calcite being less soluble than aragonite)-and porosity could favour binding of this material with bone [66]. An investigation showed that calcium phosphate-based bioceramics have been synthesized by using eggshell-derived raw materials and phosphoric acid at different mixing ratios. The mineral composition of cockle shell is almost similar to that of coral and it was suggested that cockle shell has the potential to be used as a material for orthopaedic applications [53, 65].

The shell formation process in molluscs is a promising model for the development of bioinspired ceramics for a wide variety of applications in fields as varied as adaptive surface coatings, corrosion inhibition, hybrid composite materials, and more. The hybrid nature of the mollusc shell in terms of both mineral and organic components makes it an ideal bioceramic materials candidate: it is porous, it is a result of sophisticated structure, and exhibits exceptional flexural, fracture strength, and toughness [66].

Cowrie shells of West African origin are currently being evaluated for possible applications in orthopaedics as well as dental materials. The preliminary results have shown that these shells have great potential as biomedical materials.

Soil

It can be deduced from Table 1 that most of the calcium phosphates are inorganic, and they do exist as minerals. It was based on this that one of the authors believed that there is the possibility of obtaining these materials from nature. The investigation of some naturally occurring clays show that they contain some of these minerals, and also the fact that some natural practitioners employed a special type of clay in bone fracture and spinal management. Currently, work is on-going on the possibilities of using anthill as a biomedical material.

BIOPOLYMER

Polymers play a central role both in the natural world and in modern industrial economies. Some natural polymers such as nucleic acids and proteins carry and manipulate essential biological information, while other polymers like polysaccharides—that is nature's family of sugars—provide fuel for cell activity and serve as structural elements in living systems [67]. The advantages of synthetic polymers include predictable properties, batch-to-batch uniformity, and they can be tailored easily but they are too expensive. The growing reliance on synthetic polymers has also raised a number of environmental and human health concerns. It was concluded that the focus should now be on natural polymers which are inherently biodegradable and can be promising candidates to meet different requirements [67, 68]. Emerging applications of biopolymers range from packaging, industrial chemicals, medical implant devices, to computer storage media. In this paper, the utilization of natural polymers biomedical materials will be considered. The major problems yet to be overcome with natural starting materials are their propensity for calcification and eventual biodeterioration [69]. Some of the disadvantages and advantages associated with natural biopolymers are as listed in Table 2. A study stated that polymers from natural sources are particularly useful as biomaterials, given their similarity to the extracellular matrix (ECM) other polymers in the human body [70]. Natural polymers include both ECM proteins derivatives (e.g., collagen) and some materials derived from plants and seaweed among others [71]. Biopolymers have wide applications as implantable biomaterials, controlled disease carriers or scaffold for tissue engineering.

Table 2: The advantages and disadvantages of natural polymers [72]

S/N	Advantages	Disadvantages
1	No problem with toxicity or foreign body response	A major issue is immunological reaction. Body's immune system recognizes foreign material and tries to destroy it
2	Can function biologically at molecular level, not just macroscopic level	High natural variability
3	If desired, natural degradation can occur in the body via natural enzymes. Can also add cross-links to make less degradable	Structurally more complex than traditional materials. Technological manipulation is more elaborate

Throughout history, humans have relied extensively on biological materials like wool, leather, silk, and cellulose, and these materials are natural polymers. Polymers are substances composed of repeating structural units that are linked together to form long chains [67].

Economics and Applications

Biopolymers are a diverse and versatile class of materials that have potential applications in virtually all sectors of the economy; typical examples are adhesives, absorbents, lubricants, soil conditioners, cosmetic, drug delivery vehicles, textiles, high-strength materials, and even computational switching devices [67]. The structure and some functions of natural polymers are as shown in Figure 1 [72]. It is reported that commercially available biopolymers are quite expensive but their application in specialized fields such as medical materials justified the relatively high costs. Also, the commercialization difficulties facing biopolymers in many ways resemble the problems confronting other emerging technologies such as photovoltaic cells and fuel cells. In this study, the utilization of biopolymers from natural products in the area of medicine is reviewed.

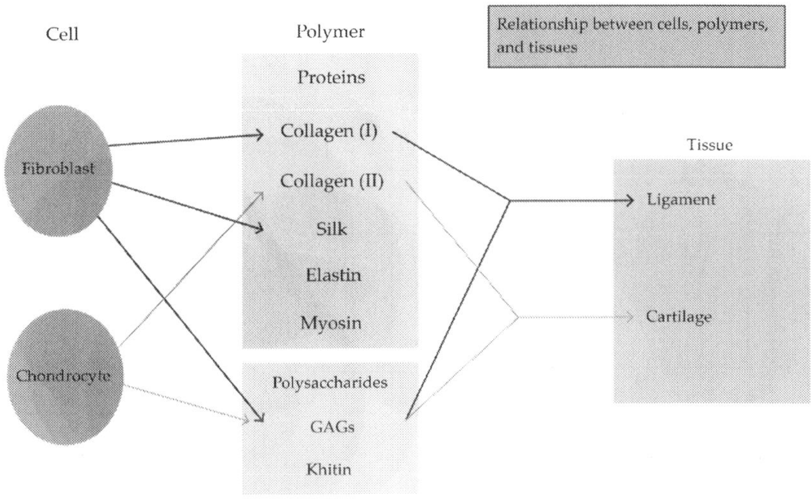

Figure 1: Natural polymers: structure and function.

Demand

It was forecast that natural polymer demand is expected to grow by 5.9% annually to \$3.6 billion in 2010, reaching £1.7 billion. It was also stated that the threats to further growth include mature applications and variable supplies for products due to climatic and political uncertainties [73]. Polymers are regarded as a class of giant molecules consisting of discrete building blocks linked together to form long chains. Simple building blocks are called monomers, while the more complicated building blocks are sometimes referred to as "repeat units."

Generally, biopolymers fall into two principal categories:

- polymers that are produced by biological systems such as micro-organisms, plants and animals, in short natural products;

- polymers that are synthesized chemically but are derived from biological starting materials such as aminoacids, sugars, and natural fats or oils. This can sometimes be referred to as synthetic natural product based polymers.

In this study, the emphasis is only on the first category of biopolymers. Exhaustive classifications are dealt with and their potential applications. Many types of polymers are widely used in biomedical devices that include orthopaedic, dental, soft tissue, and cardiovascular implants. Polymers represent the largest class of biomaterials, and they may be derived from natural sources, or from synthetic organic processes. The wide variety of natural polymers relevant to the field of biomaterials includes plant materials such as cellulose, sodium alginate, and natural rubber; animal materials like tissue-based heart valves and sutures, collagen, glycosaminoglycans, heparin and hyaluronic acid; and other natural materials, deoxyribonucleic acid (DNA), the genetic material of all living creatures [74]. Table 3 lists the various types of naturally occurring biopolymers and defines them on the basis of the chemical structure of their monomeric units and indicates the functions that these polymers serve in living organisms.

Table 3: Biopolymers found in nature and their functions [67]

Polymer	Monomer	Function(s)
Nuclei acids (DNA and RNA)	Nucleotides	Carriers of genetic information, universally recognized in all organisms
Proteins	α-aminoacids	Biological catalysis (enzymes), growth factors, receptors, structural materials (wool, leather, silk, hair, and connective tissue); hormones (insulin); toxins; antibodies
Polysaccharides (Carbohydrates)	Sugars	Structural materials in plants and some higher organisms (cellulose and chitin); energy storage materials (starch and glycogen); molecular recognition (blood types); bacterial secretions
Polyhydroxylalkanates	Fatty acids	Microbial energy reserve materials
Polyphenols	Phenols	Structural materials in plants (lignin); soil structure (peat and humics); plant defense mechanisms (tannins)
Polyphosphates	Phosphates	Inorganic energy storage materials
Polysulfates	Sulfates	Inorganic energy storage materials

Natural polymers were viewed to offer the following properties: they are often identical, very similar to macromolecular substances which the biological environment is prepared to recognize and deal with metabolically. As a result, the problems of toxicity and simulation of chronic inflammatory reaction, which provoked by many synthetic polymers are suppressed. Furthermore, there is capability for designing those biopolymers at molecular rather than the macroscopic level, equal biopolymers are degradable. However biopolymers suffer from immunogenicity, elaborate technological manipulation, decomposition or modification and natural variability in structure from the sources. These divergent factors have produced some unique biomedical materials with outstanding performance. Table 2 shows the general properties of certain natural polymers [75]. A comprehensive overview of the family of natural polymers is listed in Table 4. In this study the broad classification, presented in Table 5, will be adopted for biopolymers and they are grouped into proteins, polysaccharides, and polynucleotides. Polynucleotides are used for purposes such as nanostructural materials and assembling as well as for their electrical properties, but this is are not currently of interest in the field of biomaterials. Hence, this paper will be limited to proteins and polysaccharides.

Table 4: General properties of certain natural polymers [67]

Polymer	Incidence	Physiological Functions
(A) Protein		
Silk	Synthesized by anthropods	Protective cocoon
Keratin	Hair	Thermal insulation
Collagen	Connective tissues (tendon, skin)	Mechanical support
Gelatin	Partly amorphous collagen	Industrial product
Fibrinogen	Blood	Blood clotting
Elastin	Neck ligament	Mechanical support
Actin	Muscles	Contraction motility
Myosin	Muscles	Contraction motility
(B) Polysaccharides		
Cellulose	Plants	Mechanical support
Amylose	Plants	Energy reservoir
Dextran	Synthesized by bacteria	Matrix for growth of organism

Chitin	Insects and Crustaceans	Provides shape and form
Glycosaminoglycans	Connective tissues	Contribution to mechanical support
(C) Polynucleotides		
Deoxyribonucleic acids (DNA)	Cell nucleus	Direct protein biosynthesis
Ribonuclei acids (DNA)	Cell nucleus	Direct protein biosynthesis

Table 5: A Snapshot of the biopolymer family*

Polyesters	Polysaccharides (fungal)	Lipids/surfactants
Polyhydroxyal-kanoates	Pullulan	Acetoglycerides, waxes
Polylactic acid	Elsinan	Surfactants, emulsan
	Yeast glucans	
Proteins		Polysaccharides (bacterial)
Silks	Polysaccharides (plant/algal)	Xanthan
Collagen/gelatine	Starch (amylose/amylo-pectin)	Dextran
Elastin	Cellulose	Gellan
Resilin	Agar	Levan
Adhesives	Alginate	Curd lan
Polyaminoacids	Carrageenam	Polygalactosamine
Soy, zein, wheat glutein,	Pectin	Cellulose (bacterial)
casein, and serum albumin	Konjac	
	Various gums (guar)	Specialty Polymers
Polyphenols		Shellac
Lignin	Polysaccharides (animal)	Poly-γ-glutamic acid
Tannin	Chitin/chitosan	Natural rubber
Humic acid	Hyaluronic acid	Synthetic polymers from natural fats and oils (nylon from castor oils)

*Adapted from Herdman (1993) from p20.

Proteins

Proteins are polymers that are composed of amino acids. The specific amino acids used and the sequence of amino acids in a protein polymer chain are determined by the corresponding deoxyribonucleic acid (DNA) template. Proteins are also referred to as polypeptides, these are complex copolymers composed of up to 20 different amino acids building blocks. There are virtually a limitless number of proteins that can be formed from these 20 monomers. Proteins can contain a few hundred amino acid units or thousands of units. Protein polymers are unique in that the sequence of the monomers in the polymer chain is predetermined by the template-specific reamer of the polymerization process [14, 67].

Many proteins are of commercial interest because of their catalytic (enzymatic) or pharmaceutical properties. Nature has provided a vast array of proteins whose principal function is to form structural materials in living organisms. Some of the more familiar protein materials include wool, leather, silk, and gelatine. Also, elastomeric proteins occur in a wide range of biological systems where they have evolved to fulfil precise biological roles. The best known include proteins in vertebrate muscles and connective tissues, such as titin, elastin, and fibrillin, and spider silks. While some other examples are byssus and abductin from bivalve, molluscs, resilin from anthropods, and gluten from wheat [67, 76].

Gelatin

This is obtained through a controlled denature of the fibrous insoluble protein, collagen which is the major component of skin, bone, and connective tissue. It is characterized by having no antigenicity in comparison to its precursor. Gelatin is widely used as scaffold for tissue engineering and also has been frequently used in medicine as a wound dressing, and as an adhesive, absorbent pad for surgical use [77, 78].

Collagen

Collagen is an example of natural material which exists mostly in fibril form; it has a characteristic triple-helix structure and is the most prevalent protein in the animal world. It is the most abundant of all proteins found in mammals typically accounting for more than 30 percent of body protein. It forms a significant component of connective tissue such as bone, tendons, ligaments, and skin. There are at least 10 different types of collagen in the body. The arrangement of collagen fibres is arranged parallel to one another to give a structure with the tensile strength of a light steel wire. In skin, where strength and flexibility are required, collagen fibres are randomly oriented and woven together like felt [16, 67]. Collagen is a protein that acts as a structural support in a wide range of tissues including skin, bone, tendons, ligaments, cartilage, blood vessels, and nerves. It is usually implanted in a sponge form that does not have significant mechanical strength or stiffness. It has shown good promise as a scaffold for neotissue growth and is commercially available as a product for wound healing. Injectable collagen is widely used for the augmentation or buildup of dermal tissue for cosmetic reasons. In particular, collagen and its degradation products are often used for the attraction of fibroblasts in vivo during wound repair, fracture healing, and embryogenesis [16, 78, 79].

Silk

Silk is natural fibrous protein which is spun by Lepidoptera larvae such as silkworms, spiders, scorpions, mites, and flies [80]. Several attempts to produce silk by bacteria, yeast and plants by inserting genes in them gave insoluble silk proteins, which slumped inside cells [13].

It is well known that silk fibres are composed of at least two main proteins: sericin and fibroin. Silk fibroin shows excellent physical and chemical properties. Silk without sericin showed higher stability than others. Also, silk fibroin can be prepared in various forms; gel, powder, film, matrix, or fibre depending on the applications [80, 81]. Silk has always been a material of great fascination. Spiders can process silk protein into a material that has a tensile strength 16 times greater than that of nylon and a very high degree of elasticity. Silkworm silk is about 2 or 3 times greater than that of nylon. Silk also possesses the ability to super contract especially when they are put into liquid [67]. The breaking stress and strain for different silks are presented in Table 6 [13].

Table 6: Stress and strain of different types of silk*

Silk type	Stress (Gpa)	Stress (Gpa)
Bombyx mori cocoon silk	1.1	0.24
Nephila claripe MAS	1.75	0.15
Nephila maculate dragline	1.1	0.46

Arameus serratus framesilk	0.81	0.24
Araneus diadematus radial silk	1.2	0.40
A. serratus viscid silk	1.0	2.00
A. diadematus viscid silk	1.4	4.76

*Adapted from Ivanova (2005) [13].

It is observed that the most interesting silk for engineering is viscid silk of Adameus Diamantus since it has greater stress and strain at breaking point. The strength and toughness of silk is known to be remarkable, and some of the mechanical properties of biodegradable materials are apresented in Table 7 [11].

Table 7: Mechanical properties of biodegradable materials*

Source of biomaterial	UTS (MPa)	Modulus (GPa)	Strain (%) at breakage
Bombyx mori silk (with sericin)	500	5–12	19
Bombyx mori silk (without sericin)	610–690	15–17	4–16
Bombyx mori silk	740	10	20
Collagen	0.9–7.4	0.0018–0.046	24–68
Cross-linked collagen	47–72	0.4–0.8	12–16
Polylactic acid	28–50	1.2–3.0	2–6

*Adapted from Cao and Wang (2009) [11].

Generally, silk has been investigated for use as biomedical resource due to its unique properties which include nontoxicity, biocompatibility, and biodegradability. It must be noted that it is quite hard and expensive to produce silk from these natural sources in large quantities. Silk has been used as a biomaterial in various forms such as films, membranes, gels, sponges, powders, and scaffolds. They have been commercially applied as scaffolds, vascular prostheses, structural implants, nets, and sutures among others [30, 31, 67].

Keratins

The structural integrity and solubility of keratin as well as its natural biocompatibility, controllable biodegradability, and bioactivity makes it an ideal material medical polymer. A proposal was made that keratin extracts from hair and wool could be used as platform technology to make a new family of biomaterials used for biomedical applications such as wound healing and bone regeneration, scaffold for tissue engineering, and coatings for medical devices. This proposal was a challenge to the long-standing notion that these animal-derived proteins would not be compatible with human biological systems. It was proved that the carefully extracted keratin molecule did not elicit an adverse biologic response [82]. These are a class of biomaterials that can be derived by extraction of proteins from human hair. They have haemostatic characteristics, and it was hypothesized that a keratin hydrogel having the ability to absorb fluid and bind cells may be an effective haemostat [83]. Also the usage of keratins to mediate a robust nerve regeneration response in part through activation of Schwann cells was reported [84]. They also suggested that keratins derived from human hair are neuroinductive and can facilitate an outcome comparable to autograft in a nerve injury model.

Elastin

The best known and most widely distributed protein elastor is elastin. It is responsible for the elasticity of the aorta and skin of mammal, and is also present in the ligamentum nuchae which is involved in raising the heads of grazing hoofed animals. Elastic fibres are elastic, load-bearing protein polymers found in connective tissue such as ligaments. This rubber-like material responds to changes in temperature and is able to convert chemical energy into mechanical energy. As a consequence, the material could be used as a replacement for ligament tissue, blood vessels, or any other tissues requiring the contractile properties of elastin [67, 76].

Natural Rubber Latex

It was recently claimed that the natural rubber latex, extracted from Hervia brasilliens, is a strong candidate for use as a biomaterial. This biologically compatible material can be used in contact or inside the human body, and it has been shown that natural rubber latex performs a biological action that accelerates the healing process, being a powerful stimulator of cicatrisation. It is well known that dried latex, in the rubber form, presents high resistance; high elasticity; and is an easy-shaping material, which increases its appeal for use in the fabrication of vascular prostheses [85, 86].

Polysaccharides

Polysaccharides are inexpensive, natural biopolymers which are widely used as raw materials in several industries and being a natural compound, most polysaccharides are easily biodegradable. 75% of all organic material on earth is present in the form of polysaccharides [87, 88]. Polysaccharides are polymers or macromolecules composed of simple sugars, and they have two principal functions. Some such as cellulose serve as structural materials in living systems. Polysaccharides can be utilized as materials in medicine which include wound management, tissue engineering, drug delivery systems, and haemostatic devices among others. It can be classified as listed in Table 8.

Table 8: Types of polysaccharides

Types	Examples
Bacterial	Xanthan, dextran, gellan, and cellulose
Fungal	Pullulan, elsinan, yeast, and glucan
Plant	Starch, alginate, and tannin
Animal	Chitin, chitosan, hyaluronic acid, lignin, and tannin

Starch

Starch is one of the most abundant and cheap polysaccharides. Natural starch occurs in a granular form, and it is a principal carbohydrate storage product of plants. It is found in cereal and tuber plants such as maize and potatoes, respectively [67, 87, 89]. The content of amylose and amylopectin in starch varies and largely depends on the starch source. Most often, starch consists of about 30% amylose [a linear α-(1–4) glucan] and 70% amylopectin (dendritically branched version) [89–92]. The structure of starch is shown in Figure 2, and it is adapted from Nair and Laurencin [93]. The ratio of amylose and amylopectin in the starch may affect the starch behaviour during processing and the properties of the end product. As the amylose content increases, it will also increases the crystallinity of starch based products and this resulted in texture firming. It must be emphasized that starch is an established and widely used biodegradable polymer. Also starch possesses some properties which enable it to be compounded with other biopolymers with resultant improved products. Starch is insoluble in cold water, but it is very hygroscopic and binds water reversibly. It is renewable, biodegradable, inexpensive, and as such can play an important role in the medical field. They are studied in several biomedical applications ranging from bone replacement implants, bone cements to drug delivery systems and tissue scaffolds [94, 95].

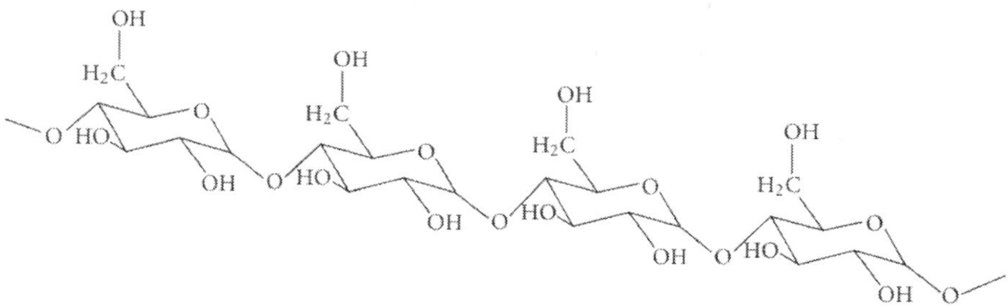

Figure 2: Structure of starch.

Chitin and Chitosan

Chitin is a nitrogen-containing polysaccharide, related chemically to cellulose, and it is insoluble in most solvents. Controlled deacetylation is used to produce chitosan which is a derivative with approximately 50% free amine. Chitins are regarded as the second most abundant natural polymer after cellulose and there are three possible sources of chitin as raw materials. They are isolation from traditional shellfish sources such as crabs and shrimps; harvesting of fungal mycelia from bioreactor processes, a typical example is mushroom; synthesizing from monomeric/dimeric units using chemical and/or enzymatic strategies. However in this paper, the third source of obtaining chitin will be neglected because it involves chemical synthesis. Recently, studies have shifted from the traditional shellfish to exciting and novel shell materials like eggshell and cowry shell. This may be attributed to the environmental pollution and the concern for the amount of residual protein [96–98].

Chitosan is currently obtained by the deacetylation of chitin-9poly-β-(1–4)-N-acetyl-D-glucosamine). Chitosan (1,4 linked 2-amino-2-deoxy-β-D-glucan) comprises of glucosamine and N-acetylglucosamine. The latter is the water soluble chitin called chitosan [99], which is a moiety of glycosaminoglucans (GAGs). The structure of chitosan and chitin is shown in Figure 3 [93]. Meanwhile, chitin and chitosan are regarded as natural resources waiting for a market and sourced from the canning industry especially the crustaceans [100]. They stated that there are four chronological steps by which chitosan can be processed from crustacean shells. The steps are as shown in Figure 4, and they are, namely, deproteination, demineralization, decolouration, and deacetylation. Chitin and chitosan are biologically stemmed aminopolysaccharides which exhibit multiple bioactivities, for example, low toxicity, biocompatibility, biodegradable, antimicrobial, and wound-healing properties. Moreover, chitosan elicits minimal foreign body reaction. Chitosan favours both soft and hard tissue regeneration [80, 101].

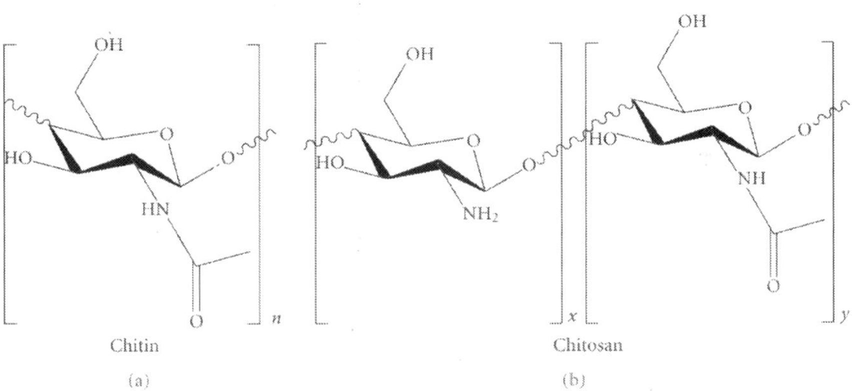

Figure 3: Structure of chitin (a) and structure of chitosan (b).

Crustacean shells ⟶ size reduction ⟶ protein separation ⟶ (NaOH) ⟶ washing demineralization (HCl) ⟶ washing and dewatering ⟶ decolouration ⟶ chitin ⟶ deacetylation (NaOH) ⟶ washing and dewatering ⟶ chitosan

Figure 4: The processing stages in producing chitosan from crustacean shells.

Chitins have various biofunctionalities including antithrombogenic, haemostatic, immunity enhancing, and wound healing. Research has shown that chitin and chitosan are nontoxic and nonallergenic, so the body does not reject these compounds as foreign invaders. Biocompatibility, biodegradability, and adsorption properties of chitin and its derivatives are much higher than synthetically substituted cellulose. The native chitin molecule has strong inter- and intra-molecular hydrogen bonds with partial N-deacetylation [102]. Chitin is a biocompatible and biodegradable polymer which demonstrates bacteriostatic and analgesic effects, which in addition, shorten the time of wound healing and the rebuilding of connective tissue. The only disadvantage of natural chitin application is its insolubility in the common solvents generally available, and the enormous related difficulties connected with chitin processing [103].

The chitin family of polymers is being widely used in various applications such as medicine, manufacturing, agriculture, and waste treatment. In the biomedical area, chitosan and its derivatives have been successfully used in wound dressings, drug delivery systems, and as materials for tissue engineering. They have been reported to be a promising candidate as a scaffold material for engineered human tissue such as skin, cartilage, and bone due to its biocompatibility, and resorbability. Chitosan is used as an intraocular lense material because of its oxygen permeability and it has also been found to expedite blood clotting. It is also being evaluated for use in the bioremediation of toxic phenolic compound [67, 104, 105].

Hyaluronic Acid

Hyaluronic acid (HA) is a natural polysaccharide polymer belonging to the same class of compounds as starch and cellulose. The substance is found naturally in the extracellular matrix of skin, cockscomb, cartilage, vitreous humor, and other body tissue and plays a role in the movement and proliferation of cells. It also occurs as an extracellular polysaccharide in a variety of bacteria [67, 106]. HA was discovered in 1934, and it is a long unbranched polysaccharide chain, composed of repeating twin sugar units. Due to the high density of negative charges along the polymer chain, HA is very hydrophilic and adopts highly extended random coil conformations. It is extremely flexible, has a high viscosity, and structure is shown in Figure 5 [67, 93]. HA consists of 2-acetamide-2-deoxy-α-D-glucose and β-D-gluconic acid residues linked by alternate (1,3) and (1,4) glycoside bonding and has the high capacity of lubrication, water sorption, and water retention, and influences several cellular functions such as migration, adhesion, and proliferation. HA is water-soluble, and must be cross-linked or otherwise modified to form a scaffold [71, 78].

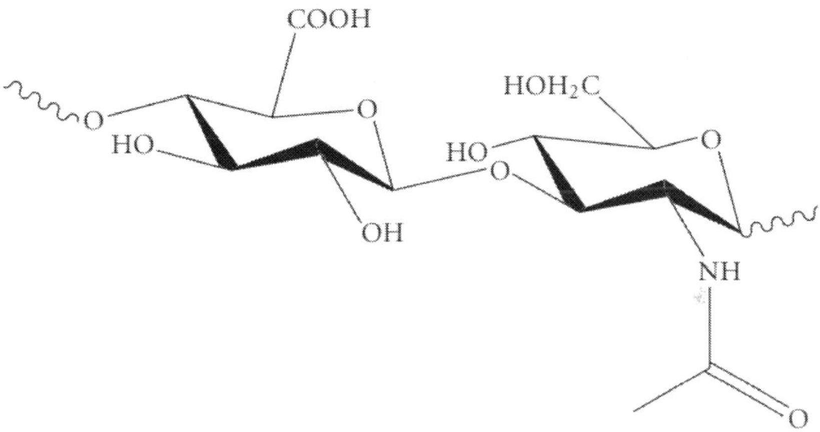

Figure 5: Structure of hyaluronic acid.

It is an extremely attractive polymer material because it is a natural product that degrades into simple sugars. Recent biomedical applications of hyaluronic acid include scaffolds for wound healing and tissue engineering, as well as ophthalmic surgery, arthritis treatment, and as a component in implant materials. It is also used as a lubricating fluid in joints and serves as a regulator in the lymphatic system. Based on its in vivo functions, HA has been adapted to commercial compounds for treating a number of tissue-related conditions, including osteoarthritis of the joints, and facial wrinkles and folds of the skin [67, 78, 107].

Pullulan

Pullulan is water-soluble polysaccharide produced outside the cell by several species of yeast, most notably Aureobasidium Pullulans. It is a linear polymer made up of monomers that contain three glucose sugars linked together [67, 108]. It is claimed that pullulan is a nonionic exopolysaccharide of fungal origin and is currently exploited in the food industry due to its many unique characteristics [109]. It is nontoxic, nonimmunogenic, nonmutagenic, and noncarcinogenic; hence there are attempts to explore pullulan for various biomedicals [110].

Pullulan acts as a plasma extender without undesirable side effects. After metabolic turnover, it is completely excreted, and it was reported that pullulan to be used as plasma blood expander should have a molecular weight of about 60 kDa. Pullulan is biodegradable, impermeable to oxygen, nonhygroscopic, and nonreducing. Recently, the role of polysaccharides in developing controlled drug delivery systems has increased significantly, and pullulan is gaining lot of attraction for this application [111]. Tissue engineering requires scaffolds or artificial extracellular matrix (ECM) that can accommodate cells and regulate their growth leading to three-dimensional tissue regeneration. It was concluded that heparin-conjugated pullulan material can thus be used for the proliferation of vascular endothelial cells and to inhibit the proliferation of smooth muscle cells (SMCs) [108]. Pullulan has very important applications in surface modification of polymeric materials so that they can be made more blood compatible and bioinert. Pullulan being adhesive is also tried for wound dressing applications. The thermal stability and elastic properties of pullulan allow it to be utilized in many different ways [109, 112, 113].

BIOCOMPOSITE

Naturally, animals and plants synthesize biocomposites with high strength consisting of fibrous biopolymers. A classic example is cellulose which consists of whisker like microfibrils that are synthesized and deposited in a definite manner which imposes high strength. Living tissues are composites themselves with a number of levels of hierarchy. Reinforcement of polymer with nanosized particles or fibres is a promising technique that is capable of yielding materials with enhanced performance but without involvement of expensive synthesis procedures [103]. Bone is regarded as a biocomposite which comprises of mainly nanohydroxyapatite crystallite orienting along collagen fibres [77].

There is no really adequate definition of a composite material; however, there are three criteria that have to be satisfied before a material can be said to be a composite. Firly, the constituents have to be present in reasonable proportions and secondly, the constituent phases must have different properties and as such ultimately leading to properties that are superior and possibly unique in some specific respects, to the properties of the individual components. Lastly, the processing technique must be (in such a way that) able to be carried out under controlled conditions. There are several means by which composites can be classified and in this paper, we are interested in natural composite materials. Some typical examples of this class are wood, bone, muscle, and other tissues. Bone is a composite material composed primarily of organic fibres, small inorganic crystals, water, and fats. The proportions of these components will vary with the type of bone, animal species, and age, but typically about 35% of the dry, fat-free weight of bone is the organic fibre, collagen. Collagen is a fibrous protein and located around the outside of the collagen fibres are small rod-like crystals of hydroxyapatite with dimensions of the order of $5 \times 5 \times 50$ nm. Therefore at this microscopic level we have a hydroxyapatite-reinforced collagen composite and this may be considered to be the basic "building block" of bone [114, 115].

It must be realised that to classify composite within the context of this work is severely restricted. This is due to the fact that only combinations of biopolymers and bioceramics are possible. Thermodynamically, metals are stable in combined state as ores (i.e., oxides, sulphides, and carbonates), and they rarely exist in their pure form; hence only precious metals are available as natural materials for biomaterials application. This paper will consider combination of biopolymer with biopolymer, biopolymer with bioceramics; there have been little or no reports on the application of two bioceramics as biocomposites.

Biopolyme-Biopolymer

Biopolymers occur abundantly in nature, and they have several applications as materials in medicine. Chitosan is regarded as a biodegradable and nontoxic hydrophilic polysaccharide with excellent mucoadhesive and permeation-enhancing properties. Alginate is the name given to a family of linear polysaccharides found in brown algae and is composed of guluronic and mannuronic units [106]. Meanwhile, there has been increasing interest in the study of alginate-chitosan microparticles as carriers for controlled release of proteins and drugs due to its biocompatible, biodegradable, and mucoadhesive properties. It was shown that the stability of alginate-chitosan capsules depended strongly on the amount of chitosan bound to the capsules [116]. The superporous hydrogel foams containing chitosan and gelatine have a lot of interesting food and medical applications. The superporous structure and mechanical properties are especially proper for scaffold preparation [105].

Chitosan/gelatine hydrogels are being developed for tracheal epithelia. A company in Italy is reported to have developed a series of modified hyaluronate esters by adding hydrophobic moieties to the carboxyl groups so as to control degradation, and it is marked as tissue engineering polymers which are being applied to bone growth and cartilage [71]. An investigation concluded that the artificial dermal skin is composed of gelatine and polysaccharides such that hydraluronic acid and β-glucan will be useful to promote wound healing [78]. While a combination of flexible protein and rigid polysaccharide results in a series of biomimetic chitosan/gelatine based biomaterials covering surface modifier and nonviral vector for gene therapy. They hold promising prospect in tissue engineering [77].

Cellulose-based hydrogels are advantageous over cellulose sponges and fabrics as their bulk chemistry can be easily modified. Independent investigations have reported the synthesis of novel biomimetic hydrogels, based on crosslinking cellulose derivatives with hyaluronic acid [117, 118]. Although mainly proposed as postoperative adhesion barriers, such hydrogels also show potential as scaffolds for regenerative medicine, with a tunable degradation rate. Indeed, the presence of hyaluronic in the cellulose network provides enzyme-sensitive degradation sites, whose density in the bulk of the hydrogel can be easily controlled [119].

The blending of polymers with starch under controlled conditions leads to copolymerization that in turn results in high-molecular polymers with thermoplastic properties. Though, the mixing or blending needs special machinery, such as an extruder, the products (polymer blend) can be handled as easily as conventional plastic resin [94]. Although starch such as that of cassava is used as a polymer, other polymers are frequently used in the blend. Polycaprolactone blend is the most commonly used polymer/starch blend because of its low-melting temperature (T_m) and high susceptibility to amylase and lipase hydrolyses [120, 121]. It must be noted that starch as polymer has been marketed under various names, and most patents describe a generic starch, this includes cassava [100].

Natural polymers such as chitosan and cellulose, and their derivatives are inherently biodegradable, and exhibit unique properties. Starch and chitosan are abundant naturally occurring polysaccharides. Both of them are cheap, renewable, non-toxic and biodegradable [122]. Bourdoom and Chinnan (2008) concluded that the starch/chitosan blend exhibits good mechanical properties, while water barrier properties, and miscibility of biodegradable blend films are affected by the ratio of starch and chitosan.

One problem associated with starch-based blend is that starch and many polymers are nonmiscible which leads to the mechanical properties of the starch/polymer blends generally becoming poor. Thus, chemical strategies are taken into consideration in order to improve the properties [123]. Starch-based biodegradable polymers have some advantages for use as medical materials such as good biocompatibility, biodegradable, and its degradation products are non-toxic, proper mechanical properties and degradation as requirement. Starch-based biodegradable polymers have been widely investigated in bone tissue engineering.

Starch-based biodegradable bone cements can provide immediate structural support and degrade from the site of application. In addition, starch-based biodegradable polymers can also be used as bone scaffold and when they are in the form of microsphere or hydrogel they are suitable for drug delivery systems. There is no need for surgical removal of the device after drug depletion. The unique properties such as hydrophilicity, permeability, biocompatibility, are to some extent similar to soft biological systems, all these make starch-based hydrogel to be useful for various biomedical applications [124–129]. Biodegradable polymers such as poly (lactic acid), poly (glycolic acid), and their respective copolymers have been used in several drug delivery systems. However, few attempts have been make to use starch-based polymers in these types of applications; despite being well known that they are biodegradable materials; they have been proposed in several works to be used as biomaterials [130–133].

Biopolymer-Bioceramics

A composite made of bioceramic-polymer composite was introduced, and it was based on the concept that cortical bone itself comprises an organic matrix reinforced with a mineral component. The material they developed has been used as an orbit implant for orbital floor fractures and volume augmentation, also it is now being used in middle ear implants, commercialized under the trade name HAPEX [134, 135].

Polymers such as alginates showing low toxicity and high biocompatibility and hydroxyapatite have been reported with osteoinductive or osteoconductive potential. Composites binding these properties could be a good choice for the development of new materials for medical application. This composite is considered to be biocompatible and partially resorbable [136].

The association of bioceramics and fibrin sealants may develop the clinical applications of bone substitutes. The physical, chemical, and biological properties of both bioceramics and fibrin glue may be cumulated for preparing advance bone substitutes. The ideal bone substitute should be biocomposite, biodegradable at the expense of bone growth and mouldable, with sufficient mechanical properties to fill and restore bone defects [137]. These biological properties can be tailored for both bioceramics and fibrin glue by changing the composition; porosity and network cross-linking. The positive effects of the combination of fibrin sealant with ceramic biomaterials have essentially been observed in clinical studies. Fibrin sealants improve the surgical handling of biomaterials and widen their field of application in bone surgery. Furthermore, the future development of this composite may be in combination with bone growth factors [138].

One of the naturally occurring biocomposites that has been deployed as biomedical materials is nacre. It is secreted only by the molluscan classes, Gastropoda, Bivalvia, Cephalopoda, and to a minor extent, Tryblidiida. Nacre has a lamellar structure consisting of alternating tablets of aragonite and organic interlamellar membranes which have a core of β-chitin surrounded by acidic proteins [139]. It is now clear that the sequence of nacre formula involves the secretion of interlamellar membranes separated by a liquid rich in silk fibroin such that subsequently the liquid is replaced with mineral. This pattern is the same for the bivalves and gastropods, so too for the other nacre-secreting molluscs, although this is yet to be determined. There are, however, structural differences between bivalve and gastropod nacre [140].

Nacre is by far the most intensively studied nonhuman organomineral biocomposite. It has a high proportion, approximately 5% of organic matter (proteins and polysaccharides), the mineral fraction being exclusively in the form of aragonite. The aragonites work of fracture is estimated to be as high as 3000 times that of inorganic aragonite but this figure is considered to be lower in reality. Nacre has found possible applications in the biomedical industry due to their superior biomechanical properties [140]. Nacre is able to form a tight bond of bone without soft and fibrous tissue formation. It is gradually and centripetally dissolved but not resorbed by cells because it is not porous [141, 142].

APPLICATION OF NATURAL PRODUCTS IN BIOMATERIALS

There are various uses of natural products in materials for medical applications, some of which are bioelectrodes, dental implantation, orthopaedic applications, adhesive and sealant, ophthalmological applications, intraocular lens implants, burn dressings, and skin substitutes among others. Table 9 lists some of the natural biomaterials and their areas of application [10]. There are three principal market segments for natural biomaterials: wound management products, drug delivery systems (DDSs), and tissue engineering. In this paper, all these three will be discussed and another important area of application will be highlighted. Biotextiles have utilized natural product-based proactive agents such as chitosan, natural dyes, neem extract, and other herbal products for antimicrobial finishing of textile substrates [143]. Biotextiles also provide a good platform to demonstrate the limitations and challenges of natural products as materials in medicine.

Table 9: Some applications of natural biomaterials

Application	Natural biomaterial
Artificial heart valves	Bovine pericardium, Intact porcine aortic valves
Hernia repair devices	Porcine small intestinal submucosa, porcine urinary bladder mucosa, Porcine dermal grafts
Sutures	Catgut (porcine or bovine intestinal wall) and porcine dermal grafts
Skin repair/wound care	Dermal allograft, porcine small intestinal submucosa, and porcine dermal grafts
Vascular prostheses	Bovine ureter, porcine small intestinal submucosa, and Ovine arteries
Urethral repair	Porcine bladder

Breast reconstruction	Dermal allograft
Ligament repair	Dermal allograft, porcine small intestinal submucosa, and fetal bovine skin
Spinal fusion/bone healing	Bone allograft

Wound Management

This segment is regarded to grow at an astronomical rate, and it represents a sizeable portion of biomedical materials. Hyaluronic acid is an extremely attractive polymer material used in wound-healing preparations, aids tissue formation and repair, provides a protective matrix for reproductive cells, serves as a regulator in the lymphatic system, and acts as a lubricating fluid in joints. Zein, which is a major storage protein of corn, has been used widely as an adhesive, fibre, cosmetic powder, and ink [67]. Hide glue derived from gelatin has been in used for centuries. Lastly, mention must also be made of burn dressings methods like autografts, allografts from cadavers, and xenografts of which domestic swine is the most commonly used as a temporary wound closure material [14].

Drug Delivery System

The discovery of new drugs has been the major thrust in drug research, and although this trend will continue for a while, there is an increasing emphasis being placed on the development of novel drug delivery systems. The use of biopolymer materials for drug delivery can minimize tissue reaction and allow drugs to be administered in nonconvectional ways. The use of biopolymers in these formulations has thus far been restricted to a narrow set of applications [67]. Natural polysaccharides due to their outstanding merits have received more and more attention in the field of drug delivery systems. In particular, polysaccharides seem to be the most promising materials in the preparation of nanometric carriers. In nature, polysaccharides have various resources from algal origin (e.g., alginate), plant origin (e.g., pectin and guar gum), and microbial origin (e.g., chitosan and chondroitin). Among various polysaccharides, chitosan is the early one to be used to prepare nanoparticles [144, 145]. Also, the main applied form of albumin for biomaterials is pharmaceutical microspheres, which has been extensively investigated for drug targeting to various organs and tissues [146, 147].

Tissue Engineering

One key area of research gaining significant attention over the past several years is tissue engineering. This technology combines an engineered scaffold, or 3-dimensional structure with living cells. These scaffolds can be constructed of various materials and made into different shaped depending on the desired application [14]. Many naturally occurring scaffolds can be used as biomaterials for tissue engineering purposes, and one of the typical examples is hydroxyapatite which is obtained from coral or animal bone. Proteins are one of the important candidates for tissue engineering. They provide cell support for anchorage and adherence through cellular growth and development [148, 149]. Currently available proteins for application in tissue engineering include fibrin, collagen, zein, silk fibroin, keratin, casein, and albumin. A more speculative application is the use of biopolymers as scaffolding in the formation of new cartilage in the body. With the advantage of biocompatibility, biopolymers are likely to be used in many more novel orthopaedic applications [67].

Medical Fibres and Biotextiles

The application of fibres and biotextiles as components for implantable devices is widespread and covers all aspects of medicine and healthcare. They are used as drapes and protective apparel such as protective surgical gowns, operating room curtains, masks, and shoe covers. Textiles can also be used as topical and percutaneous applications in areas like bandages, wound coverings, and nappies. Lastly, three key applications of biotextiles in general surgery are sutures, haemostatic devices, and hernia repair meshes. Catgut is a natural collagen-based suture material obtained from ovine intestine, which is cross-linked and cut into narrow strips, and it is one of the first bioabsorbable fibres used in surgery. Silk and collagen are two natural fibres that have been widely used in medicine and multiple applications. Silk is inexpensive and is considered to be the gold standard in suture-handling characteristics. Collagen sutures are primarily used in microsurgery or ophthalmic surgery. Cotton was and still is commonly employed for bandages, surgical sponges, curtains, and surgical apparel, and in surgical gowns [14].

Antimicrobial textiles help to reduce effectively the ill effects associated with microbial growth on textile material. The use of natural products such as chitosan and natural dyes for antimicrobial textile materials has been reported. Plant products comprise the major segment of natural antimicrobial agents; sericin, natural dyes, tannins, and aloe vera among others. Aloe vera also possesses antifungal and antibacterial properties, which can be exploited for medical textile applications, such as wound dressing, suture, and bioactive textiles. The major challenges in application of natural products for textile application are that most of these biomaterials are complex mixtures of several compounds, and also the composition varies in different species of the same plant. The activity and composition also vary depending on their geographical location, age, and method of extraction. The availability of these products in bulk quantities, their extraction, isolation, and purification to get standardized products are other challenges in their applications [143].

CONCLUSION

Natural products have found many applications in the field of pharmaceuticals and medicine; healthcare and nutritional supplements; agrochemicals among others. The paper looks at the recent research in natural biomaterials towards applications in various biomedical fields. The purpose of this paper is to highlight information available on the various forms of natural products for biomaterials as well as to highlight the applications of natural biomaterials. The properties, demand, and economic importance of different natural biomaterials are discussed. Three major areas of intervention of natural biomaterials were discussed. It is advocated that future work should be focussed on the methods of reducing the major disadvantages like poor immunogenic response, variability, and the technological processing techniques. The poor immunogenic response is attributed to the presence of antigenic determinants which can be reduced by either chemical modification or standardized natural material sourcing. There is a need to develop a well-tested production and processing route for producing the biomaterials from the natural products as this will reduce the effects of variability among others.

In this paper, an attempt has been made to increase the understanding of the utilization of natural products in the biomedical field. This will attract the attention of specialists in the study of natural products and biomaterials.

REFERENCES

1. J. P. A. Nicolai and G. Rakhorst, Introduction: Biomaterials in Modern Medicine, 2008, http://www.worldscibooks.com/etextbooks/6562/6562.chap01.pdf.
2. G. Heness and B. Ben-Nissan, "Innovative bioceramics," Materials Forum, vol. 27, pp. 104–114, 2004.
3. D. F. Williams, Definition in Biomaterials, Elservier, Amsterdam, The Netherlands, 1987.
4. A. F. Von Recum and M. LaBerge, "Educational goals for biomaterials science and engineering: prospective view," Journal of Applied Biomaterials, vol. 6, no. 2, pp. 137–144, 1995.
5. L. Sedel, "Biomaterials: Medical Viewpoint," 2004, http://www.atsmp.whut.edu.cn/resources/pdf/4078.pdf.
6. J. P. Brown, M. Bide, and M. Phaneuf, "Cellular encapsulation into porous alginate fibers," Tech. Rep. M04-CL13, National Textile Center Research Briefs—Materials Competency, 2005.
7. L. M. Zimmerman and I Veith, Great Ideas in the History of Surgery, Norman Publishers, New York, NY, USA, 1st edition, 1993.
8. DSM, Corporate Communications, A Brief History of Biomedical Materials, 2009, http://www.dsm.com/en_US/cworld/public/medical downloads/.
9. R. H. Doremus, "Review—bioceramics," Journal of Materials Science, vol. 27, no. 2, pp. 285–297, 1992.
10. J. C. Coburn and A. Pandit, "Chapter 13: development of naturally-derived biomaterials and optimization of their biomechanical properties," in Topics in Tissue Engineering, supplement 13, pp. 1–23, 2007.
11. Y. Cao and B. Wang, "Biodegradation of silk biomaterials," International Journal of Molecular Sciences, vol. 10, no. 4, pp. 1514–1524, 2009.
12. K. McLean, "Evaluation of Biomaterial Performance," 2006, http://www.csiro.au/.
13. N. Ivanova, "Biomaterials: Silk as a Natural Biomaterial," 2005, http://www.sinc.sunysb.edu.edu/.

14. B. D. Ratner, A. S. Hoffmann, F. J. Schoen, and J. C. Lemons, Biomaterials Science: An Introduction to Materials in Medicine, Elsevier Academic Press, London, UK, 2nd edition, 2004.
15. P. Heffernan, "Section 3: discovery research measure (2007–2013), marine biodiscovery/biotechnology research programme," in Sea Change—A Marine Knowledge, Research and Innovation Strategy for Ireland (2007–2013), Marine Institute, Rinville Oranmore, Galway, Ireland, 2006.
16. ASMI, "Overview of biomaterials and their use in medical devices," in Handbook of Materials for Medical Devices, 2003, http://asmiinternational.org/bookstore.
17. P. Fratzl, "Biomimetic materials research: what can we really learn from nature›s structural materials?" Journal of the Royal Society Interface, vol. 4, no. 15, pp. 637–642, 2007.
18. R. Schoental, "Toxicology of natural products," Food and Cosmetics Toxicology, vol. 3, no. 4, pp. 609–620, 1965.
19. G. A. Holt and A. Chandra, "Herbs in the modern healthcare environment—an overview of uses, legalities, and the role of the healthcare professional," Clinical Research and Regulatory Affairs, vol. 19, no. 1, pp. 83–107, 2002.
20. C. B. Spainhour, "Natural products," in Drug Discovery Handbook, pp. 11–72, John Wiley & Sons, Inc., New York, NY, USA, 2005.
21. B. B. Jarvis, "Evolution of metabolic pathways," in The Role of Natural Products in Evolution, J. T. Romeo, Ed., vol. 34, pp. 1–24, 1st edition, 2000.
22. http://www.thefreedictionary.com/Natural+products+chemistry.
23. http://en.wikipedia.org/wiki/Natural_product.
24. K. Nakanishi, "An historical perspective of natural products chemistry," in Comprehensive Natural Products Chemistry, vol. 8, pp. 21–48, 1999.
25. B. R. O›Keefe, "Biologically active proteins from natural product extracts," Journal of Natural Products, vol. 64, no. 10, pp. 1373–1381, 2001.
26. J. D. Connolly, "Natural products from around the world," Revista Latinoamericana de Quimica, vol. 25, no. 2, pp. 77–85, 1997.

27. K. H. Lee, "Anticancer drug design based on plant-derived natural products," Journal of Biomedical Science, vol. 6, no. 4, pp. 236–250, 1999.
28. G. W. Qin and R. S. Xu, "Recent advances on bioactive natural products from Chinese medicinal plants," Medicinal Research Reviews, vol. 18, no. 6, pp. 375–382, 1998.
29. V. J. Ram and S. Kumari, "Natural products of plant origin as anticancer agents," Drug News and Perspectives, vol. 14, no. 8, pp. 465–482, 2001.
30. C. Acharya, B. Hinz, and S. C. Kundu, "The effect of lactose-conjugated silk biomaterials on the development of fibrogenic fibroblasts," Biomaterials, vol. 29, no. 35, pp. 4665–4675, 2008.
31. C. Acharya, S. K. Ghosh, and S. C. Kundu, "Silk fibroin protein from mulberry and non-mulberry silkworms: cytotoxicity, biocompatibility and kinetics of L929 murine fibroblast adhesion," Journal of Materials Science, vol. 19, no. 8, pp. 2827–2836, 2008.
32. N. Minoura, M. Tsukada, and M. Nagura, "Physico-chemical properties of silk fibroin membrane as a biomaterial," Biomaterials, vol. 11, no. 6, pp. 430–434, 1990.
33. N. Minoura, S. I. Aiba, M. Higuchi, Y. Gotoh, M. Tsukada, and Y. Imai, "Attachment and growth of fibroblast cells on silk fibroin," Biochemical and Biophysical Research Communications, vol. 208, no. 2, pp. 511–516, 1995.
34. M. Santin, A. Motta, G. Freddi, and M. Cannas, "In vitro evaluation of the inflammatory potential of the silk fibroin," Journal of Biomedical Materials Research, vol. 46, no. 3, pp. 382–389, 1999.
35. L. Meinel, R. Fajardo, S. Hofmann et al., "Silk implants for the healing of critical size bone defects," Bone, vol. 37, no. 5, pp. 688–698, 2005.
36. G. M. Cragg and D. J. Newman, "Natural product drug discovery in the next millenium," Pharmaceutical Biology, vol. 39, supplement 1, pp. 8–17, 2001.
37. A. M. Burja, B. Banaigs, E. Abou-Mansour, J. Grant Burgess, and P. C. Wright, "Marine cyanobacteria — a profile source of natural products," Tetrahedron, vol. 57, no. 46, pp. 9347–9377, 2001.

38. D. J. Faulkner, "Marine natural products," Natural Product Reports, vol. 17, no. 1, pp. 7–55, 2000.
39. D. J. Faulkner, "Highlights of marine natural products chemistry (1972–1999)," Natural Product Reports, vol. 17, no. 1, pp. 1–6, 2000.
40. D. J. Faulkner, "Marine natural products," Natural Product Reports, vol. 19, no. 1, pp. 1–48, 2002.
41. G. M. König and A. D. Wright, "Marine natural products research: current directions and future potential," Planta Medica, vol. 62, no. 3, pp. 193–211, 1996.
42. K. Liberra and U. Lindequist, "Marine fungi—a prolific resource of biologically active natural products?" Pharmazie, vol. 50, no. 9, pp. 583–588, 1995.
43. J. K. Volkman, "Australasian research on marine natural products: chemistry, bioactivity and ecology," Marine and Freshwater Research, vol. 50, no. 8, pp. 761–779, 1999.
44. B. Haefner, "Drugs from the deep: marine natural products as drug candidates," Drug Discovery Today, vol. 8, no. 12, pp. 536–544, 2003.
45. N. Fusetani, "Introduction," in Drugs from the Sea, N. Fusetani, Ed., Karger, Basel, Switzerland, 2000.·
46. A. Kelecom, "Chemistry of marine natural products: yesterday, today and tomorrow," Anais da Academia Brasileira de Ciencias, vol. 71, no. 2, pp. 249–263, 1999.
47. L. L. Hench, "Bioceramics: from concept to clinic," Journal of the American Ceramic Society, vol. 74, no. 7, pp. 1487–1510, 1991.
48. L. L. Hench and S. Best, "Ceramics, glasses and glass-ceramics," in Biomaterials Science: An Introduction to Materials in Medicine, B. D. Ratner, A. S. Hoffmann, F. J. Schoen, and J. E. Lemons, Eds., Elsevier Academic Press, London, UK, 2004.
49. S. H. Gupta SH, "Mineralized tissues," Tech. Rep., Department of Biomaterials, Biomaterials, Max Planck Institute of Colloids and Interfaces, Postdam, Germany, 2005-2006.
50. P. Fratzl, "Bone material quality and Osteoporosis research," in Research in the Department of Biomaterials, Biomaterials, Max

Planck Institute of Colloids and Interfaces, Biannual Report (2005-2006), P. Fratzl, Ed., Postdam, Germany, 2006.

51. P. Fratzl, The Bone Quality Book—A Guide to Factors Influencing Bone Strength, Materials Properties: Mineral Crystals, Excerota Medica, Amsterdam, The Netherlands, 2006, Edited by Dempster D., Felsenberg D. and Van Der Geest S.

52. F. C. M. Driessens, "Formation and stability of calcium phosphate in relatuion to the phase composition of the mineral in calcified tissue," in Bioceramics of Calcium Phosphate, K. DeGroot, Ed., CRC Press, Boca Raton, Fla, USA, 1983.

53. C. Balázsi, F. Wéber, Z. Kövér , E. Horváth, and C. Németh, "Preparation of calcium-phosphate bioceramics from natural resources," Journal of the European Ceramic Society, vol. 27, no. 2-3, pp. 1601–1606, 2007.

54. P. Chattopadhyay, S. Pal, A. K. Wahi, L. Singh, and A. Verma, "Synthesis of crystalline hydroxyapetite from coral (Gergonacea sp) and cytotoxicity evaluation," Trends in Biomaterials and Artificial Organs, vol. 20, no. 2, pp. 139–142, 2007.

55. G. Guillenium, J. Patat, J. Fournie, and M. Chetail, "The use of coral as a bone graft substitute," Journal of Biomedical Materials Research, vol. 21, no. 5, pp. 557–567, 1987.

56. M. Richard, E. Aguado, G. Daculsi, and M. Cottrel, "Ultrastructural and electron diffraction of the bone-ceramic interfacial zone in coral and biphasic calcium phosphate implants.," Calcified Tissue International, vol. 62, no. 5, pp. 437–442, 1998.

57. C. Demers, C. R. Hamdy, K. Corsi, F. Chellat, M. Tabrizian, and L. Yahia, "Natural coral exoskeleton as a bone graft substitute: a review," Bio-Medical Materials and Engineering, vol. 12, no. 1, pp. 15–35, 2002.

58. J. Chou, B. Ben-Nissan, A. H. Choi, R. Wuhrer, and D. Green, "Conversion of coral sand to calcium phosphate for biomedical applications," Journal of the Australian Ceramic Society, vol. 43, no. 1, pp. 44–48, 2007.

59. J. Hu, J. J. Russell, R. Vago, and B. Ben-Nissan, "Production and analysis of hydroxyapatite from Australian corals via hydrothermal

process," Journal of Materials Science Letters, vol. 20, no. 1, pp. 85–87, 2001.

60. R. Z. LeGeros, "Properties of osteoconductive biomaterials: calcium phosphates," Clinical Orthopaedics and Related Research, no. 395, pp. 81–98, 2002.
61. Y. Barbotteau, J. L. Irigaray, and J. F. Mathiot, "Modelling by percolation theory of the behaviour of natural coral used as bone substitute," Physics in Medicine and Biology, vol. 48, no. 21, pp. 3611–3623, 2003.
62. F. Roux, D. Brasnu, B. Loty, B. Georges, and G. Guillenimin, "Madreporic coral: a new bone graft substitute for cranial surgery," Journal of Biomedical Materials Research, vol. 69, no. 4, pp. 510–513, 1988.
63. E. Arnaud, C. Morieux, M. Wybier, and M. C. De Vernejoul, "Potentiation of transforming growth factor (TGF-β1) by natural coral and fibrin in a rabbit cranioplasty model," Calcified Tissue International, vol. 54, no. 6, pp. 493–498, 1994.
64. R. A. Yukna, "Clinical evaluation of coralline calcium carbonate as a bone replacement graft material in human periodontal osseous defects," Journal of Periodontology, vol. 65, no. 2, pp. 177–185, 1994.
65. A. J. Awang-Hazmi, A. B. C. Z. Zuki, M. M. Noordim, A. Jalila, and Y. Norimah, "Mineral composition of the cockle (Anadara Granosa) shells of west coast of peninsular malaysia and it's potential as biomaterial for use in bone repair," Journal of Animal and Veterinary Advances, vol. 6, no. 5, pp. 591–594, 2007.
66. A. B. Rodríguez-Navarro, C. CabraldeMelo, N. Batista et al., "Microstructure and crystallographic-texture of giant barnacle (Austromegabalanus psittacus) shell," Journal of Structural Biology, vol. 156, no. 2, pp. 355–362, 2006.
67. R. C. Herdman, "U.S. Congress, office of technology assessment, biopolymers: making materials nature's way," Tech. Rep. OTA-BP-E-102, US Government Printing Office, Washington, DC, USA, 1998, 1998, http://www.fas.org/ota/reports/9313.pdf.

68. D. R. Lu, C. M. Xiao, and S. J. Xu, "Starch-based completely biodegradable polymer materials," Express Polymer Letters, vol. 3, no. 6, pp. 366–375, 2009.
69. R. E. Baier, "Advanced biomaterials development from natural products," Journal of Biomaterials Applications, vol. 2, no. 4, pp. 615–626, 1988.
70. R. L. Reis, Natural-Based Polymers for Biomedical Applications, Woodhead Publishing Limited, Abinghton, UK, 2008.
71. L. G. Griffith, "Biomaterials," WTEC Panel Report on Tissue Engineering Research, WTEC, 2003.
72. T. J. Lujan, "Biomaterials science ME 491: lecture 5: natural polymers," 2010, http://www.biomechresearch.org/staff/lujan/teaching/Biomtrl_5%20Natural%20Polymers.pdf.
73. "Natural Polymers US Industry Study with Forecasts to 2010 and 2015," 2007, http://www.bharatbook.com.
74. S. L. Cooper, S. A. Visser, R. W. Hergenrother, and N. M. K. Lamba, "Polymer," in Biomaterials Science: An Introduction to Materials in Medicine, B. D. Ratner, A. S. Hoffman, F. J. Schoen, and J. E. Lemons, Eds., pp. 67–79, Elsevier, London, UK, 2nd edition, 2004.
75. I. V. Yannas, "Natural Materials," in Biomaterials Science: An Introduction to Materials in Medicine, B. D. Ratner, A. S. Hoffman, F. J. Schoen, and J. E. Lemons, Eds., Elsevier, London, UK, 2nd edition, 2004.
76. P. R. Shewry, A. S. Tatham, and A. J. Bailey, Elastomeric Proteins: Structures, Biomechanical Properties, and Biological Roles, Cambridge University Press, 2003.
77. K. Yao, J. Mao, J. Yin et al., "Chitosan/gelatin network based biomaterials in tissue engineering," Biomedical Engineering Applications Basis and Communications, vol. 14, no. 3, pp. 115–121, 2002.
78. S. B. Lee, H. W. Jeon, Y. W. Lee et al., "Artificial dermis composed of gelatin, hyaluronic acid and (1–3), (1–6), -β—Glucan," Macromolecular Research, vol. 11, no. 5, pp. 368–374, 2003.

79. J. Wood, "Tissue-like constructs made in minutes," Materials Today, vol. 8, no. 12, p. 22, 2003.
80. H. J. Jin, J. Park, R. Valluzzi, P. Cebe, and D. L. Kaplan, "Biomaterial films of bombyx mori silk fibroin with poly (ethylene oxide)," Biomacromolecules, vol. 5, no. 3, pp. 711–717, 2004.
81. S. Prasong, S. Yaowalak, and S. Wilaiwan, "Characteristics of silk fiber with and without sericin component: a comparison between botnbyx tnori and philosamia ricini silks," Pakistan Journal of Biological Sciences, vol. 12, no. 11, pp. 872–876, 2009.
82. A. Mueller, Engineering Biomedical Materials from Corn and Wool, Alcht, Chemical Engineering Progress (CEP), 2005.
83. T. Aboushwareb, D. Eberli, C. Ward et al., "A keratin biomaterial gel hemostat derived from human hair: evaluation in a rabbit model of lethal liver injury," Journal of Biomedical Materials Research, vol. 90, no. 1, pp. 45–54, 2009.
84. P. Sierpinski, J. Garrett, J. Ma et al., "The use of keratin biomaterials derived from human hair for the promotion of rapid regeneration of peripheral nerves," Biomaterials, vol. 29, no. 1, pp. 118–128, 2008.
85. W. F. P. Neves-Junior, M. Ferreira, M. C. O. Alves et al., "Influence of fabrication process on the final properties of natural-rubber latex tubes for vascular prosthesis," Brazilian Journal of Physics, vol. 36, no. 2, pp. 586–591, 2006.
86. F. Mrue, J. Coutinho-Netto, R. Ceneviva, J. J. Lachat, J. A. Thomazini, and H. Tambelini, "Evaluation of the biocompatibility of a new biomembrane," Materials Research, vol. 7, no. 2, pp. 277–283, 2004.
87. E. S. Stevens, Green Plastics, an Introduction to the New Acience of Biodegradable Plastics, Princeton University Press, Princeton, NJ, USA, 2002.
88. A. Atala and D. J. Mooney, Synthetic Biodegradable Polymer Scaffolds, Birkhäuser, Boston, Mass, USA, 1997.
89. A. R. Talja, Preparation and characterization of potato starch films plasticized with polyols, Ph.D. thesis, University of Helsinki, Helsinki, Finland, 2007.
90. A. K. Sugih, Synthesis and properties of starch based biomaterials,

Ph.D. thesis, University of Groningen, Groningen, The Netherlands, 2008.

91. C. G. Biliaderis, "Structure and phase transition of starch in food systems," Food Technology, vol. 46, no. 6, pp. 98–109, 1992.

92. J. C. Vengal and M. Srikumar, "Processing and study of novel lignin-starch and lignin-gelatin biodegradable polymeric films," Trends in Biomaterials and Artificial Organs, vol. 18, no. 2, pp. 237–241, 2005.

93. L. S. Nair and C. T. Laurencin, "Polymers as biomaterials for tissue engineering and controlled drug delivery," Advances in Biochemical Engineering/Biotechnology, vol. 102, pp. 47–90, 2006.

94. A. P. Marques, Biofunctionality and immunocompatibility of starch-based biomaterials, Ph.D. thesis, Departmento engenharia polimerios, Universidale Do Minho, Escola de Engenharia, 2004.

95. K. Sriroth, R. Onollakup, K. Piyachomkwan, and C. G. Oates, "Biodegradable plastics from cassava starch in Thailand," 1999, http://www.ciat.cgiar.org/Paginas/index.aspx.

96. E. Khor, "Chitin: a biomaterial in waiting," Current Opinion in Solid State and Materials Science., vol. 6, no. 4, pp. 313–317, 2002.

97. E. Khor, "Chitin and chitosan as biomaterials: going forward based on lessons learnt," Journal of Metals, Materials and Minerals, vol. 15, no. 1, pp. 69–72, 2005.

98. J. F. Louvier-Hernandez, R. A. Mauricio-Sanchez, G. A. Camacho-Bragado, G. Luna-Barcenas, and R. B. Gupta, "Chitin nanofibrous three-dimensional scaffold prepared by supercritical antisolvent precipitation," in Proceedings of the 8th Inter American Congress of Electron Microscopy, La Hanaba, Cuba, September 2005.

99. R. Naznin, "Extraction of chitin and chitosan from shrimp (Metapenaeus monoceros) shell by chemical method," Pakistan Journal of Biological Sciences, vol. 8, no. 7, pp. 1051–1054, 2005.

100. P. K. Dutta, J. Duta, and V. S. Tripathi, "Chitin and chitosan: chemistry, properties and applications," Journal of Scientific and Industrial Research, vol. 63, no. 1, pp. 20–31, 2004.

101. M. M. Saad, "Chelating ability of the chitosan-glucan complex from aspergillus niger NRRL595 biomass recycling in citric acid

production," Research Journal of Agric & Biological Sciences, vol. 2, no. 3, pp. 132–136, 2006.

102. S. Hirano and T. Midorikawa, "Novel method for the preparation of N-acylchitosan fiber and N-acylchitosan-cellulose fiber," Biomaterials, vol. 19, no. 1–3, pp. 293–297, 1998.

103. A. Blasinka, I. Krucinska, and M. Chrzanowski, "Dibutyrylchitin nonwoven biomaterials manufacturing using electrospinning method," Fibres and Textiles in Eastern Europe, vol. 4, no. 48, pp. 51–55, 2004.

104. R. Shelma, W. Paul, and C. P. Sharma, "Chitin nanofibre reinforced thin chitosan films for wound healing application," Trends in Biomaterials and Artificial Organs, vol. 22, no. 2, pp. 107–111, 2008.

105. M. Beran, M. Urban, L. Adamek, L. Jandusik, and J. Speracek, "Applications of mushroom chitosans in medical biomaterials," 2007, http://www.vupp.cz/.

106. S. Honary, M. Maleki, and M. Karami, "The effect of chitosan molecular weight on the properties of alginate/chitosan microparticles containing prednisolone," Tropical Journal of Pharmaceutical Research, vol. 8, no. 1, pp. 53–61, 2009.

107. "Natural polymers US industry study with forecasts to 2010 and 2015," 2007, http://www.bharatbook.com.

108. Y. M. Tsujisaka and M. Tsuhashi, "Pullulan," in Industrial Gums: Polysaccharides and Their Derivatives, R. Whistkler and J. N. BeMiller, Eds., pp. 447–460, Academic Press, San Diego, Calif, USA, 1993.

109. M. R. Rekha and C. P. Sharma, "Pullulan as a promising biomaterials for biomedical applications: a perspective," Trends in Biomaterials and Artificial Organs, vol. 20, no. 2, pp. 116–121, 2007.

110. US Congress, "Office of technology assessment, biopolymers: making materials nature's way," Tech. Rep. OTA-BP-E-102, US Government Printing Office, Washington, DC, USA, 1993.

111. U. Hasegawa, S. M. Nomura, S. C. Kaul, T. Hirano, and K. Akiyosh, "Nanogel-quantum dot hybrid nanoparticles for live cell imaging," Biochemical and Biophysical Research Communications, vol. 331, no. 4, pp. 917–921, 2005.

112. H. Hasuda, O. H. Kwon, I. K. Kang, and Y. Ito, "Synthesis of photoreactive pullulan for surface modification," Biomaterials, vol. 26, no. 15, pp. 2401–2406, 2005.
113. Y. Ito and M. Nogawa, "Preparation of a protein micro-array using a photo-reactive polymer for a cell-adhesion assay," Biomaterials, vol. 24, no. 18, pp. 3021–3026, 2003.
114. D. Hull and T. W. Clyne, An Introduction to Composite Materials, Cambridge University Press, Cambridge, Mass, USA, 2nd edition, 1981.
115. F. L. Matthews and R. D. Rawlings, Composite Materials Engineering and Science, Chapman & Hall/CRC, London, UK, 1994.
116. O. Gåserød, A. Sannes, and G. Skjåk-Bræk, "Microcapsules of alginate-chitosan. II. a study of capsule stability and permeability," Biomaterials, vol. 20, no. 8, pp. 773–783, 1999.
117. A. Sannino, S. Pappadà, M. Madaghiele, A. Maffezzoli, L. Ambrosio, and L. Nicolais, "Crosslinking of cellulose derivatives and hyaluronic acid with water-soluble carbodiimide," Polymer, vol. 46, no. 21, pp. 11206–11212, 2005.
118. T. Ito, Y. Yeo, C. B. Highley, E. Bellas, C. A. Benitez, and D. S. Kohane, "The prevention of peritoneal adhesions by in situ cross-linking hydrogels of hyaluronic acid and cellulose derivatives," Biomaterials, vol. 28, no. 6, pp. 975–983, 2007.
119. A. Sannino, C. Demitri, and M. Madaghiele, "Biodegradable cellulose-based hydrogels: design and applications," Materials, vol. 2, no. 2, pp. 353–373, 2009.
120. Y. Tokiwa, A. Iwamoto, and M. Koyama, "Development of biodegradable plastics containing polycaprolactose and/or starch," American Chemical Society, Division of Polymeric Materials, vol. 63, pp. 742–746, 1990.
121. Y. Tokiwa, T. Ando, T. Suzuki, and T. Takeda, "Biodegradation of synthetic polymers containing ester bonds," Proceedings of the ACS Division of Polymeric Materials, vol. 62, pp. 988–992, 1990.
122. M. Zhai, L. Zhao, F. Yoshii, and T. Kume, "Study on antibacterial starch/chitosan blend film formed under the action of irradiation," Carbohydrate Polymers, vol. 57, no. 1, pp. 83–88, 2004.

123. D. R. Lu, C. M. Xiao, and S. J. Xu, "Starch-based completely biodegradable polymer materials," Express Polymer Letters, vol. 3, no. 6, pp. 366–375, 2009.

124. S. C. Mendes, R. L. Reis, Y. P. Bovell, A. M. Cunha, C. A. Van Blitterswijk, and J. D. De Bruijn, "Biocompatibility testing of novel starch-based materials with potential application in orthopaedic surgery: a preliminary study," Biomaterials, vol. 22, no. 14, pp. 2057–2064, 2001.

125. A. P. Marques, R. L. Reis, and J. A. Hunt, "The biocompatibility of novel starch-based polymers and composites: in vitro studies," Biomaterials, vol. 23, no. 6, pp. 1471–1478, 2002.

126. H. S. Azevedo, F. M. Gama, and R. L. Reis, "In vitro assessment of the enzymatic degradation of several starch based biomaterials," Biomacromolecules, vol. 4, no. 6, pp. 1703–1712, 2003.

127. L. F. Boesel, J. F. Mano, and R. L. Reis, "Optimization of the formulation and mechanical properties of starch based partially degradable bone cements," Journal of Materials Science, vol. 15, no. 1, pp. 73–83, 2004.

128. A. V. Reis, M. R. Guilherme, T. A. Moia, L. H. C. Mattoso, E. C. Muniz, and E. B. Tambourgi, "Synthesis and characterization of a starch-modified hydrogel as potential carrier for drug delivery system," Journal of Polymer Science, A, vol. 46, no. 7, pp. 2567–2574, 2008.

129. N. A. Peppas, P. Bures, W. Leobandung, and H. Ichikawa, "Hydrogels in pharmaceutical formulations," European Journal of Pharmaceutics and Biopharmaceutics, vol. 50, no. 1, pp. 27–46, 2000.

130. K. J. Zhu, L. Xiangzhou, and Y. Shilin, "Preparation, characterization and properties of polylactide (PLA)-poly(ethylene glycol) (PEG) copolymers. A potential drug carrier," Journal of Applied Polymer Science, vol. 39, no. 1, pp. 1–9, 1990.

131. C. S. Pereira, A. M. Cunha, R. L. Reis, B. Vázquez, and J. San Román, "New starch-based thermoplastic hydrogels for use as bone cements or drug-delivery carriers," Journal of Materials Science, vol. 9, no. 12, pp. 825–833, 1998.

132. C. Bastiol, "Starch-polymer composites," in Degradable Polymer-Principles and Applications, p. 112, Chapman & Hall, London, UK, 1995.

133. R. L. Reis, A. M. Cunha, and M. J. Bevis, "Using non-conventional processing routes to develop anisotropic and buiodegradable composites of starch-based thermoplastic reinforced with bone-like ceramics," Medical Plastics and Biomaterials Magazine, vol. 4, pp. 46–50, 1997.

134. W. Bonfield, M. D. Grynpas, A. E. Tully, J. Bowman, and J. Abram, "Hydroxyapatite reinforced polyethylene—a mechanically compatible implant material for bone replacement," Biomaterials, vol. 2, no. 3, pp. 185–186, 1981.

135. W. Bonfield, "Composite biomaterials, in bioceramics," in Proceedings of the 9th International Symposium on Ceramics in Medicine, T. Kukubo, T. Nakamura, and F. Miyaji, Eds., Pergamon, Otsu, Japan, 1996.

136. C. L. Jardelino, D. R. Gomes, I. I. Castro-Silva, M. H. Rocha-Leao, A. M. Rossi, and J. M. Granjeiro, "Tissue behaviour in response to alginate-hydroxyapatite-capsule containing membrane," in Proceedings of the 11th International Conference on Advanced Materials (ICAM ‹09), Rio de Janeiro, Brazil, September0 2009.

137. T. W. Bauer and G. F. Muschler, "Bone graft materials: an overview of the basic science," Clinical Orthopaedics and Related Research, vol. 371, pp. 10–27, 2000.

138. W. K. Czaja, D. J. Young, M. Kawecki, and R. M. Brown, "The future prospects of microbial cellulose in biomedical applications," Biomacromolecules, vol. 8, no. 1, pp. 1–12, 2007.

139. Y. Levi-Kalisman, G. Falini, L. Addadi, and S. Weiner, "Structure of the nacreous organic matrix of a bivalve mollusk shell examined in the hydrated state using Cryo-TEM," Journal of Structural Biology, vol. 135, no. 1, pp. 8–17, 2001.

140. A. G. Checa, J. H. E. Cartwright, and M. G. Willinger, "The key role of the surface membrane in why gastropod nacre grows in towers," Proceedings of the National Academy of Sciences of the United States of America, vol. 106, no. 1, pp. 38–43, 2009.

141. S. Camprasse, G. Camprasse, M. Pouzol, and E. Lopez, "Artifical dental root made of natural calcium carbonate (bioracine)," Clinical Materials, vol. 5, no. 2–4, pp. 235–250, 1990.

142. G. Atlan, N. Balmain, S. Berland, B. Vldal, and E. Lopez, "Reconstruction of human maxillary defects with nacre powder: histological evidence for bone regeneration," Comptes Rendus de l›Academie des Sciences—Serie III, vol. 320, no. 3, pp. 253–258, 1997.

143. M. Joshi, S. W. Ali, and R. Purwar, "Ecofriendly antimicrobial finishing of textiles using bioactive agents based on natural products," Indian Journal of Fibre and Textile Research, vol. 34, no. 3, pp. 295–304, 2009.

144. V. R. Sinha and R. Kumria, "Polysaccharides in colon-specific drug delivery," International Journal of Pharmaceutics, vol. 224, no. 1-2, pp. 19–38, 2001.

145. M. Veerapandian and K. Yun, "The state of the art in biomaterials as nanobiopharmaceuticals," Digest Journal of Nanomaterials and Biostructures, vol. 4, no. 2, pp. 243–262, 2009.

146. F. Iemma, U. G. Spizzirri, F. Puoci et al., "pH-Sensitive hydrogels based on bovine serum albumin for oral drug delivery," International Journal of Pharmaceutics, vol. 312, no. 1-2, pp. 151–157, 2006.

147. C. Y. Gan, L. H. Cheng, E. T. Phuah, P. N. Chin, A. F. M. AlKarkhi, and A. M. Easa, "Combined cross-linking treatments of bovine serum albumin gel beadlets for controlled-delivery of caffeine," Food Hydrocolloids, vol. 23, no. 5, pp. 1398–1405, 2009.

148. O. A. Diak, A. Bani-Jaber, B. Amro, D. Jones, and G. P. Andrews, "The manufacture and characterization of casein films as novel tablet coatings," Food and Bioproducts Processing C, vol. 85, no. 3, pp. 284–290, 2007.

149. K. Yamauchi, H. Hojo, Y. Yamamoto, and T. Tanabe, "Enhanced cell adhesion on RGDS-carrying keratin film," Materials Science and Engineering C, vol. 23, no. 4, pp. 467–472, 2003.

CHAPTER 4

PROGRESS IN BIO-BASED PLASTICS AND PLASTICIZING MODIFICATIONS

Tizazu Mekonnen[a], Paolo Mussone[a], Hamdy Khalil[b] and David Bressler[a]

[a]Biorefining Conversions and Fermentations Laboratory, Department of Agricultural, Food and Nutritional Sciences, University of Alberta, Edmonton, AB T6G 2P5, Canada

[b]The Woodbridge Group, 8214 Kipling Avenue, Woodbridge, ON, Canada L4L 2A4

ABSTRACT

Over the coming few decades bioplastic materials are expected to complement and gradually replace some of the fossil oil based materials. Multidisciplinary research efforts have generated a significant level of technical and commercial success towards these bio-based materials. However, extensive application of these bio-based plastics is still challenged by one or more of their possible inherent limitations, such as poor processability, brittleness, hydrophilicity, poor moisture and gas barrier, inferior compatibility, poor electrical, thermal and physical properties. The incorporation of additives such as plasticizers into the biopolymers is a common practice to improve these inherent limitations. Generally, plasticizers are added to both synthetic and bio-based polymeric materials to impart flexibility, improve toughness, and lower the glass transition temperature. This review introduces the most common bio-based plastics and provides an overview of recent advances in the selection and use of plasticizers, and their effect on the performance of these materials. In addition to plasticizers, we also present a perspective of other emerging techniques of improving the overall performance of bio-based plastics. Although a wide variety of bio-based plastics are under development, this review focuses on plasticizers utilized for the most extensively studied bioplastics including poly(lactic acid), polyhydroxyalkanoates, thermoplastic starch, proteinaceous plastics and cellulose acetates. The ongoing challenge and future potentials of plasticizers for bio-based plastics are also discussed.

INTRODUCTION

Plastics are amorphous organic solid polymers covering a wide range of polymerization products suitable for the manufacture of diversified products. Worldwide annual plastics production is estimated to surpass 300 million tons by 2015 (ref. 1) representing trillions of dollars in terms of global economic returns.[2] Plastics are highly valued materials because of their low cost and extraordinary versatility and they constitute the largest petroleum application second only to energy.[3] Among the many applications of plastics, packaging accounts for almost one-third of their use followed by construction and consumer products.[4]The materials science community has been striving for decades to generate bio-based plastics to substitute or complement conventional synthetic plastics based on exclusively petroleum feedstock. According to current estimates, the global production of bioplastics is expected to grow at an annual rate of up to 30% in the coming decade to reach 3.5 million tonnes in 2020.[5]

Bioplastics may also be bio-based (*i.e.* polymer derived from renewable feedstock) and biodegradable (*i.e.* polymer that can return to nature).[6] Biodegradability and compostability depend on the chemical structure rather than the feedstock source. According to the US Department of Agriculture (USDA), bio-based products are defined as commercial or industrial goods (other than feed or food) composed in whole or in significant part of biological products.[7] Thus, synonymous use of the terms bio-based plastic and biodegradable plastic is not correct. Some of the most commonly known bio-based plastics in today's marketplace in terms of production and renewability are poly(lactic acid) (PLA), polyhydroxyalkanoates (PHAs), starch plastics, cellulose esters and protein based plastics (Fig. 1). Other bio-based plastics, such as bio enriched polyurethane manufactured using modified vegetable oils, polyethylene monomers derived from the dehydration of bio-ethanol, polypropylene monomers derived from dehydration of bio-butanol and poly(ethylene terephthalate) monomers produced *via* fermentation, catalytic pyrolysis or gasification of biomass,[8] that have at least partial sourcing from plants constitute emerging technologies expected to make a significant market impact.

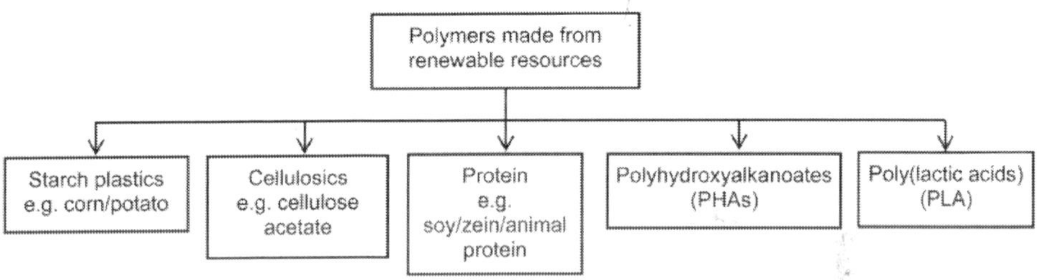

Figure 1: Major bio-based plastics and their production routes.[6]

Bio-based plastics could overcome the sustainability issues and environmental challenges posed by the production and disposal of synthetic plastics. However, the large scale commercial deployment of bio-based plastics to replace conventional plastic materials remains challenged by several factors. Some of the challenges are attributed to the relatively poor performance, variability of properties of the feedstock associated with location and the time of harvest, high production cost and lack of infrastructure. Recent development in bacteria synthesized plastics (PHAs) and the utilization of nature's own building blocks such as proteins, fats, carbohydrates, lignin, *etc.* obtained from agricultural feedstock and agricultural industry wastes constitute a major progress towards bio-based plastics in the last decade.

Plasticizers have long been known for their effectiveness in enhancing the flexibility of synthetic plastics such as polyvinyl chloride (PVC) and epoxy resins. New types of plasticizers compatible with bio-based plastics are being developed. For technical and economic reasons, polymer additives are a large and increasingly significant component of the polymer industry.[9] Among the additives, plasticizers constitute about one third of the global additive market,[10] with a worldwide consumption of over 4.6 million metric tonnes in 2003 (ref. 11) and over 6.4 metric tonnes in 2011.[12] Generally, plasticizers are small, relatively non-volatile, organic molecules that are added to polymers to reduce brittleness, impart flexibility, and improve toughness, reducing crystallinity, lowering glass transition and melting temperatures.[13,14] Plasticization reduces the relative number of polymer–polymer contacts thereby decreasing the rigidity of the three-dimensional structure thereby allowing deformation without rupture.[15] Consequently, plasticizers improve processability, flexibility, durability and in some cases reduce the cost of polymers.[16,17]

The use of plasticized polymers in pharmaceutical applications ranging from packaging materials or auxiliary substances in conventional dosage forms to membranes or matrices modifying and controlling the drug release characteristics in therapeutic systems has been reported in the literature.[16,18,19] The processing behavior, such as film formation and coating dispersion, and properties of polymers in various applications are greatly improved by adequate choice of plasticizer type and quantity.[16,20]Generally, the choice of these plasticizers to be used as modifiers of plastics is limited by the required safety, environmental favorability, chemical and physical property that dictate their miscibility, processing temperature and required flexibility towards the target application.[17]

The risk of leaching out of certain plasticizers, such as phthalates during storage or end-user application, constitutes a major safety risk.[21–24] This coupled with other shortcomings (*e.g.* toxicity, poor compatibility) limits some plasticizers from application in the medical, pharmaceutical and food packaging fields. The ideal plasticizer significantly lowers the glass transition temperature (T_g), is biodegradable, nonvolatile, and nontoxic, and exhibits minimal leaching or migration during use or aging.

Recent advances in bio-based plastics are spurred by factors such as public concern over the depletion of petroleum based raw materials, the desire of manufacturing companies to develop more sustainable raw material sources, the improvement in properties as well as cost competitive relationship of bioplastics.[25,26] As these bio-based plastic industries continuously grow, the demand for new types of plasticizers with new characteristics, performance and other additives that are compatible with the bioplastics also grows in the same direction.[27] In the realm of developing packaging materials from bio-based materials, a high ductility at room temperature is required and thus, there is no tolerance for the polymer film tearing or cracking when subjected to stresses during package manufacturing or use.[28] Moreover, increase in the utilization of plasticized polymers for biomedical and pharmaceutical application,[16] the search for safer plasticizers for commodity plastics such as poly(vinyl chloride)[29–31] and efforts to produce renewable and biodegradable plasticizers[29,30] constitute an additional motive for the recent development of new plasticizers. This review briefly reports recent progress in the development of plasticizers utilized for bio-based plastics, and their influence on the performance of bio-based plastics.

PLASTICIZATION MECHANISM

Two types of plasticizers are defined in polymer science: internal and external.[22,23,32] Internal plasticizers are part of the polymer molecules, co-polymerized into the polymer structure, grafted or reacted with the original polymer thereby making the polymer chains more difficult to fit and compact with each other closely.[23] They soften polymers by lowering the glass transition temperature (T_g) and reducing the elastic modulus.[22] External plasticizers, on the other hand, are low volatility molecules added to interact with polymers and produce swelling without chemical reaction. Internal molecular forces between plasticizer molecules and between a plasticizer and a polymer such as dispersion forces, induction forces, dipole–dipole interaction, hydrogen bonds are important in external plasticization.[23]

Several theories have been proposed to explain the mechanism and action of plasticizers on polymers. Among those theories, the following plasticizing mechanisms have been widely accepted to describe the effect of plasticizers on polymeric networks:[33–36] (a) the lubricity theory: this theory is similar to metal parts lubrication by oil. The plasticizer acts as a lubricant to reduce friction and facilitates polymer chain mobility past one another, consequently lowering deformation; (b) the gel theory: this theory extends the lubricity theory and suggests that a plasticizer disrupts and replaces polymer–polymer interactions (hydrogen bonds, van der Waals or ionic forces, *etc.*) that hold polymer chains together resulting in reduction of the polymer gel structure and increased flexibility; and (c) the free volume theory: for any polymeric material the free volume is defined as the internal space available in a polymer for the movement of chains. Free volume is usually described as the difference between the observed volume at absolute zero and the volume measured at a selected temperature. Rigid resins are characterized by limited free volume whereas flexible resins have relatively large amounts of free volume. Plasticizers increase the free volume of resins and also maintain the free volume after the polymer–plasticizer mixture post processing is cooled down. The free volume theory explains the effect of plasticizers in lowering the glass transition temperature.

Although these theories are widely accepted and utilized in the selection of plasticizer for polymers, Shtarkman and Razinskaya[35] stressed the limitation of the current plasticization theories. According to these authors,[35] the plasticization theories are limited and not feasible for plasticizer selection for the following reasons: (1) direct studies of the plasticization mechanism is lacking and (2) the existing plasticization theories have limited predicting capability and are limited to only particular cases. For this purpose, the authors[35] suggested the necessity of a compatibility–efficiency–property study that takes into account the structure of the polymeric system to select a specific plasticizer rather than relying on the theories.

The aforementioned plasticization theories/mechanisms were developed for synthetic plastics, particularly PVC. Limited attention has been devoted to developing new theories/mechanisms or improving established theories to explain the plasticization mechanism of the newly developed and emerging bio-based plastics. The complex nature of some of the biological feedstock macromolecules makes bio-based plastics radically different from the common repeating monomer based synthetic polymers. Hence, renewed efforts are required to investigate other more explanatory plasticization possibilities and theories.

PLASTICIZATION OF BIO-BASED PLASTICS

Poly(lactic acid) plastics

Poly(lactic acid) (PLA) is one of the most promising innovative plastics for various end-use applications. This polymer is thermoplastic, renewable, biodegradable and biocompatible, a set of highly attractive attributes for pharmaceutical, biological and medical applications.[37,38] The raw material of PLA, L-lactic acid, can be produced by fermentation of renewable sugar resources such as starch and other polysaccharides.[39,40] Moreover, PLA exhibits a remarkable balance of performance properties comparable to traditional thermoplastics[39] processed using conventional plastic processing techniques. From a physical property standpoint it is often loosely compared to polystyrene.[37] Similar to polystyrene, standard grade PLA has high modulus and strength.[37,41] Moreover, the degradation products of polylactides are nontoxic which enhances practical applications in biomedicine.[42] PLA is currently being commercialized for a wide spectrum of technologically important fields and applications by companies such as Cargill and Dow Chemicals.[38]

PLA belongs to the family of aliphatic polyesters commonly made from lactic acid (2-hydroxypropionic acid) building block shown in Fig. 3. The synthesis of lactic acid into high-molecular weight PLA can follow two different routes of polymerization,[40,43] as depicted in Fig. 4. The monomer lactic acid is condensation polymerized to yield a low-molecular weight, brittle, glassy polymer in the first route, which, for the most part, is unusable unless external coupling agents are used to increase the molecular weight of the polymer.[40] The second route of producing PLA is to collect, purify, and ring-open and polymerize lactide to yield high molecular weight (average $M_w > 100\,000$) PLA.[40,43,44]

The combination of the chiral lactic acid monomers (Fig. 2) or the depolymerization of low molecular weight PLA (Fig. 3) could give rise to distinct forms of polylactides. These polylactides are poly(L-lactide) (or LL-lactide), poly(D-lactide) (or DD-lactide), poly(LD-lactide) (or *meso*-lactide) as shown in Fig. 4 or a mixture of L-and D-lactides, called racemic lactide (*rac*-lactide).[38,46,47] While the D- and L- lactides are optically active, *meso*- is not (Fig. 4).[46] Highly crystalline PLA can be obtained with low D content (<2%), fully amorphous PLA on the other hand can be obtained with high D content (>20%).[48] Semi-crystalline PLA is obtained with 2 to 20% of D content.[2] The amount and stereosequence of these lactides in the polymer backbone give rise to a wide range of molecular weights. These changes as a result impact the melt behavior, thermal, mechanical, optical properties, barrier properties and biological properties of PLA.[49,50]

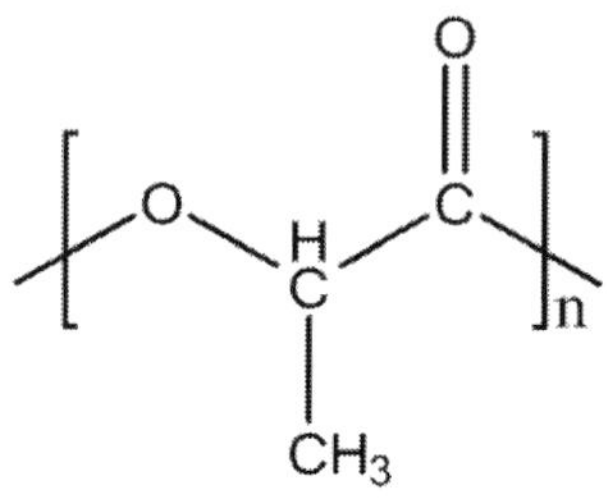

Figure 2: Basic structure of PLA.

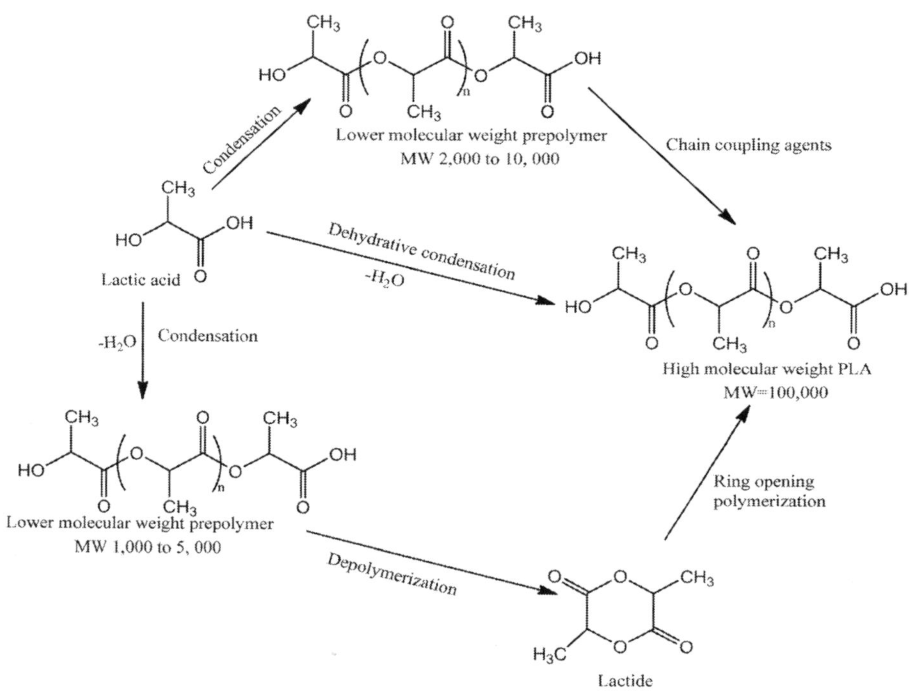

Figure 3: Synthesis methods for high molecular weight PLA.[44,45]

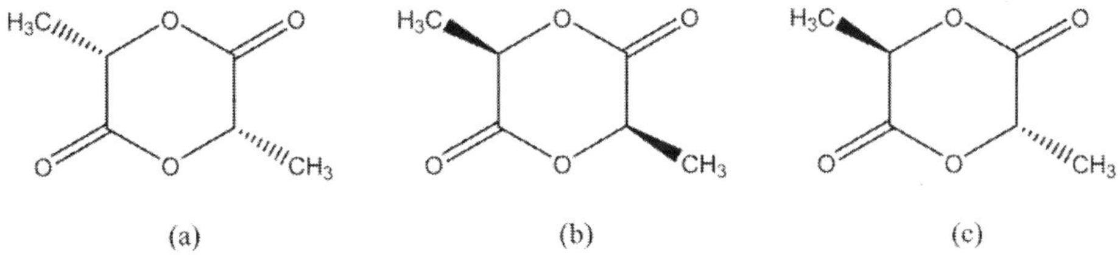

Figure 4: Chemical structures of dimeric (a) D-lactide, (b) L-lactide and (c) *meso*-lactide.

PLA is brittle, with relatively poor impact strength and low thermal degradation temperature limiting its applicability.[39,40]Relatively poor strength, coupled with its hydrophobicity, semi-crystalline properties, limited thermal processability, lack of reactive functional groups along the polymer backbone and high cost constitute the majority of its limitation in wide industrial and medical applications.[39,45] Accordingly, to compete with the low-cost and flexible commodity polymers and upgrade the PLA performance, considerable research effort is being carried out. These attempts include modifying PLA with plasticizers, blending with other polymers,[51] copolymerization, and incorporation of fillers.[37,45,52]

Plasticizers for poly(lactic acid).

Low molecular weight compounds such as oligomeric lactic acid, glycerol, triacetin, and low molecular weight citrates,[17] partial fatty acid esters[53,54] are common plasticizers of PLA. A large number of investigations have also been reported on blending PLA with various polymers as plasticizers, for example, thermoplastic starch (TPS),[51]poly(ethylene oxide),[55,56] poly(ethylene glycol)(PEG),[55] poly(ε-caprolactone),[55–57] poly(vinyl acetate),[58] poly(hydroxy butyrate),[59]cellulose acetate,[60] poly(butylene succinate),[61] and poly(hexamethylene succinate),[61] to improve its flexibility. Most of the resulting plasticized PLA materials exhibited better impact resistance, increased deformation at break and improved resilience.Table 1 reports common monomeric and polymeric plasticizers and their plasticization effects on the glass transition temperature and mechanical property of PLA.

Table 1: Mechanical and thermal performance of PLA plasticized with different modifiers

Modifiers	M_W (g mol^{-1})	Conc. (wt%)	T_g (°C)	E^a (MPa)	ε^a (%)	σ^a (MPa)
PLA[54,55]	137 000	100	59	1720	7	51.7
Triethyl citrate[54]	276	20	32.6		382	12.6
Tributyl citrate[54]	360	20	17.6		350	7.1
Acetyl triethyl citrate[54]	318	20	30		320	9.6
Acetyl tributyl citrate[54]	402	20	17		420	9.2
Poly(oxyethylene)[55]	10 000	21	31	320	7	49
Poly(ε-caprolactone)[55]	10 000	20	35	961	25	19
Glycerol[51]	92.09	20	53		—	—
PEG monolaurate[51]	400	20	21		142	—
Plasticized TPS[51]	—	25	—		2.9	30.2
PEG[62]	1500	10	34.3	1750	150	15.1
PEG[62]	1500	20	23.2	1460	150	14.6

^{a} Tensile modulus (E), tensile stress at yield (σ), and elongation at break (ε).

The results in Table 1 show that all citrates at 20% concentration reduced the glass transition temperature and improved the flexibility while reducing the tensile strength of the PLA control. A significant improvement of elongation at break was achieved at the expense of tensile strength. Ljungberg and Wesslen[63] also investigated the use of triacetin, tributyl citrate, triethyl citrate, acetyl tributyl citrate, acetyl triethyl citrate as potential plasticizers of PLA and reported a drastic lowering of the glass transition temperature of PLA at concentrations as low as 15% resulting in a homogeneous and flexible film. However, it was reported that the migration of citrates onto the film surfaces during aging, especially the low molecular weight citrates, was a major challenge.[17] This issue could be addressed by increasing the molecular weight. For instance, Ljungberg and Wesslen[17]transesterified tributyl citrate (M_W 360 g mol^{-1}) with diethyl glycol that resulted in two oligomeric plasticizers with higher molecular weights (M_W 4500 g mol^{-1} and 63 600 g mol^{-1}). The investigation of the effects of these oligomers on thermo-mechanical and aging properties of PLA shows that both oligomers did not lower the T_g as greatly as monomeric citrates. Among the two oligomeric plasticizers, a relatively larger reduction in T_g was achieved by the oligomer with the lower molecular weight.

Similarly, molecular weight variation, concentration and the presence of polar amide groups of plasticizers can positively interact with PLA chains, affecting the compatibility between PLA and the plasticizer and controlling elongation and morphological stability that result in leaching during aging or use.[17,63] The plasticizer with lower molecular weight that resulted in lower T_g of PLA may also facilitate the migration of the plasticizer from the bulk of the material compared to the higher molecular weight plasticizer.[17] The effect of triacetin (0–30%) and tributyl citrate (0–25%) loading on PLA was studied[63] and an almost linear decrease in T_g with the increase of plasticizer content was observed.

Other methods of improving PLA performance.

New PLA toughening strategies using citrate family (tributyl *O*-acetylcitrate and tributyl citrate) to improve the PLA ductility were reported by Hassouna *et al.*[64] This strategy involves grafting of tributyl citrate onto neat and maleic anhydride modified PLA with tributyl *O*-acetylcitrate.[62] The maleation of PLA was carried out by reactive extrusion with the aim of incorporating hydroxyl functional groups into the PLA. The neat PLA and hydroxyl functionalized PLA were then copolymerized with tributyl citrate that already contains a hydroxyl functional group. Such toughening was shown to drastically decrease the T_g of PLA. However, the grafting reaction of tributyl citrate into anhydride-grafted-PLA revealed a shift of PLA T_g toward higher values compared to neat PLA grafted with tributyl citrate. After six months of aging, no phase separation was observed and no major leaching phenomenon was noticed in both cases. These observations indicated that the mobility restriction as a result of hydrogen bonding occurring between PLA and tributyl citrate as well as the grafting reaction of tributyl citrate into anhydride-grafted-PLA diminished the leaching phenomena.

Maglio *et al.*[55] studied the copolymerization of PLA with poly(ε-caprolactone) (PCL) and poly(oxyethylene) (PEO) to improve the brittleness and reduce T_g. The copolymers obtained as a result exhibited high elongations at break and as a result much lower tensile moduli than the PLA pure polymer (Table 1). The T_g of the copolymers was also much lower than that of the original PLA. A recent study by Hassouna *et al.*[62] investigated grafting of poly(ethylene glycol) (PEG) onto PLA through reactive extrusion to develop plasticized PLA. It was shown that the T_g and modulus were invariably reduced. In all cases, both T_g and elastic modulus were dependent on the content of PEG grafted onto the PLA. The *in situ* reactive grafting of PEG onto PLA exhibited a marked T_g reduction than the blending option. Other plasticizers of PLA reported in the literature include epoxidized soybean oil,[65] ionic liquids,[66] mixed plasticizers,[67] *etc.*

In summary, it can be pointed out that several studies have demonstrated that plasticizers play a significant role in determining the performance properties of PLA plastics. Plasticizers can solve most of the problems that occur during processing or in final use. New characteristics of PLA observed during plasticization may also pave the way to novel applications. The limitations of the currently studied PLA plasticizers include leaching during use, lack of thermal stability, need for offering more ductility and more performance, biocompatibility issues, cost, need for high percentage loading to lower price of PLA, need for a bio-based plasticizer that reduces the overall carbon footprint, *etc.*

Polyhydroxyalkanoates (PHAs)

Microbial-produced PHAs are fully biodegradable bio-polyesters produced by a wide variety of microorganisms for internal carbon and energy storage as part of their survival mechanism.[68,69] PHAs, also known as poly(4-alkan-2-oxelanones) according to IUPAC naming, have attracted much attention recently as alternative polymeric materials that can be produced from renewable and biowaste resources. PHAs, with the general structure shown in Fig. 5, vary widely in their structure and properties (flexibility, crystallinity, melting temperature, *etc.*), depending on the producing microorganisms, the conditions of biosynthesis and the type of carbon source.[70] PHAs are piezoelectric, perfectly isotactic/optically active and biocompatible thermoplastic polyesters amenable to melt-processing into various final forms.[71–74] High molecular weight PHAs have attracted considerable attention as potential replacements for non-degradable commodity plastics (*e.g.* polyethylene and polypropylene), as well as biodegradable and biocompatible biomaterials for implant purposes.[72,75]

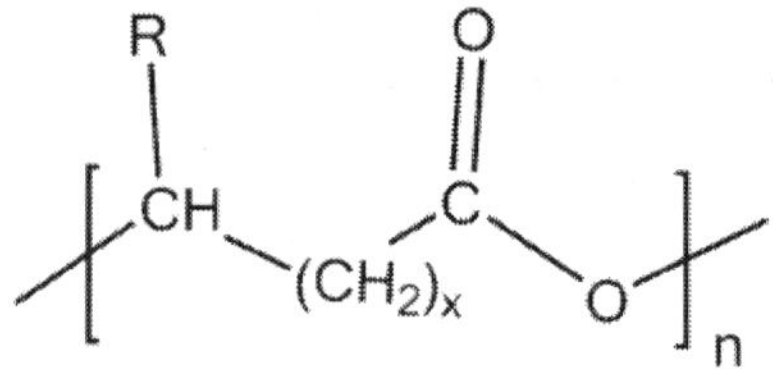

Figure 5: General structure of PHAs. *n* = 1 (R = hydrogen, poly(3-hydroxypropionate); R = methyl, poly(3-hydroxybutyrate); R = ethyl,

poly(3-hydroxyvalerate); R = propyl, poly(3-hydroxyhexanoate); R = pentyl, poly(3-hydroxyoctanoate); R = nonyl, poly(3-hydroxydodecanoate)), *n* = 2 (R = hydrogen, poly(4-hydroxybutyrate)), and *n* = 3 (R = hydrogen, poly(5-hydroxyvalerate)).

Poly(3-hydroxybutyrate) (PHB), R = methyl, being the first among the isolated PHAs, is the most extensively studied PHA produced in nature in the presence of excess carbon by bacteria as storage granules providing food, energy and reducing power.[76,77] This polymer and its copolymer with polyhydroxyvalerate to make poly(3-hydroxybutyrate-*co*-3-hydroxyvalerate) (PHBV) are at present the only known PHAs relevant for practical applications.[78] PHB and PHBV are completely biodegradable in the environment and can be extruded, moulded and spun on conventional plastic processing equipment.[71,79] These features make these polymers an ideal candidate for the production of biodegradable packaging materials and other disposable articles.[79] However, the industrial scale production of PHB/PHBV is hindered by roadblocks. Thermal processing is challenging because of their relative low decomposition temperatures near their melting points, pronounced brittleness, very low deformability and susceptibility to a rapid thermal degradation.[78,80] Furthermore, the current cost of production of PHB is high compared to other high-volume synthetic plastics.[70,78] Because of its limited thermal stability, the melt flow index changes rapidly with time and its volatile decomposition products need to be handled safely. PHB's slow crystallization rates also lead to tacky products (*e.g.* fibres, films, *etc.*).[70] Its copolymer with valerate (PHBV) has overall better properties, especially regarding improved toughness with an acceptable loss of strength and modulus.[73] However, the present large-scale production cost of PHBV remains higher than that of PHB.[78]

The toughness and processability of PHB can be improved by incorporation of the hydroxyvalerate (HV) monomers in the bacterial fermentation process.[81] While PHBV with a high HV content has high flexibility, low crystallinity, and low crystallization rate, it compromises the yield strength and Young›s modulus of PHB, which can result in rubbery materials, meanwhile, it increases the cost of materials.[81] Various approaches, such as use of nucleating agents, plasticizers and agents that promote crystallization of the polymer, modification of the polymer structure, blending, *etc.*, have been carried out to overcome the processing and product difficulties and other shortcomings.[70,73,78,82]

Plasticizers for PHAs.

The use of monomeric and polymeric plasticizers of PHA lowers the glass transition temperature and the melting points, allowing processing at lower temperature and avoiding thermal degradation.[79,82] In addition, plasticizers improve both toughness and softness of the polymer by decreasing its crystallinity, weakening the intramacromolecular bonding and facilitating conformational changes.[83] The plasticization of PHBV using soybean oil, epoxidized soybean oil, epoxidized linseed oil, dibutyl phthalate, polyester plasticizer (Lapol 108), triethyl citrate, acetyl tributyl citrate and polyethylene glycol has been described in the literature[82–88] as shown in Table 2.

Table 2: Effect of plasticizing modifiers on the glass transition temperature and mechanical properties of PHBV

Sample	Modifier concentration (wt%)	M_w (g mol^{-1})	T_g (°C)	ε (%)	TS (MPa)
PHBV[83]	100	680 000	−6.6	6	43.1
Soy oil[83]	20	814.3	−3.4	3	33.7
Epoxidized soy oil[83]	20	874.2	−19.0	7.2	22.1
Dibutyl phthalate[83]	20	278.2	−28.5	10	11.7
Triethyl citrate[83]	20	276.1	−30.0	10	10.9
Epoxy soyate[88]	20	—	—	7.51	13.8
Soy oil[88]	10	—	—	5.09	18.7
Epoxidized linseed oil[88]	10	—	—	7.46	19.6

It is observed from Table 2 that the plasticizers induced depression of glass transition temperature and improvement in the elongation at break with all the plasticizing additives used, with the exception of triglyceride soy oil. From Park and Choi's study,[83] triethyl citrate was the most effective plasticizer in terms of reduction of the glass transition temperature as well as in terms of improvement of the impact strength and elongation. The difference in the effectiveness of these plasticizers can be attributed to the variation in the combined effect of chemical structure, molecular weight compatibility or solubility of the plasticizer with the polymer.[83] On the other hand, studies of impact strength and elongation properties by Seydibeyoglu *et al.*[88] showed that functionalized oils such as epoxy soyate are much more effective than triglycerides of epoxidized soybean or linseed oil. This might be due to the better reactivity of epoxy soyate than the counterpart triglycerides owing to its lower molecular size and simple molecular structure.

The use of low molecular weight, biodegradable and non-toxic compounds as plasticizing additives such as dibutyl sebacate (DBS), dioctyl sebacate (DOS), polyethylene glycol (PEG), Laprol503 (L503), Laprol5003 (L5003), and a nonpolar polymer polyisobutylene (PIB) with concentration up to 50 wt% was investigated to improve the deformative characteristics of PHB.[89] These plasticizers were completely compatible with the polymer and formed a monophase system in mixtures of up to 15–20 wt%. Conversely, when the concentration was beyond 20 wt% the system becomes considerably weak, because of overloading. The majority of the plasticizers examined by the author cause a considerable decrease in crystallization temperature and improvement of mechanical properties. Other plasticizers reported in the literature include dodecanol, lauric acid, tributyrin, and trilaurin.[90]

Other methods of toughening.

Blending of PHB or PHBV-based materials with polymers such as poly(butylene succinate),[81] poly(ethylene succinate),[81,91] polyethylene,[92] polypropylene,[92] poly(ε-caprolactone),[93] poly(lactic acid),[93,94] *etc.* has been extensively examined, and improvements of mechanical and thermal properties were reported. However, in most of these studies poor interfacial adhesion and phase separation of the PHB (V) and the other polymer blends was reported. These limitations could be improved through the use of a compatibilizing agent, functionalization, chain extension, controlled chemical crosslinking and optimizing the process conditions.

Ma *et al.*[81] used a free radical initiator (dicumyl peroxide) to induce compatibilization and partial crosslinking between PHB and PBS. The resulting compatibilized blends were shown to have a smaller particle size, improved interfacial adhesion and consequently resulted in improved tensile strength, impact toughness and elongation at break. Sadi *et al.*[95] evaluated the compatibilization efficiency of polypropylene/PHB blends with copolymers such as poly(propylene-g-maleic anhydride), poly(ethylene-*co*-methyl acrylate), poly(ethylene-*co*-glycidyl methacrylate), and poly(ethylene-*co*-methyl-acrylate-*co*-glycidyl methacrylate). Their study showed that poly(propylene-g-maleic anhydride), having the strongest adhesion between the phases, was the most efficient in terms of improving the mechanical performance of the blend.

Optimization of the processing conditions, taking the relationship among structure, composition, and polymer properties into account, is of particular importance as well.[96] High shearing rates during process operations such as extrusion (10–100 000 S^{-1}) and injection molding (1000–100 000 S^{-1})[97] in addition to the heating applied during processing are expected to change the molecular weight and as a result its performance. For instance, Yamaguchi *et al.*[98] reported 25–30% molecular weight reduction of PHB through shearing of PHB at 180 °C, at a shear rate of 6.3 s^{-1} within 5 min interval in addition to an order-of-magnitude decrease in shear viscosity.

Thermoplastic starch (TPS)

Starch, a polysaccharide of granular structure, is one of the most attractive feedstock for the development of biodegradable polymers because it is relatively inexpensive, abundant and renewable. Starch plays an important role both in the development of the commercialized bio-based plastics[99] and in the bio-ethanol industry. The role of starch in the development of bio-based plastics could be in the development of (a) thermoplastic starch (TPS) plastics where TPS is obtained through direct modification of starch, (b) poly(lactic acid) where its feedstock (lactic acid) is originated from starch derived sugars fermentation and (c) PHAs where starch derived sugars are used as a carbon source for the microorganisms producing PHAs.

Starch is composed of two homopolymers of D-glucose: the linear (1,4)-linked α-D-glucan amylase, typically constituting about 30% of starch depending on the source of starch, and a highly branched (1,6)-linked α-D-glucan amylopectin (Fig. 6). Commonly, amylopectin takes part in the formation of a crystalline structure and amylose does not.[100] Virgin starch is brittle and difficult to process into articles due to its relatively high glass transition and melting temperatures. The T_g of virgin dried starch is estimated to be approximately 240 °C,[101] which is above the starting point of its thermal degradation (about 220 °C).[102] High T_g and brittleness of starch are mainly caused by the presence of strong inter- and intra-molecular hydrogen bonds between the starch macromolecules.[101] Furthermore, TPS polymers based solely on starch are extremely water sensitive[103] and can suffer from significant molecular weight change during processing (extrusion or injection molding).[104] These drawbacks limit the possible shapes that can be imparted to the materials into films with adequate mechanical properties[105] and thus of limited practical value. Therefore, starch must be modified to breakdown the crystalline granules, decrease the T_g and melting temperature (T_m) either by incorporating plasticizers,[106] blending with other polymers,[51,52] chemical modification or combinations before they can be processed into plastics.[107]

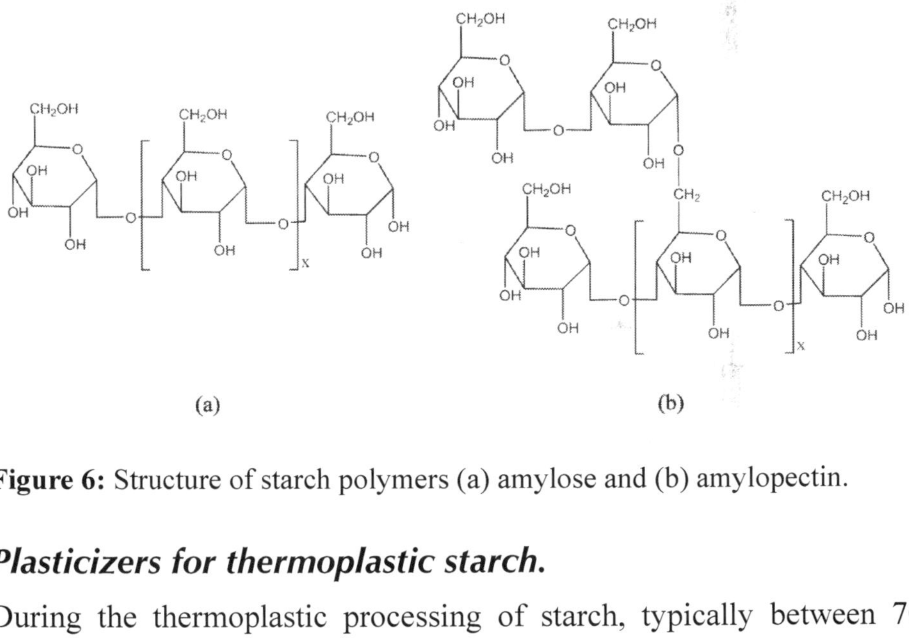

Figure 6: Structure of starch polymers (a) amylose and (b) amylopectin.

Plasticizers for thermoplastic starch.

During the thermoplastic processing of starch, typically between 70 and 90 °C in the presence of a plasticizer (*e.g.* water), a semicrystalline granule of starch is transformed into a homogeneous material with hydrogen bond cleavage between starch molecules. This process, called gelatinization, leads to loss of crystallinity[106] and is associated with the loss of double helices as well as with the loss of lamellar and long range crystalline structure.[108] Plasticizers penetrate starch granules and destroy the inner hydrogen bonds of starch, and eliminate starch–starch interactions because they are replaced by starch–plasticizer interactions.

The plasticized moldable thermoplastic material, called TPS, is fit for injection molding, extrusion or blow molding similar to other synthetic thermoplastic polymers.[109]

There are several substances used as plasticizers for the preparation of thermoplastic starch. Some of the most studied and reported TPS plasticizers in the literature include polyols such as glycerol, glycol, sorbitol, xylitol, maltitol, ethylene glycol, propylene glycol, butanediol;[107,110,111] sucrose, fructose, mannose,[112] fatty acids (such as myristate or palmitate),[113] *etc.* It is also necessary to note that water is a good plasticizer of starch. However, the use of water alone as a plasticizer is not preferable because the resulting product will be brittle when equilibrated with ambient humidity[110] and due to volatilization of water. Glycerol, a classical plasticizer of starch, is perhaps the most widely studied and used polyol plasticizer of TPS. This is because of its low cost, nontoxicity (for food and biomedical application) and high boiling point (292 °C). Moreover, the hydrolysis and/or transesterification of lipids (triglycerides) into fatty acids for the biodiesel industry produce glycerol as a by-product. Utilizing such by-products provides glycerol with an additional market driver in addition to the opportunity of improving the economics of both the biodiesel and the bioplastic industries. Nonetheless, glycerol is known to leach out during aging and humidity exposure, a major limitation for large scale applications.

The properties of plasticized starch can be tuned by changing the temperature of processing, water content and the properties and amount of plasticizers. For instance, Yu *et al.*[101] reported that the elongation at break of the thermoplastic starch is significantly improved by plasticization with glycol, glycerol and hexylene glycol. In addition, the thermal properties of plasticized starch are a function of water and plasticizer content.[110,114] The source of starch is also important for the property of TPS. This is because starches from various sources have different amylose/amylopectin ratios, molecular weights, molecular weight distributions and granular size crystallinity (Fig. 7). This as a result influences the gelatinization and glass transition temperatures[115,116] that are directly correlated with the thermoplasticity of the TPS. The effect of various plasticizers at different concentrations on the gelatinization temperatures, thermal stability, and glass transition temperature has been studied and reported in the literature. Some of the plasticizers and their effects are reviewed and shown in Table 3.

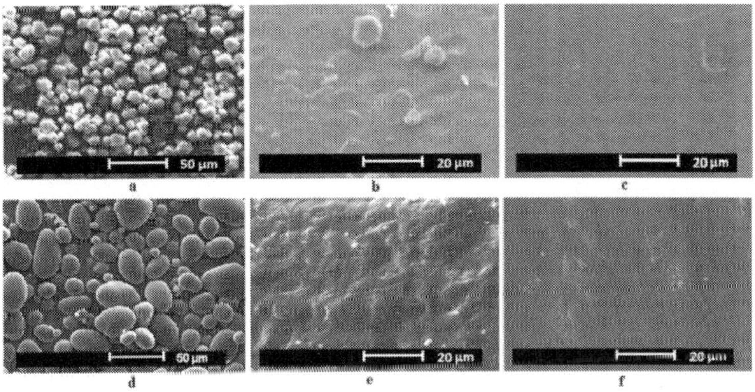

Figure 7: Scanning electron microscopy micrograph of (a) native corn starch granules, (b) ethylenebisformamide (25%) plasticized TPS, (c) ethylenebisformamide (30%) plasticized TPS (adapted from: ref. 126 John Wiley & Sons, copyright © 2006) and (d) native potato starch.

Table 3: Common plasticizers for some of the common starch sources

Starch source	Plasticizer	Plasticizer concentration (wt%)	Gelatinization onset (°C)	Gelatinization peak/ conclusion (°C)/T_g
Wheat starch in the presence of water[108]	Glycerol	65	74.7	91.5[a]
	Sorbitol	65	73	92[a]
	Diglycerol	65	90	115[a]
Sago starch[117]	Starch (control)	0	123.7	157.2[b]
	Glycerol	30, 40, 50	149, 152, 141	169, 175, 164[b]
	Sorbitol	30, 40, 50	124, 126, 122	158, 151, 155[b]
	Sorbitol : glycerol (1 : 1)	30, 40, 50	120, 118, 142	150, 147, 176[b]
Corn starch[118,119]	Glycerol,[118,119] PLA,[120] poly(butylene adipate-*co*-terephthalate)[120]	—	—	—
Potato starch[111,121]	Glycerol–xylitol	40	—	−66.4[c]
	Glycerol–sorbitol	40	—	−69.3[c]
	Xylitol–sorbitol	40	—	−44.1[c]
Rice starch[122–124]	Glycerol	20, 25, 30, 35		
	Poly(ethylene glycol)	3, 6, 9		
	Sorbitol	30, 35, 40, 45		
	Others(formamide,[123] urea,[124] propylene and triethylene glycol[124])	10–30		

[a] Gelatinization conclusion. [b] Gelatinization peak. [c] T_g.

Abdorreza *et al.*[117] showed that the type and concentration of plasticizers govern the heat sealability as well as the seal strength of sago starch based films. The same authors showed that sorbitol-plasticized films exhibited significantly better heat sealability than did the glycerol type. However, the highest seal strength was obtained with a combination of sorbitol and glycerol. The effect of starch gelatinization in the presence of high molecular weight polyol plasticizers and water was also studied under static and dynamic conditions by Taghizadeh and Favis.[108] Their investigation showed that glycerol and sorbitol exhibited similar gelatinization temperatures, while an ascending T_g was observed from glycerol to diglycerol and polyglycerol attributed to the viscosity and molecular weight increase and hydroxyl bond density diminution of the latter two plasticizers.

Other plasticizers such as urea, formamide, combinations of urea and formaldehyde,[125] used with thermoplastic corn starch at different concentrations were also reported. Property evaluation by Ma *et al.*[125] showed that mixtures of urea (20 wt%) and formamide (10 wt%) plasticized TPS exhibited better thermal stability, water resistance and better mechanical properties than conventional glycerol plasticized TPS. According to Ma *et al.*,[125] the reasons behind such property improvement with the urea–formamide mix plasticizer could be due to the formation of more stable and stronger hydrogen bonds with the hydroxyl groups of starch molecules than with glycerol. Yang *et al.*[126,127] reported ethylenebisformamide, synthesized from methyl formate and ethylenediamine, as a novel and effective plasticizer of corn starch and potato starch. Ethylenebisformamide was shown to be effective in destroying the crystalline morphology of the native starch granule and conversion into a homogeneous phase TPS through plasticization and extrusion under shear and pressure. The morphology of the native crystalline starch and the homogeneous plasticized starch at 25% and 30% ethylenebisformamide loading was studied by scanning electron microscopy (SEM)[126,127] and shown in Fig. 7.

The SEM study (Fig. 7) clearly showed that the action of ethylenebisformamide and temperature processing (in this case extrusion) resulted in destruction of the crystalline native starch granules (Fig. 7a and d) morphology to form a continuous phase of TPS having a different crystallinity as further confirmed by X-ray diffraction crystallography.[126,127] The effect of plasticizer loading had also an effect on the continuity of the plasticized TPS phase. Higher concentrations resulted in more uniform phases for the studied loading range. Possible hydrogen bond formation between ethylenebisformamide and starch[126] during plasticization is shown in Fig. 8 below. The hydrogen bonds formed can be stronger than the intra and intermolecular bonds in starch, and as a result corn and potato starch were effectively plasticized with ethylenebisformamide.[126,127] In general, besides the plasticizer type, the quantity of plasticizer used and the processing method applied also affect the physical, thermal and mechanical properties of the resulting starch based bioplastics. For example, Flores *et al.*[128] studied and reported the effect of different gelatinization and drying techniques on the performance of glycerol plasticized starch films. The authors[128] finding shows that gelatinization and drying techniques used to obtain TPS films affected network characteristics that as a result determines the changes in physical properties potentially affecting the film performance as well.

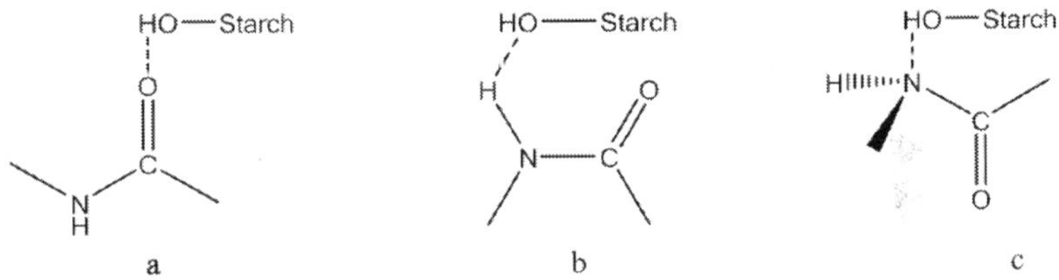

Figure 8: Possible hydrogen bonds between ethylenebisformamide and starch.[126]

In recent studies, the use of novel multifunctional ionic liquid plasticizers such as 1-allyl-3-methylimidazolium chloride[109,129]and 1-butyl-3-methyl imidazolium chloride as a plasticizer[130] and a compatibilizing agent[131] of starch has been reported. Ionic liquids, organic salts that are liquid at ambient temperature, are gaining interest because of their unique properties including non-volatility, non-flammability, low viscosity, chemical and electrochemical stability.[132] These liquids (examples of structures are shown in Fig. 9) have strong hydrogen bond forming abilities with starch owing to their high concentration of chloride ions. TPS plasticized using 1-butyl-3-methylimidazolium chloride shows less hygroscopicity and a much higher elongation at break in the rubbery state than the control glycerol-plasticized TPS samples.[130] The potential application of ionic liquids plasticized starch as solid biopolymer electrolytes was also reported by Wang *et al.*[129] This paves the way for a wide variety of potential applications of TPS bioplastics such as antistatic plastics, electronic shielding, biosensor, and environmentally sensitive membranes.

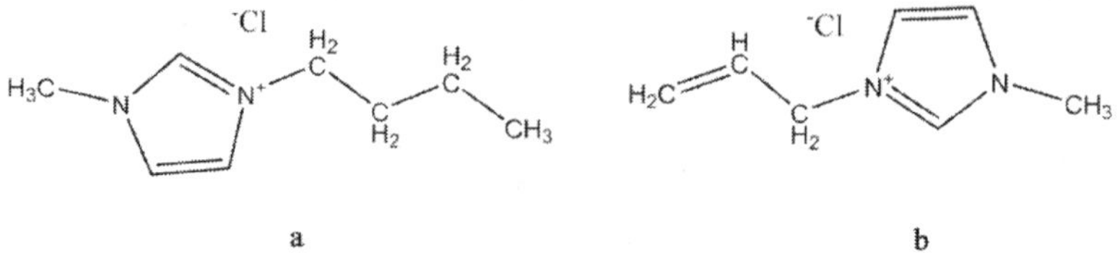

Figure 9: Chemical structure of ionic liquids 1-allyl-3-methylimidazolium chloride and 1-butyl-3-methyl imidazoliumchloride.

Other methods of improving performance of TPS.

Chemical modification can also be an effective method of improving the processability and product performance of TPS. For example, hydroxylation,[133,134] acylation,[135] oxidation[136] and acetylation[136,137] of starch by substituting the ester or ether groups for the hydroxyl were reported to improve the processing behavior, hydrophobicity and mechanical properties. The synthesis of thermoplastic starch acetate with a high degree of substitution through acetylation of starch is one of the common chemical modifications of starch.[137,138] The starch acetates were shown to have higher thermal stability and hydrophobicity due to the reduction of the hydroxyl group with the acetylation. Some of the chemical modifications may reduce biodegradability, biocompatibility and generate some toxic chemical by-product during synthesis.[139] On the other hand, some of the chemically modified starch (*e.g.* starch acetate) could be biocompatible and safe enough to be used in tissue engineering and other medical applications.[138]

Other toughening modifications recently reported in the literature to improve performance and overall economics of TPS based polymers include blending of TPS with protein,[131] PVA,[140] polycaprolactone, polyhydroxybutyrate, polymethacrylate, polystyrene mostly in the presence of urea and polyol plasticizers.[103,120] Surface modifications such as polymeric surface coating,[109] chemical and photo crosslinking of TPS and blends were also shown to reduce surface hydrophilic characteristics and improve water resistance, increase the tensile strength and Young's modulus while decreasing the elongation at break.[140–142]In summary, starch based plastics have grown to represent a major portion of the biodegradable polymer market. The commercial success of TPS polymers is hugely affected by the source, safety, quality, cost and functionality of plasticizers. Selective plasticization/toughening methods of TPS provide an attractive base for developing starch polymers that can be used as biodegradable and renewable packaging materials, environmentally sensitive membranes, and in biomedical and pharmaceutical applications such as drug and protein carriers, tissue engineering applications, *etc.*

Protein based plastics

Proteins are renewable, biodegradable and optically active natural[143] polymers produced by animals, plants and bacteria. Until recently proteins have been utilized exclusively in the food industries. Recent studies on non-food uses of agricultural feedstock initiated an interest in protein based plastics as well. Due to the continuous and cohesive matrix forming ability of proteins, various proteins of both plant and animal origin have received attention for the production of biodegradable plastics, edible films and sheets.[144,145] Furthermore, microencapsulating agents and active compounds in pharmaceutical applications,[146,147]adhesives, blend and composite materials,[142,148] wound dressing[149] and bionanocomposites[142,150] can be produced. Protein based biomaterials can also promote tissue regeneration, such as new bone growth,[149] integrate into blood clots and stimulate collagen deposition, and stimulate cells to produce new tissue, with no need for expensive growth factors.[151] Plant proteins that can be used for bio-based plastics include soy protein,[150] corn zein,[33] wheat protein,[152,153] cottonseed protein,[154] sunflower protein,[155]*etc.* Animal proteins such as blood meal,[156] gelatine and collagen,[157] keratin and feather quill,[158] egg protein,[159] whey protein,[155] meat and bone meal[160,161] can also be used as a feedstock of such bio-based plastics (Table 4).

Table 4: Examples of plasticizers, animal and plant derived protein biomass, and the mechanical properties of the biopolymers made

Plasticizer	Protein studied	TS (MPa)	*E* (MPa)	ε (%)	Ref.
Urea (20%)	Blood meal	12.3	608	—	156
	Wheat gluten	—	—	—	165
Diethyl tartrate (30%)	Feather quill	19.2	1267.9	1.6	158
	Chicken fibers	19.0	907.9	3.3	166
Dibutyl tartrate (30%)	Corn zein	20	1325	—	167

Glycerol (30%)	Corn zein	19.3	620	—	167
	Wheat gluten	6.7	51	118	168
	Soy protein	13.8	250.5	177.5	169
	Feather quill	15.2	380.5	13.6	158
	Chicken feathers	15.7	332.3	8.5	166
	Peanut proteins	8.0	147.0	63.0	168
	Sunflower protein	8.5	—	140	170
Ethylene glycol (30%)	Chicken feathers	17.8	354.0	43.8	166
	Sunflower protein	8.7	—	23	170
	Feather quill	16.8	321	64.9	158
Propylene glycol (30%)	Chicken feathers	22.3	811.2	7.6	166
	Sunflower protein	7.2	—	63	170
	Feather quill	20.5	529.5	11.2	158
	Soy protein	4.5	108.4	8.5	169
Oleic acid (20%)	Gelatin	54	2500	2.9	171
Sorbitol (20%)	Gelatin	52	1997	4.4	171
Mannitol (20%)	Gelatin	57	2250	4.5	171

Proteins are interesting biomaterials based on 20 amino acids which confers a wide range of functional and film-forming properties as a function of various extrinsic or intrinsic conditions such as plasticizer type and concentration.[162] The major drawback of protein-based plastics, with the notable exception of keratin, is their sensitivity towards relative humidity.[163] For example, Zheng *et al.*[163] reported that soy protein sheets submerged in water for 20 h absorbed up to 180% water. In addition, protein films and coatings are often quite stiff and brittle due to extensive intermolecular interactions between protein chains through hydrogen bonding, electrostatic forces, hydrophobic bonding and disulfide cross-linking.[34] Thus, thermoplastic processing of proteins into bio-based plastics is usually accompanied by plasticization and/or other form of modification for the successful development of useful proteinaceous biopolymers. Plasticizers can reduce the aforementioned chain-to-chain interaction and induce flexibility, moisture resistance and ease of processability.

Thermosetting protein plastics processing, on the other hand, occurs through chemical crosslinking that involves the formation of covalent bond bridges between protein chains by using a crosslinking agent. The crosslinkers chiefly target the reaction between themselves and protein functional groups – such as primary amines, carboxyl, hydroxyl, and sulfhydryls – of amino acid residues to provide mechanical strength and moisture resistance.[160,161] Protein-polymer grafting is another method of producing a protein based biomaterial usually with complementary advantages of each component. Thermoplastic processing, which involves melting a polymer followed by shaping and cooling, is the most widely adopted method for the production of protein-based bioplastics.

Plasticizers of protein based plastics.

Plasticizers of proteins are generally added to the protein matrix during thermoplastic processing such as extrusion or injection molding to improve processability, reduce brittleness and modify the properties of the final structure.[164] Plasticizers added to the protein resin or compound usually consist of low molecular weight, low volatility substances, which mainly compete for hydrogen bonding and electrostatic interactions with the protein chains to produce swelling.[34]Table 4 below summarizes common plasticizers for different protein biomass sources that have been used in the thermoplastic processing of proteins in the literature.

Similar to most other bio-based plastics, the composition, size, and shape of plasticizers influence the mechanical, physical, thermal, moisture permeability and aging behavior of proteinaceous plastics.[158,166,170,171] Orliac *et al.*[170] demonstrated that sunflower protein isolate films plasticized with different polyalcohols, such as glycerol, ethylene glycol, propylene glycol, polyethylene glycols, and polypropylene glycols, exhibited high mechanical properties, and good moisture impermeability to the level that it can be used for agricultural mulching. Cao *et al.*[171] compared the plasticizing effect of polyethylene glycol (PEG) with different molecular weights (300, 400, 600, 800, 1500, 4000, 10 000, 20 000) on gelatin films. The result showed that PEG with lower molecular weight gave better plasticizing effect (higher elongation), lower water vapor permeability and better visual effect. An increase in molecular weight of PEG in contrast induced an increase in the tensile strength, elastic modulus and a decrease in the elongation of gelatin films. Polar groups (–OH) along plasticizer chains are believed to develop polymer–plasticizer hydrogen bonds replacing the polymer–polymer interactions in biopolymer films.[172]

Thus, hydrogen bonding ability of PEGs was affected by factors such as the number of hydroxyl groups per mole, molecular size, solubility and polarity that will explain the observed variation. Recent studies by Ullah *et al.*[158,166] also demonstrated that the variation in hydrogen bonding interactions between plasticizers (glycerol, diethyl tartrate, propylene glycol and diethyl tartrate) and keratin from poultry feather quills and poultry feather fiber could be responsible for the variation in plasticization efficacy. The best mechanical properties, transparency, flowability, and processability were observed in the case of ethylene glycol plasticized keratin quill and keratin feather, conceivably because of the formation of strong hydrogen bonding between the ethylene glycol and quill keratin.

Proteins are hydrophilic materials and as such they need to be coupled with adequate plasticizers to reduce the water absorbance of the corresponding plastics. Therefore, extensive attempts to improve moisture barrier properties are being conducted.[173–175] The introduction of hydrophobic materials such as lipids, long chain fatty acids and waxes incorporated into protein films has shown promising results.[173,176,177] For instance, Sohail *et al.*[173] studied and reported the moisture barrier property improvement of protein biopolymers (casein films), as a result of wax incorporation in the film formation and surface wax coating. The wax application on moisture barrier properties was more efficient in wax-coated casein films than wax incorporated biopolymers. While both wax-coating and incorporation improved the flexibility of the films at the expense of tensile strength reduction, the wax incorporated polymers exhibited better flexibility than the coated ones. Pommet *et al.*[176]likewise reported an improvement in the water vapor permeability of gluten protein films with the use of saturated fatty acids with an even number of carbons from 6 to 18 (C6 : 0 : hexanoic acid, C8 : 0 : octanoic acid, C10 : 0 : decanoic acid, C12 : 0 : lauric acid, C14 : 0 : myristic acid, C16 : 0 : palmitic acid, C18 : 0 : stearic acid).

Shellhammer and Krochta[178] studied the effect of lipid type and amount on the plasticization of whey protein biopolymer using beeswax, candelilla wax, carnauba wax and a high melting fraction of anhydrous milk fat. According to the authors, an increase in lipid level decreased the strength of the biopolymers. Among the studied lipids, candelilla wax incorporation provided the weakest films, followed by beeswax, milk fat, and carnauba wax. Furthermore, a positive correlation between water vapor permeability of the lipids and the lipid plasticized protein plastics explains the increment in water vapor permeability of some of the biopolymers. Fabra *et al.*[177] reported the formation of bilayer structures by saturated fatty acids in sodium caseinate film forming solution that led to water vapor permeability improvement. The self-association of saturated fatty acid molecules occurs to form bilayers of different sizes in the film forming dispersions, and these laminar structures grow and persist in the dried film. The crystal formations as a result greatly limit water vapor permeability and yield rigid nonflexible films that show opacity and low gloss. According to the same authors,[177] unsaturated fatty acids such as oleic acid did not form laminar structures due to the double bond while it provokes a synergic plasticizing effect with water that seriously increased the water vapor permeability and film flexibility at intermediate relative humidity levels.[179]

The synergetic effects of using mixed glycerol (polar) and oleic acid (amphiphilic) plasticizers on sodium caseinate[179] and zein protein biopolymers[180] were also reported recently. According to Ibragimo *et al.*,[181] the plasticization obtained by glycerol is structural (inter-packet) and that of oleic acid is molecular (intra-packet). These two different molecules with different plasticization mechanisms provide the possibility for their interaction during film formation. The combination of these two plasticizers in zein films exhibited synergy and as a result a change in tensile strength (highest at 3 : 1 ratio of oleic acid to glycerol), decrease in glass transition temperature and change in microscopic molecular structure were observed.[180]

Tummala *et al.*[182] reported the use of glycerol, sorbitol and their blend to plasticize and compatibilize soy-protein and polyester amide, and compared their influence on the performance of the resulting biopolymers. While sorbitol plasticized soy-polyester amide plastics were more rigid, with a higher tensile modulus and tensile strength and thermal stability, glycerol plasticized soy-polyester amide plastics had the highest impact strength. The blend of the two on the other hand provided an intermediate tensile strength and modulus. Other types of protein biopolymer modifications reported in the literature include blending of gelling agents such as agar, agargel, phytagel,[183] incorporation of nanoclays,[184]*etc.* A recent study by Kim and Netravali[183] demonstrated that the blending of gelling agents with soy protein significantly improved the mechanical, thermal stability resistance of the protein biopolymers. This is because of the possible formation of interpenetrating network (IPN) structures between the gelling agents and the protein with a high degree of intermolecular interactions.[183]

Other methods of improving performance of protein based plastics.

The most commonly utilized techniques for polymer modifications besides plasticization are blending, grafting, crosslinking, and composite formation, which are all multicomponent systems.[185,186] While blending is the physical mixing of multiple polymers to obtain the requisite properties, grafting and crosslinking are among the major irreversible methods of chemically modifying polymer properties. Grafting is a method by which a monomer or a polymer is covalently attached onto another polymer chain. Crosslinking on the other hand is the joining of two or more molecules or molecular chains through a covalent bond by another monomer or polymer called crosslinking agent. Grafting of polymers to protein chains results in a new class of proteinaceous biomaterials comprising natural and synthetic building blocks that are important in diverse fields of application including drug delivery, biotechnology, nanotechnology and nanobiotechnology.[187,188] Materials produced as a result of covalent attachment of synthetic polymers to proteins have the potential to synergistically merge the advantages of proteins and synthetic polymers. The hybrid materials may keep the chemical structure, diverse functionalities, stability, solubility, biocompatibility and biocompostability of proteins while keeping the stability and processability of synthetic polymers.[189,190]

The graft polymerization of styrene on soy protein isolate,[191] 2-hydroxyethyl methacrylate on soy protein,[192] poly(ethylene glycol) on soy protein,[193] polycaprolactone on zein,[194] waterborne polyurethane on soy protein,[195] poly(ethylene oxide) diglycidyl ether on wheat protein,[196] methyl methacrylate, ethyl methacrylate and butyl methacrylate on camelina meal[197] has been widely reported. Wu *et al.*[194] reported that the grafting of polycaprolactone onto zein protein resulted in a dramatic flexibility improvement, while the strength remained constant. Moreover, the glass transition temperature and melting temperature were also shown to decrease due to the plasticizing effect of polycaprolactone on the protein. Kurniawan *et al.*[196] also showed that the chemical modification of wheat protein based biopolymers with poly(ethylene oxide) diglycidyl ether resulted in the formation of a different network structure of the biopolymer with an improved flexibility, and improved mechanical performance.

Chemical crosslinking modification of protein with various agents to improve the mechanical, thermal and moisture resistance of the resulting biopolymers is another technique that has been widely studied.[160,161,196,198,199] Chemical crosslinking of proteins usually depends on the availability of particular chemicals that are capable of reacting with the specific kinds of functional groups that exist in proteins. The most extensively used chemical crosslinking agents of proteins include aldehydes (formaldehyde, glutaraldehyde, glyoxal, benzaldehyde),[161,200,201] carbodiimide, maleic anhydride, hydroxysuccinimide, *etc.*[202–204] Most of the studies show that crosslinking improved the tensile strength, tensile modulus, and moisture and solvent resistance, while the flexibility is reduced.[160,161] In summary, protein-based plastics can be easily modified through plasticization; grafting or crosslinking due to the presence of several functional groups provides protein-based plastics great promise in a wide range of applications. Further research into plasticizer/modification technique selection that combines the characteristics of the different protein feedstock with performance is necessary if protein based plastics are to achieve their full commercial potential.

Cellulose acetate

Cellulose is an abundant, renewable, and biodegradable natural polymer that constitutes the skeletal part of plants. It is a homogeneous polysaccharide formed by repeating connection of D-glucose building blocks (Fig. 10), with an average degree of polymerization of 1500 to 3000 depending on the source.[205] Industrial materials are being developed from cellulose and its derivative over a broad range of application because of its abundance, environmental and biocompatibility benefits, relatively low cost and ease of modification. Cellulose by itself is poorly soluble in common solvents and is not melt processable as it decomposes before it undergoes melt flow.[206] Nonetheless, the chemistry of cellulose opens the way to various chemical modifications, such as esterification and etherification give entry into a broad variety of products including coatings, base for photographic films, filters, pharmaceutics, fragrances, polymer additives, membranes and building materials.[207–209] The most industrially relevant and oldest biodegradable cellulose ester derivative is cellulose acetate.[208] Cellulose acetate and mixed cellulose esters, such as cellulose diacetate, cellulose triacetate, cellulose acetate propionate, and cellulose acetate butyrate, are all commercially available materials.[208] These thermoplastic materials are usually synthesized through esterification of cellulose,[210] where other substituent groups replace the hydroxyl groups of cellulose (Fig. 10). For instance, cellulose acetate is the product of esterification reaction between cellulose and acetic anhydride in the presence of sulfuric acid catalyst to form fully acetylated cellulose triacetate, followed by partial hydrolysis to remove acid catalyst and produce a degree of substitution in the polymer that yields the desired working properties.[211] Other methods of cellulose ester synthesis, such as transesterification of cellulose with vinyl esters under catalysis, are reported in the literature recently.[212]

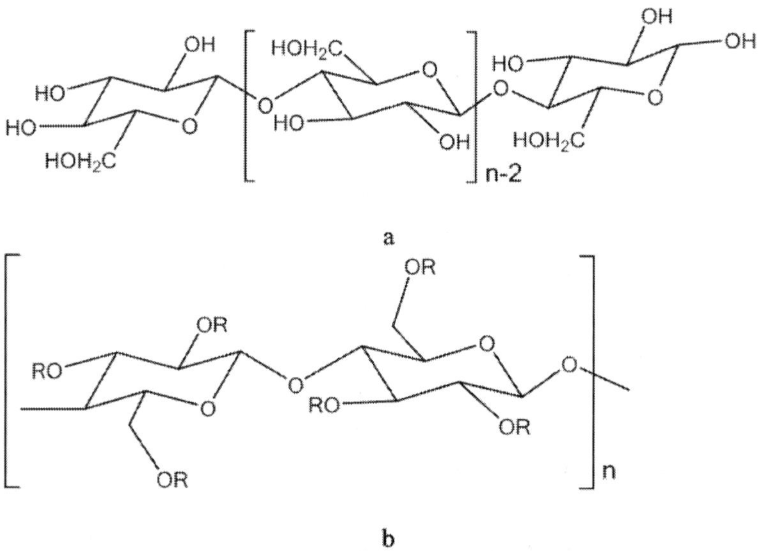

Figure 10: Schematic representation of the molecular structure of (a) cellulose (*n*-degree of polymerization) and (b) cellulose ester (R-functional group for each type of cellulose ester).

Raw materials such as cotton, recycled paper, wood cellulose, and sugarcane are used in making cellulose ester biopolymers in powder form.[210] Cellulose ester powders combined with plasticizers and additives are extruded to produce various grades of commercial cellulosic plastics in pelletized form. Of great interest as potential biodegradable plastics are also long chain aliphatic acid esters of cellulose.[206,213] These cellulose esters are characterized by stiffness, moderate heat resistance, high moisture vapor transmission, grease resistance, clarity and appearance, and moderate impact resistance.[206] The presence of polar functional groups in the cellulose acetate chain offers an additional advantage of affinity to solvents including plasticizers and lithium ions for the development of polymer electrolytes.[214] Nevertheless, owing to the high viscosity and elevated glass transition temperature, cellulose acetate derivatives themselves are not processable as a thermoplastic.[209] In an effort to modify its properties and facilitate processing, cellulose acetate is modified through plasticization by various aliphatic and aromatic esters,[215,216] chemically modified through grafting onto the polysaccharide backbone, and modification by forming polymer blends.[208,216,217]

Plasticizers for cellulose acetate.

The thermal and rheological properties of polymers are crucial factors for thermoplastic processability of polymers. A broad processing window without thermal degradation of polymer is necessary to adjust the required rheological behavior of the polymers. Cellulose acetate is characterized by high glass transition and melting temperatures. As such the addition of processing aids such as plasticizers is required to improve its rheological behavior or thermal processability in the polymer melt. The most common plasticizers for cellulose acetate plastics reported in the literature include diethyl phthalate, dimethyl phthalate, triphenyl phosphate, ethylhexyl adipate, flexol, triacetin, glycerol triacetate, triethylene, glycol dipropionate and a wide variety of other plasticizers.[218–220]

Zepnik *et al.*[221] have recently studied the effect of plasticizer type and concentration on cellulose acetates using benzoate, acetates, phosphate and citrates based plasticizers. An increase in plasticizer concentration resulted in significant broadening of the thermoplastic processing window due to a strong decrease in glass transition temperature. It was thus possible to tune the rheology, melt strength and thermoplastic processing cellulose acetate by changing the plasticizer concentration. On the other hand molecular size, chemical structure, and solubility variation of plasticizers were shown to influence its compatibility, and ultimately the efficacy. It is generally agreed that plasticizers that have higher thermodynamic compatibility with the base polymer cause better plasticization than those with limited compatibility. The selection of an efficient plasticizer for cellulose esters was suggested by Fridman and Sorokina[219] who developed a set of criteria for efficient plasticization of cellulose acetate. An efficient plasticizer should take into account the compatibility of components, temperature durability and mechanical properties during processing and service time of the final polymer.

The efficiency of a plasticizer depends also on the loading concentration. Fig. 11 and 12 show the effect of one of the common cellulose acetate plasticizer (diethyl phthalate) concentration on the thermal and mechanical properties of cellulose acetate (drawn from tabulated data reported by Fridman and Sorokina[219]). As the plasticizer concentration increases, a reduction in glass transition temperature was observed and hence a significantly lower processing temperature is needed, while a substantial thermal stability drop resulted in the cellulose ester plastic (Fig. 11). On the other hand, an increase in plasticizer concentration resulted in an increment of impact strength (Fig. 12) and elongation at break accompanied by a drop in tensile strength. Similar trend of cellulose acetate stiffness and toughness properties was observed by Mohanty *et al.*[210] upon increasing triethyl citrate plasticizer concentration. In summary, the type of the plasticizer and optimum plasticizer concentration are key parameters to reduce the processing temperature without compromising the stability and other performances of the plastic.

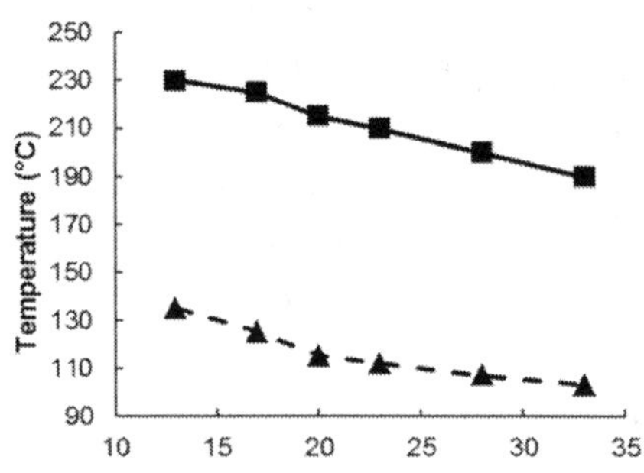

Figure 11: Influence of diethyl phthalate concentration on T_g (▲) and processing temp (■) based on the data from Fridman and Sorokina.[219]

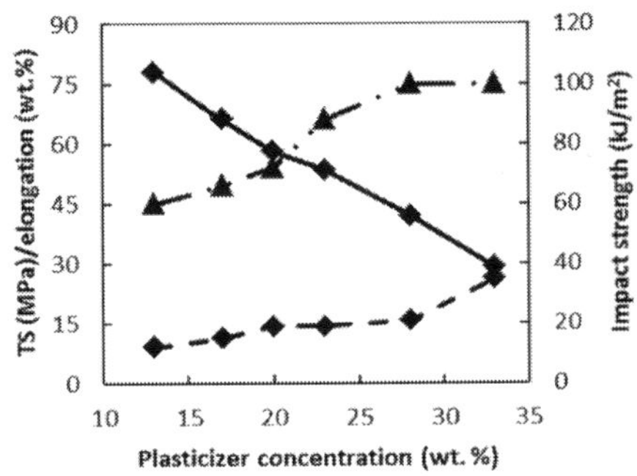

Figure 12: Influence of diethyl phthalate concentration on impact strength (▲), elongation (■) and tensile strength (TS) (♦) based on the data from Fridman and Sorokina.[212]

While cellulose acetate or its degradation products are safe, some of its common plasticizers such as phthalates, triacetin, glycerin, polyethylene glycol are associated with high toxicity, relatively high diffusion and water solubility.[222,223] For example, deterioration of cellulose acetate as a result of migration or evaporation of plasticizers, reaction of plasticizers with other chemicals in their surroundings to form other products has been documented in the literature.[218] As a result of such deterioration, not only unsafe plasticizers and plasticizer reaction products are released to the environment, but materials developed from cellulose ester became prone to cracking, warping, discoloration, exudation, shrinkage and powdering as they age.[224]

To mitigate these safety issues in addition to awareness of green technology and government legislations, several mitigation efforts are being conducted. These efforts include the development of safe, more stable and more compatible, bio-based, and functional plasticizers. Sugar based plasticizers, such as sorbitan,[216,225] polyoxyethylene sorbitan mono palmitate,[225]polyoxyethylene sorbitan monostearate,[216] are also reported. The use of an ionic liquid plasticizer synthesized from choline chloride and urea, such as deep eutectic solvent (DES), has been recently reported as a safe and novel plasticizer of cellulose acetate.[226] These plasticizers have the high solvating potential of crystalline cellulose acetate, and are less expensive, non-toxic and biodegradable in addition to their large electronegativity and delocalization of charge that enables them to positively influence ionic conductivity of cellulose ester.[226,227] Other ionic solvent cellulose ester plasticizers reported in the literature include 1-allyl-3-methylimidazolium chloride,[228] and other ionic liquids based on methylimidazolium and methylpyridinium cores with allyl-, ethyl-, or butyl-side chains.[229]

Other techniques of improving performance of cellulose acetate.

Chemical modification of cellulose acetate through grafting and crosslinking with other polymers is the other widely reported strategy for improving the performance of cellulose acetate.[208,217,230] Graft copolymerization of polymers generally offers an attractive and versatile means of imparting a variety of functional groups to a polymer.[185] Cellulose acetate grafting is a process aimed at the introduction of branches of synthetic polymers along the main polysaccharide chain to confer specific additional properties to the former without modifying its intrinsic characteristics.[231] Grafting offers the opportunity to combine the best properties of two or more polymers in one physical unit. Moreover, by varying parameters such as the degree of polymerization, polydispersities of the main chain and the side chains, graft density, and the distribution of the grafts, a more polymeric material with more valuable properties may result.[217] Grafting of cellulose acetate with other polymers such as poly(lactic acid),[232,233] poly(methyl methacrylate),[231,234]poly(hydroxybuty rate-*co*-valerate),[235] caprolactone,[32] polystyrene,[236] poly(ethylene glycol) and poly(hydroxybutyrate)[237] with new and improved performance features including better conductivity, thermal stability, and other marked advantages with better control and property tuning capability has been reported in the literature.

CONCLUDING REMARKS

The rapid technological development of bio-based plastics, such as PLA, polyhydroxyalkanoates, (PHA), bio-based epoxy resin and bio-based PE, has yet to be translated into significant market impact, primarily due to high production cost and performance limitations. Plasticizers are important additives and performance enhancers of polymers. As such they can augment the processability of most of these bio-based plastics, and constitute a significant opportunity and at the same time barrier in the applications of bio-based plastics in various fields of applications. Moreover, the performance, safety, biodegradability, economics and functional utilization of bio-based plastics are strongly dependent on the performance of the incorporated plasticizers. This review highlights that selective use of a plasticizer at an optimum concentration allows controlling the balance between the processability, strength, modulus, toughness and other important mechanical properties of plastics. Such a selection is usually conducted based on reasons for application of plasticizer, mechanism of plasticizer action, interaction and effect of the plasticizer on other additives besides the biopolymer, concentration range depending on the application.

In addition to the sanction of some common plasticizers (*e.g.* phthalates) for various applications, due to environmental and health concern as a result of migration and leaching during aging or use of plastics, newer concerns are emerging with regard to the effect of plasticizers in maintaining the renewability and biodegradability of bio-based plastics. The development and utilization of bio-based plasticizers such as polyols, fatty acids and fatty acid derivatives, epoxidized soy oil, ester amides, citrates and ionic liquids, such as methylimidazolium chloride and deep eutectic solvent are widely reported to tackle such issues. These new forms of plasticizers offer new dimensions of plasticizer selection that provide additional functionality (*e.g.* electric conductivity by ionic liquids) to the bio-based plastics and some of the others provide high safety to be used even in edible food packaging applications. While most of the fundamental mechanism, physico–chemical interaction or rule of thumb in the selection of a suitable plasticizer were established for the synthetic plastics (mainly for PVC), there are hardly no newer theories/mechanisms for the relatively new bio-based plastics. As a result, most of the current plasticization investigations are conducted under the assumption that the mechanisms developed for PVC would also be valid for the bio-based plastics. A fundamental understanding of the plasticization mechanism of bio-based materials is essential, along with their similarity and difference with PVC, if they are to reach their full potential and success.

Besides the usual purpose of improving flexibility and processability, research is progressing in areas such as the search and modification of plasticizers that impart additional functions of flame retardancy, optical quality, electric conductivity or insulation, thermoxidative stability, chemical and temperature (high and low) resistance in demanding environments; reactive plasticizers that provide chemical integrity, gas and moisture impermeability improvement, provide or improve biodegradability and biocompatibility to the polymers are under investigation. The migration of some of the current bio-based plasticizers through either volatilization or mass transfer to a liquid or solid in contact poses another challenge. These and the other challenges led to the search for newer types of plasticizers and alternative methods of improving the processability and overall performance. Some of the alternative methods to plasticization include molecular orientation, physical and reactive blending, chemical crosslinking and grafting of the bio-based polymers with other polymers to tailor the ultimate product properties.

ACKNOWLEDGEMENTS

Tizazu H. Mekonnen, Paolo G. Mussone and David C. Bressler are grateful to PrioNet Canada, Alberta Prion Research Institute, Alberta Livestock and Meat Agency, and the Biorefining Conversion Network at the University of Alberta for the financial support they received while they were researching and preparing this work.

REFERENCES

1. R. U. Halden, *Annu. Rev. Public Health*, 2010, 31, 179–194.
2. M. M. Reddy, S. Vivekanandhan, M. Misra, S. K. Bhatia and A. Mohanty, *Prog. Polym. Sci.*, 2013 DOI:10.1016/j.progpolymsci.2013.05.006 .
3. B. Gervet, *The use of crude oil in plastic making contributes to global warming*, Lulea University of Technology, Sweden, 2007 Search .
4. M. McCoy, *Chem. Eng. News*, 1998, 76, 15–18 Search .
5. L. Shen, E. Worrell and M. Patel, *Biofuels, Bioprod. Biorefin.*, 2010, 4, 25–40.
6. M. M. Reddy, M. Misra and A. K. Mohanty, *Chem. Eng. Prog.*, 2012, 108, 37–42 .

7. USDA Agriculture, ed. USDA, http://www.biocom.iastate.edu/workshop/2012workshop/presentations/pruszko.pdf, Washington, DC., 2008, accessed 21 May 2013.
8. R. Mathers, *J. Polym. Sci., Part A: Polym. Chem.*, 2012, 50, 1–15.
9. M. Rahman and C. Brazel, *Prog. Polym. Sci.*, 2004, 29, 1223–1248 .
10. I. Lemer, Plasticizers to reach $8 billion by 2004 Look Smart, Ltd, http://www.findarticles.com/cf_dls/m0FVP/9_258/65462810/p1/article.jhtml, accessed January 2010.
11. C. E. Wikes, J. W. Summers and C. A. Daniels, *PVC handbook*, Hanser Gardner Publications, Inc., 2005.
12. S. Cullen, Global Plasticizer Update, http://www.plasticsindustry.org/files/events/Stephen%20Cullen_Tuesday.pdf, accessed 28 April 2013.
13. L. Sperling, *Introduction To Physical Polymer Science*, 4th edn, 2006, pp. 1–845.
14. P. De Groote, J. Devaux and P. Godard, *J. Polym. Sci., Part B: Polym. Phys.*, 2002, 40, 2208–2218.
15. S. Varughese and D. Tripathy, *J. Elastomers Plast.*, 1993, 25, 343–357 .
16. E. Snejdrova and M. Dittrich, *Pharmaceutical Applications of Plasticized Polymers, Recent Advances in Plasticizers*, ed. Mohammad Luqman, Intech, 2012.
17. N. Ljungberg and B. Wesslen, *Polymer*, 2003, 44, 7679–7688 .
18. D. Klee and H. Hocker, *Biomedical Applications: Polymer Blends*, 1999, 149, 1–57.
19. S. Tomic, M. Micic, S. Dobic, J. Filipovic and E. Suljovrujic, *Radiat. Phys. Chem.*, 2010, 79, 643–649 .
20. Y. Zhu, N. Shah, A. Malick, M. Infeld and J. McGinity, *Int. J. Pharm.*, 2002, 241, 301–310.
21. Y. Li, C. Wang, G. Wang and Z. Qu, *J. Wuhan Univ. Technol., Mater. Sci. Ed.*, 2008, 23, 100–104 .
22. R. Sothornvit and J. M. Krochta, *Innovations in Food Packaging*, 2005, pp. 403–433.

23. E. H. Immergut and H. F. Mark, Principles of Plasticization, in *Plasticization and Plasticizer Processes*, American Chemical Society, Washington, DC, 1965.
24. A. S. Wilson, *Plasticisers: Principles and Practice*, The Institute of Materials, London, UK, 1995.
25. A. Mohanty, M. Misra and L. Drzal, *Abstracts of Papers of the American Chemical Society*, 2002, vol. 223, pp. D70–D70.
26. P. Sarnacke and W. Stephen, *Disposable Bioplastics: Consumer disposables agricultural films, a market opportunity study*, United soybean board, Omni Tech International, Ltd., Midland, Michigan, 2008 .
27. M. Vieira, M. da Silva, L. dos Santos and M. Beppu, *Eur. Polym. J.*, 2011, 47, 254–263 .
28. L. Lim, R. Auras and M. Rubino, *Prog. Polym. Sci.*, 2008, 33, 820–852 .
29. N. Gil, M. Saska and I. Negulescu, *J. Appl. Polym. Sci.*, 2006, 102, 1366–1373.
30. O. Fenollar, D. Garcia, L. Sanchez, J. Lopez and R. Balart, *Eur. Polym. J.*, 2009, 45, 2674–2684 .
31. A. Stuart, M. McCallum, D. Fan, D. LeCaptain, C. Lee and D. Mohanty, *Polym. Bull.*, 2010, 65, 589–598.
32. B. Videki, S. Klebert and B. Pukanszky, *J. Polym. Sci., Part B: Polym. Phys.*, 2007, 45, 873–883.
33. L. Di Gioia, B. Cuq and S. Guilbert, *Int. J. Biol. Macromol.*, 1999, 24, 341–350.
34. R. Sothornvit and J. Krochta, *J. Food Eng.*, 2001, 50, 149–155 .
35. B. P. Shtarkman and I. N. Razinskaya, *Acta Polym.*, 1983, 34, 514–520.
36. J. K. Sears and J. R. Darby, *The Technology of Plasticizers*, John Wiley & Dpmd, New York, NY, 1982.
37. R. Drumright, P. Gruber and D. Henton, *Adv. Biomater.*, 2000, 12, 1841–1846 .
38. B. Gupta, N. Revagade and J. Hilborn, *Prog. Polym. Sci.*, 2007, 32, 455–482 .

39. X. Hu, X. Jing, S. Sharma and A. Mudhoo, *Handbook of Applied Biopolymer Technology: Synthesis, Degradation and Applications*, 2011, pp. 291–310.
40. D. Garlotta, *J. Polym. Environ.*, 2001, 9, 63–84.
41. A. Singha, V. Thakur, A. Singha and V. Thakur, *Green Polymer Materials*, 2012, pp. 1–14.
42. J. P. Penning, H. Dijkstra and A. J. Pennings, *Polymer*, 1993, 34, 942–951.
43. J. Lunt, *Polym. Degrad. Stab.*, 1998, 59, 145–152.
44. P. Kurcok, A. Matuszowicz, Z. Jedlinski, H. Kricheldorf, P. Dubois and R. Jerome, *Macromol. Rapid Commun.*, 1995, 16, 513–519.
45. H. Liu and J. Zhang, *J. Polym. Sci., Part B: Polym. Phys.*, 2011, 49, 1051–1083.
46. L. Averous, M. Belgacem and A. Gandini, *Monomers, Polym. Compos. Renewable Resour.*, 2008, 433–450.
47. A. Gupta and V. Kumar, *Eur. Polym. J.*, 2007, 43, 4053–4074 .
48. J. Huang, M. Lisowski, J. Runt, E. Hall, R. Kean, N. Buehler and J. Lin, *Macromolecules*, 1998, 31, 2593–2599.
49. R. Auras, B. Harte and S. Selke, *Macromol. Biosci.*, 2004, 4, 835–864 .
50. P. Zhao, Q. F. Wang, Q. Zhong, N. W. Zhang and J. Ren, *J. Appl. Polym. Sci.*, 2010, 115, 2955–2961.
51. O. Martin and L. Averous, *Polymer*, 2001, 42, 6209–6219.
52. L. Averous and C. Fringant, *Polym. Eng. Sci.*, 2001, 41, 727–734 .
53. S. Jacobsen and H. Fritz, *Polym. Eng. Sci.*, 1999, 39, 1303–1310 .
54. L. Labrecque, R. Kumar, V. Dave, R. Gross and S. McCarthy, *J. Appl. Polym. Sci.*, 1997, 66, 1507–1513.
55. G. Maglio, A. Migliozzi and R. Palumbo, *Polymer*, 2003, 44, 369–375.
56. G. Maglio, M. Malinconico, A. Migliozzi and G. Groeninckx, *Macromol. Chem. Phys.*, 2004, 205, 946–950.
57. G. Maglio, A. Migliozzi, R. Palumbo, B. Immirzi and M. Volpe, *Macromol. Rapid Commun.*, 1999, 20, 236–238.

58. A. Gajria, V. Dave, R. Gross and S. McCarthy, *Polymer*, 1996, 37, 437–444.

59. E. Blumm and A. Owen, *Polymer*, 1995, 36, 4077–4081 .

60. N. Ogata, T. Tatsushima, K. Nakane, K. Sasaki and T. Ogihara, *J. Appl. Polym. Sci.*, 2002, 85, 1219–1226.

61. J. Park and S. Im, *J. Polym. Sci., Part B: Polym. Phys.*, 2002, 40, 1931–1939.

62. F. Hassouna, J. Raquez, F. Addiego, P. Dubois, V. Toniazzo and D. Ruch, *Eur. Polym. J.*, 2011, 47, 2134–2144 .

63. N. Ljungberg and B. Wesslen, *J. Appl. Polym. Sci.*, 2002, 86, 1227–1234.

64. F. Hassouna, J. Raquez, F. Addiego, V. Toniazzo, P. Dubois and D. Ruch, *Eur. Polym. J.*, 2012, 48, 404–415 .

65. Y. Xu and J. Qu, *J. Appl. Polym. Sci.*, 2009, 112, 3185–3191.

66. B. Chen, T. Wu, Y. Chang and A. Chen, *Chem. Eng. J.*, 2013, 215, 886–893.

67. Z. Ren, L. Dong and Y. Yang, *J. Appl. Polym. Sci.*, 2006, 101, 1583–1590.

68. P. Suriyamongkol, R. Weselake, S. Narine, M. Moloney and S. Shah, *Biotechnol. Adv.*, 2007, 25, 148–175 .

69. S. Park, T. Kim, M. Kim, S. Lee and S. Lim, *Biotechnol. Adv.*, 2012, 30, 1196–1206 .

70. T. G. Volova, Polyhydroxyalkanoates plastic material of the 21st century, Nova science publishers, Inc., Hauppauge, New York, 2004.

71. K. Snell and O. Peoples, *Biofuels, Bioprod. Biorefin.*, 2009, 3, 456–467.

72. G. Chen, Q. Wu, Y. Wang, Z. Zheng, X. Zhang, J. Tanaka, Y. Yu and Y. Tabata, *Asbm6: Advanced Biomaterials Vi*, 2005, 288–289, 437–440 .

73. L. Yu, K. Dean and L. Li, *Prog. Polym. Sci.*, 2006, 31, 576–602 .

74. B. Laycock, P. Halley, S. Pratt, A. Werker and P. Lant, *Prog. Polym. Sci.*, 2013, 38, 536–583 .

75. G. Chen and Q. Wu, *Biomaterials*, 2005, 26, 6565–6578 .

76. A. Anderson and E. Dawes, *Microbiol. Rev.*, 1990, 54, 450–472 .

77. H. Salehizadeh and M. Van Loosdrecht, *Biotechnol. Adv.*, 2004, 22, 261–279 .

78. I. Chodak, in *Monomers, polymers and composites from renewable resources*, Elsevier Ltd., Boston, US, 2008.

79. R. Baltieri, L. Mei and J. Bartoli, *Macromol. Symp.*, 2003, 197, 33–44.

80. I. Chodak, in *Degradable polymers*, Kluwer Academic publishers, The Netherlands, 2002.

81. P. Ma, D. Hristova-Bogaerds, P. Lemstra, Y. Zhang and S. Wang, *Macromol. Mater. Eng.*, 2012, 297, 402–410.

82. M. Erceg, T. Kovacic and I. Klaric, *Polym. Degrad. Stab.*, 2005, 90, 313–318 .

83. J. Choi and W. Park, *Polym. Test.*, 2004, 23, 455–460 .

84. M. Correa, M. Branciforti, E. Pollet, J. Agnelli, P. Nascente and L. Averous, *J. Polym. Environ.*, 2012, 20, 283–290 .

85. S. Hong, H. Hsu and M. Ye, *J. Therm. Anal. Calorim.*, 2013, 111, 1243–1250 .

86. W. V. Srubar, Z. C. Wright, A. Tsui, A. T. Michel, S. L. Billington and C. W. Frank, *Polym. Degrad. Stab.*, 2012, 97, 1922–1929 .

87. M. A. Abdelwahab, A. Flynn, B. S. Chiou, S. Imam, W. Orts and E. Chiellini, *Polym. Degrad. Stab.*, 2012, 97, 1822–1828 .

88. M. O. Seydibeyoglu, M. Misra and A. Mohanty, *Int. J. Plast. Technol.*, 2010, 14, 1–16 .

89. I. Bibers, V. Tupureina, A. Dzene and M. Kalnins, *Mech. Compos. Mater.*, 1999, 35, 357–364.

90. N. Yoshie, K. Nakasato, M. Fujiwara, K. Kasuya, H. Abe, Y. Doi and Y. Inoue, *Polymer*, 2000, 41, 3227–3234.

91. L. Miao, Z. Qiu, W. Yang and T. Ikehara, *React. Funct. Polym.*, 2008, 68, 446–457 .

92. T. Sadik, V. Massardier, F. Becquart and M. Taha, *J. Appl. Polym. Sci.*, 2013, 127, 1148–1156.

93. C. Del Gaudio, L. Fioravanzo, M. Folin, F. Marchi, E. Ercolani and A. Bianco, *J. Biomed. Mater. Res., Part B*, 2012, 100, 1883–1898.
94. P. Ma, D. Hristova-Bogaerds, J. Goossens, A. Spoelstra, Y. Zhang and P. Lemstra, *Eur. Polym. J.*, 2012, 48, 146–154 .
95. R. Sadi, R. Kurusu, G. Fechine and N. Demarquette, *J. Appl. Polym. Sci.*, 2012, 123, 3511–3519.
96. A. Tsui, Z. Wright and C. Frank, *Annu. Rev. Chem. Biomol. Eng.*, 2013, 4, 143–170.
97. D. van Krevelen and K. te Nijenhuis, Thermal Decomposition, Processing Properties, in *Properties of Polymers*, Elsevier, Amsterdam, 2009.
98. M. Yamaguchi and K. Arakawa, *Eur. Polym. J.*, 2006, 42, 1479–1486 .
99. C. Bastioli, *Starch/Staerke*, 2001, 53, 351–355.
100. N. Cheetham and L. Tao, *Carbohydr. Polym.*, 1998, 36, 277–284.
101. J. Yu, J. Gao and T. Lin, *J. Appl. Polym. Sci.*, 1996, 62, 1491–1494.
102. P. Russell, *J. Cereal Sci.*, 1987, 6, 133–145.
103. P. J. Halley, R. W. Truss, M. G. Markotsis, C. Chaleat, M. Russo, A. L. Sargent, I. Tan and P. A. Sopade, *Polymer Durability and Radiation Effects*, 2008, vol. 978, pp. 287–300.
104. A. D. Sagar and E. W. Merrill, *J. Appl. Polym. Sci.*, 1995, 58, 1647–1656.
105. F. Otey and R. Westhoff, *Ind. Eng. Chem. Prod. Res. Dev.*, 1984, 23, 284–287.
106. E. Rudnik and E. Rudnik, *Compostable Polymer Materials*, 2008, pp. 11–36.
107. J. Yu, N. Wang and X. Ma, *Starch/Staerke*, 2005, 57, 494–504.
108. A. Taghizadeh and B. D. Favis, *Carbohydr. Polym.*, 2013, 92, 1799–1808 .
109. X. Ma, P. R. Chang and J. Yu, in *Starch-Based Polymeric Materials and Nanocomposites*, CRC Press, Florida, US, 2012.
110. P. Forssell, J. Mikkila, G. Moates and R. Parker, *Carbohydr. Polym.*, 1997, 34, 275–282.

111. R. A. Talja, H. Helen, Y. H. Roos and K. Jouppila, *Carbohydr. Polym.*, 2007, 67, 288–295 .

112. Y. Zhang and J. H. Han, *J. Food Sci.*, 2008, 73, E313–E324 .

113. S. Raphaelides, G. Dimitreli, S. Exarhopoulos, G. Kokonidis and E. Tzani, *Carbohydr. Polym.*, 2011, 83, 727–736 .

114. F. Rodriguez-Gonzalez, B. Ramsay and B. Favis, *Carbohydr. Polym.*, 2004, 58, 139–147 .

115. P. Liu, L. Yu, X. Y. Wang, D. Li, L. Chen and X. X. Li, *J. Cereal Sci.*, 2010, 51, 388–391 .

116. R. A. de Graaf, A. P. Karman and L. Janssen, *Starch/Staerke*, 2003, 55, 80–86.

117. M. N. Abdorreza, L. H. Cheng and A. A. Karim, *Food Hydrocolloids*, 2011, 25, 56–60 .

118. O. V. Lopez, C. J. Lecot, N. E. Zaritzky and M. A. Garcia, *J. Food Eng.*, 2011, 105, 254–263 .

119. Q. Q. Yan, H. X. Hou, P. Guo and H. Z. Dong, *Carbohydr. Polym.*, 2012, 87, 707–712 .

120. J. Ren, H. Y. Fu, T. B. Ren and W. Z. Yuan, *Carbohydr. Polym.*, 2009, 77, 576–582 .

121. R. A. Talja, H. Helen, Y. H. Roos and K. Jouppila, *Carbohydr. Polym.*, 2008, 71, 269–276 .

122. N. Laohakunjit and A. Noomhorm, *Starch/Staerke*, 2004, 56, 348–356.

123. X. Ma and J. Yu, *J. Appl. Polym. Sci.*, 2004, 93, 1769–1773.

124. R. Shogren, C. Swanson and A. Thompson, *Starch/Staerke*, 1992, 44, 335–338.

125. X. Ma, J. Yu and F. Jin, *Polym. Int.*, 2004, 53, 1780–1785.

126. J. Yang, J. Yu and X. Ma, *Chin. Chem. Lett.*, 2006, 17, 133–136 .

127. J. Yang, J. Yu and X. Ma, *Carbohydr. Polym.*, 2006, 63, 218–223 .

128. S. Flores, L. Fama, A. Rojas, S. Goyanes and L. Gerschenson, *Food Res. Int.*, 2007, 40, 257–265 .

129. N. Wang, X. Zhang, H. Liu and B. He, *Carbohydr. Polym.*, 2009, 76, 482–484 .

130. A. Sankri, A. Arhaliass, I. Dez, A. Gaumont, Y. Grohens, D. Lourdin, I. Pillin, A. Rolland-Sabate and E. Leroy, *Carbohydr. Polym.*, 2010, 82, 256–263 .

131. E. Leroy, P. Jacquet, G. Coativy, A. Reguerre and D. Lourdin, *Carbohydr. Polym.*, 2012, 89, 955–963 .

132. P. Kubisa, *Prog. Polym. Sci.*, 2004, 29, 3–12 .

133. A. L. Chaudhary, P. J. Torley, P. J. Halley, N. McCaffery and D. S. Chaudhary, *Carbohydr. Polym.*, 2009, 78, 917–925 .

134. A. L. Chaudhary, M. Miler, P. J. Torley, P. A. Sopade and P. J. Halley, *Carbohydr. Polym.*, 2008, 74, 907–913 .

135. A. Alissandratos, N. Baudendistel, S. L. Flitsch, B. Hauer and P. J. Halling, *BMC Biotechnol.*, 2010, 10.

136. P. B. Zamudio-Flores, A. V. Torres, R. Salgado-Delgado and L. A. Bello-Perez, *J. Appl. Polym. Sci.*, 2010, 115, 991–998.

137. B. Volkert, A. Lehmann, T. Greco and M. H. Nejad, *Carbohydr. Polym.*, 2010, 79, 571–577 .

138. N. Reddy and Y. Q. Yang, *Biotechnol. Bioeng.*, 2009, 103, 1016–1022 .

139. F. Xie, E. Pollet, P. J. Halley and L. Averous, *Prog. Polym. Sci.*, 2013 DOI:10.1016/j.progpolymsci.2013.05.002 .

140. Z. Liu, Y. Dong, H. Men, M. Jiang, J. Tong and J. Zhou, *Carbohydr. Polym.*, 2012, 89, 473–477 .

141. J. Zhou, J. Zhang, Y. Ma and J. Tong, *Carbohydr. Polym.*, 2008, 74, 405–410 .

142. R. Nakamura, A. Netravali, A. Morgan, M. Nyden and J. Gilman, *Fire Mater.*, 2013, 37, 75–90.

143. V. Madison and J. Schellma, *Biopolymers*, 1972, 11, 1041–1076 .

144. S. Swain, S. Biswal, P. Nanda and P. Nayak, *J. Polym. Environ.*, 2004, 12, 35–42.

145. A. Gennadios, C. L. Weller, M. A. Hanna and G. W. Froning, *J. Food Sci.*, 1996, 61, 585–589.

146. K. Dangaran, P. Tomasula, P. Qi, M. Embuscado and K. Huber, *Edible Films and Coatings For Food Applications*, 2009, pp. 25–56.

147. A. Elzoghby, W. El-Fotoh and N. Elgindy, *J. Controlled Release*, 2011, 153, 206–216 .
148. S. Khosravi, P. Nordqvist, F. Khabbaz and M. Johansson, *Ind. Crops Prod.*, 2011, 34, 1509–1515 .
149. Z. Peles and M. Zilberman, *Acta Biomater.*, 2012, 8, 209–217 .
150. H. Tian, *J. Compos. Mater.*, 2012, 46, 427–435 .
151. M. Santin and L. Ambrosio, *Expert Rev. Med. Devices*, 2008, 5, 349–358 .
152. Y. Song, Q. Zheng and Q. Zhang, *J. Cereal Sci.*, 2009, 50, 376–380 .
153. S. Domenek, M. Morel, A. Redl and S. Guilbert, *Macromol. Symp.*, 2003, 197, 181–191.
154. H. Yue, Y. Cui, P. Shuttleworth and J. Clark, *Green Chem.*, 2012, 14, 2009–2016 .
155. C. Verbeek and L. van den Berg, *Macromol. Mater. Eng.*, 2010, 295, 10–21.
156. C. J. R. Verbeek and L. E. van den Berg, *J. Polym. Environ.*, 2011, 19, 1–10.
157. A. Denis, N. Brambati, B. Dessauvages, S. Guedj, C. Ridoux, N. Meffre and C. Autier, *Food Hydrocolloids*, 2008, 22, 989–994 .
158. A. Ullah, T. Vasanthan, D. Bressler, A. L. Elias and J. P. Wu, *Biomacromolecules*, 2011, 12, 3826–3832 .
159. A. Jerez, P. Partal, I. Martinez, C. Gallegos and A. Guerrero, *J. Food Eng.*, 2007, 82, 608–617 .
160. T. Mekonnen, P. Mussone, N. El-Daher, P. Choi and D. Bressler, *Macromol. Mater. Eng.*, 2013 DOI:10.1002/mame.201200429 .
161. N. El-Thaher, T. Mekonnen, P. Mussone, D. Bressler and P. Choi, *Ind. Eng. Chem. Res.*, 2013, 52, 4987–4993.
162. B. Cuq, C. Aymard, J. Cuq and S. Guilbert, *J. Food Sci.*, 1995, 60, 1369–1374.
163. H. Zheng, Z. Tan, Y. Zhan and J. Huang, *J. Appl. Polym. Sci.*, 2003, 90, 3676–3682.
164. V. Hernandez-Izquierdo and J. Krochta, *J. Food Sci.*, 2008, 73, R30–R39 .

165. M. Pommet, A. Redl, S. Guilbert and M. Morel, *J. Cereal Sci.*, 2005, 42, 81–91 .

166. A. Ullah and J. Wu, *Macromol. Mater. Eng.*, 2013, 298, 153–162.

167. J. W. Lawton, *Cereal Chem.*, 2004, 81, 1–5.

168. N. Reddy, L. H. Chen and Y. Q. Yang, *Ind. Crops Prod.*, 2013, 43, 159–164 .

169. X. Q. Mo and X. Z. Sun, *J. Am. Oil Chem. Soc.*, 2002, 79, 197–202.

170. O. Orliac, A. Rouilly, F. Silvestre and L. Rigal, *Ind. Crops Prod.*, 2003, 18, 91–100.

171. N. Cao, X. Yang and Y. Fu, *Food Hydrocolloids*, 2009, 23, 729–735.

172. L. Yang and A. T. Paulson, *Food Res. Int.*, 2000, 33, 563–570.

173. S. S. Sohail, B. W. Wang, M. A. S. Biswas and J. H. Oh, *J. Food Sci.*, 2006, 71, C255–C259.

174. M. Wihodo and C. I. Moraru, *J. Food Eng.*, 2013, 114, 292–302 .

175. W. Wu, N. Hettiarachchy and M. Qi, *J. Am. Oil Chem. Soc.*, 1998, 75, 845–850.

176. M. Pommet, A. Redl, M. Morel and S. Guilbert, *Polymer*, 2003, 44, 115–122.

177. M. J. Fabra, A. Jimenez, L. Atares, P. Talens and A. Chiralt, *Biomacromolecules*, 2009, 10, 1500–1507 .

178. T. H. Shellhammer and J. M. Krochta, *J. Food Sci.*, 1997, 62, 390–394.

179. M. J. Fabra, P. Talens and A. Chiralt, *Food Hydrocolloids*, 2010, 24, 384–391 .

180. H. Xu, Y. W. Chai and G. Y. Zhang, *J. Agric. Food Chem.*, 2012, 60, 10075–10081 .

181. Z. Ibragimo, A. Yulchiba, Y. V. Zelenev and K. U. Usmanov, *Vysokomol. Soedin., Ser. A*, 1973, 15, 1831–1838.

182. P. Tummala, W. J. Liu, L. T. Drzal, A. K. Mohanty and M. Misra, *Ind. Eng. Chem. Res.*, 2006, 45, 7491–7496.

183. J. T. Kim and A. N. Netravali, *Macromol. Mater. Eng.*, 2012, 297, 176–183.

184. X. S. Huang and A. N. Netravali, *Biomacromolecules*, 2006, 7, 2783–2789 .

185. A. Bhattacharya and B. N. Misra, *Prog. Polym. Sci.*, 2004, 29, 767–814 .

186. A. Bhattacharya and P. Ray, in *Polymer grafting and crosslinking*, John Wiley & Sons, Inc., Hoboken, New Jersey, 2009.

187. R. M. Broyer, G. N. Grover and H. D. Maynard, *Chem. Commun.*, 2011, 47, 2212–2226 .

188. Y. X. Hu, D. Samanta, S. S. Parelkar, S. W. Hong, Q. A. Wang, T. P. Russell and T. Emrick, *Adv. Funct. Mater.*, 2010, 20, 3603–3612.

189. J. Shu, B. Panganiban and T. Xu, *Annu. Rev. Phys. Chem.*, 2013, 64, 631–651 .

190. N. Dube, A. D. Presley, J. Y. Shu and T. Xu, *Macromol. Rapid Commun.*, 2011, 32, 344–353 .

191. D. L. Xi, C. Yang, X. Y. Liu, M. Q. Chen, C. Sun and Y. L. Xu, *J. Appl. Polym. Sci.*, 2005, 98, 1457–1461.

192. H. Zhou, P. Jiang, C. Yang, J. Q. Jiang, H. Y. Bai, M. Q. Chen and X. Y. Liu, *Acta Polym. Sin.*, 2008, 424–429.

193. R. Snyders, K. I. Shingel, O. Zabeida, C. Roberge, M. P. Faure, L. Martinu and J. E. Klemberg-Sapieha, *J. Biomed. Mater. Res., Part A*, 2007, 83, 88–97.

194. Q. X. Wu, T. Yoshino, H. Sakabe, H. K. Zhang and S. Isobe, *Polymer*, 2003, 44, 3909–3919.

195. H. F. Tian, Y. X. Wang, L. N. Zhang, C. Y. Quan and X. Z. Zhang, *Ind. Crops Prod.*, 2010, 32, 13–20 .

196. L. Kurniawan, G. G. Qiao and X. Q. Zhang, *Biomacromolecules*, 2007, 8, 2909–2915 .

197. N. Reddy, E. Q. Jin, L. H. Chen, X. Jiang and Y. Q. Yang, *J. Agric. Food Chem.*, 2012, 60, 4872–4879 .

198. N. G. Wang, L. N. Zhang and J. M. Gu, *J. Appl. Polym. Sci.*, 2005, 95, 465–473.

199. C. M. Vaz, L. A. de Graaf, R. L. Reis and A. M. Cunha, *Polym. Degrad. Stab.*, 2003, 81, 65–74.

200. S. K. Park, D. H. Bae and N. S. Hettiarachchy, *J. Am. Oil Chem. Soc.*, 2000, 77, 1223–1227 .

201. C. Marquie, *J. Agric. Food Chem.*, 2001, 49, 4676–4681 .

202. V. Tropini, J. P. Lens, W. Mulder and F. Silvestre, *Ind. Crops Prod.*, 2004, 20, 281–289 .
203. V. Tropini, J. P. Lens, W. J. Mulder and F. Silvestre, *Cereal Chem.*, 2000, 77, 333–338.
204. S. Kim, D. J. Sessa and J. W. Lawton, *Ind. Crops Prod.*, 2004, 20, 291–300 .
205. E. Malmstrom and A. Carlmark, *Polym. Chem.*, 2012, 3, 1702–1713.
206. K. Edgar, C. Buchanan, J. Debenham, P. Rundquist, B. Seiler, M. Shelton and D. Tindall, *Prog. Polym. Sci.*, 2001, 26, 1605–1688.
207. T. Heinze and K. Petzold, *Monomers, Polym. Compos. Renewable Resour.*, 2008, 343–368.
208. R. Quintana, O. Persenaire, L. Bonnaud and P. Dubois, *Polym. Chem.*, 2012, 3, 591–595 .
209. P. Zugenmaier, *Macromol. Symp.*, 2004, 208, 81–166.
210. A. K. Mohanty, A. Wibowo, M. Misra and L. T. Drzal, *Polym. Eng. Sci.*, 2003, 43, 1151–1161 .
211. H. L. La Nieve, in *Handbook of Fiber Chemistry*, 3rd edn, ed. M. Lewin, 2007, vol. 16, pp. 1–1030.
212. X. Cao, S. Sun, X. Peng, L. Zhong, R. Sun and D. Jiang, *J. Agric. Food Chem.*, 2013, 61, 2489–2495 .
213. N. Joly, R. Granet, P. Branland, B. Verneuil and P. Krausz, *J. Appl. Polym. Sci.*, 2005, 97, 1266–1278.
214. S. Ramesh and S. C. Lu, *J. Power Sources*, 2008, 185, 1439–1443 .
215. M. Yoshioka, N. Hagiwara and N. Shiraishi, *Cellulose*, 1999, 6, 193–212.
216. J. Amim, L. Blachechen and D. Petri, *J. Therm. Anal. Calorim.*, 2012, 107, 1259–1265.
217. D. Roy, M. Semsarilar, J. Guthrie and S. Perrier, *Chem. Soc. Rev.*, 2009, 38, 2046–2064 .
218. M. Schilling, M. Bouchard, H. Khanjian, T. Learner, A. Phenix and R. Rivenc, *Acc. Chem. Res.*, 2010, 43, 888–896 .
219. O. Fridman and A. Sorokina, *Polym. Sci., Ser. B*, 2006, 48, 233–236.
220. G. Wypych, *Handbook of Plasticizers*, William Andrew Inc., New York, 2004.

221. S. Zepnik, S. Kabasci, H.-J. Radusch and T. Wodke, *J. Mater. Sci. Eng. A*, 2012, 2, 152–163 .

222. A. Ghebremeskel, C. Vernavarapu and M. Lodaya, *Int. J. Pharm.*, 2007, 328, 119–129 .

223. A. Ghebremeskel, C. Vemavarapu and M. Lodaya, *Pharm. Res.*, 2006, 23, 1928–1936 .

224. J. Ballany, D. Littlejohn, R. A. Pethrick and A. Quye, Probing the factors that control degradation in museum collections of cellulose acetate artefacts. in *Historic textiles, papers, and polymers in museums*, American Chemical Society, Washington, DC, 2001.

225. J. Amim, Y. Kawano and D. F. S. Petri, *Mater. Sci. Eng., C*, 2009, 29, 420–425.

226. S. Ramesh, R. Shanti and E. Morris, *Carbohydr. Polym.*, 2013, 91, 14–21 .

227. H. R. Jhong, D. S. H. Wong, C. C. Wan, Y. Y. Wang and T. C. Wei, *Electrochem. Commun.*, 2009, 11, 209–211 .

228. S. Ramesh, R. Shanti and E. Morris, *Carbohydr. Polym.*, 2012, 87, 2624–2629 .

229. A. Pinkert, K. N. Marsh, S. S. Pang and M. P. Staiger, *Chem. Rev.*, 2009, 109, 6712–6728 .

230. G. Stiubianu, A. Nicolescu, A. Nistor, M. Cazacu, C. Varganici and B. Simionescu, *Polym. Int.*, 2012, 61, 1115–1126.

231. E. Bianchi, A. Bonazza, E. Marsano and S. Russo, *Carbohydr. Polym.*, 2000, 41, 47–53.

232. C. Yan, J. Zhang, Y. Lv, J. Yu, J. Wu, J. Zhang and J. He, *Biomacromolecules*, 2009, 10, 2013–2018 .

233. Y. Teramoto and Y. Nishio, *Biomacromolecules*, 2004, 5, 407–414 .

234. J. Zhong, X. Chai and S. Fu, *Carbohydr. Polym.*, 2012, 87, 1869–1873 .

235. C. Buchanan, S. Gedon, A. White and M. Wood, *Macromolecules*, 1992, 25, 7373–7381.

236. C. Biermann, J. Chung and R. Narayan, *Macromolecules*, 1987, 20, 954–957.

237. G. Cheng, T. Wang, Q. Zhao, X. Ma and L. Zhang, *J. Appl. Polym. Sci.*, 2006, 100, 1471–1478.

CHAPTER

5

SEMIFLEXIBLE BIOPOLYMERS IN BUNDLED ARRANGEMENTS

Jörg Schnauß [1,2], Tina Händler [1,2] and Josef A. Käs [1]

[1]Institute for Experimental Physics I, Universität Leipzig, Linnéstraße 5, Leipzig 04103, Germany

[2]Fraunhofer Institute for Cell Therapy and Immunology, Perlickstraße 1, Leipzig 04103, Germany

ABSTRACT

Bundles and networks of semiflexible biopolymers are key elements in cells, lending them mechanical integrity while also enabling dynamic functions. Networks have been the subject of many studies, revealing a variety of fundamental characteristics often determined via bulk measurements. Although bundles are equally important in biological systems, they have garnered much less scientific attention since they have to be probed on the mesoscopic scale. Here, we review theoretical as well as experimental approaches, which mainly employ the naturally occurring biopolymer actin, to highlight the principles behind these structures on the single bundle level.

Keywords: semiflexible polymers; bundles; worm-like bundle; dynamics; actin; microtubules; depletion forces; counterion condensation; crosslinkers

INTRODUCTION

Semiflexible polymers cannot be treated in the classical physical frame of flexible polymers or rigid rods due to their non-vanishing backbone stiffness. While they still show strong thermal fluctuations, they remain in an outstretched configuration, which lends them unique mechanical properties. During the last few decades, the worm-like chain (WLC) model was established as the standard description for this special class of polymers. Its defining property is the bending stiffness κ as an intrinsic material constant [1]. Many experimental studies have successfully demonstrated the applicability of this description to a broad range of biopolymers, which are mainly considered to be semiflexible. From a biological point of view, the unique mechanical properties of this polymer class allow the formation of stable networks even for low volume fractions, which provides biological material with mechanical stability while still enabling transport processes [2]. These network arrangements can be modeled via collective properties of WLCs. Within entangled networks, the WLC model even allows derivations of the scaling of the elastic plateau shear modulus with regard to polymer concentration and their according single filament stiffness [3,4,5]. Due to their biological importance and experimental accessibility, networks received a lot of scientific attention [6,7,8,9].

In contrast, the arrangement of WLCs into bundled structures received much less attention, although they play key roles in biological matter. A natural model system for semiflexible polymers is the protein actin, which can polymerize into long, semiflexible filaments. Besides forming isotropic networks, they can be assembled into bundles via a variety of effects, and filaments are either parallel or antiparallel aligned.

Thus, single bundles are anisotropic singularities in a system impeding bulk measurements such as shear rheology, which commonly relies on an isotropic, homogenous material distribution throughout the entire sample. Anisotropies do not average out and highly bias each experiment in a different way rendering controlled investigations challenging (except networks of bundles spanning the entire sample [10,11,12]). Each bundle needs to be experimentally probed on the mesoscopic scale and has to be investigated separately since bundles do not form uniformly. Actin filaments, for instance, have a polydisperse length distribution leading to length and thickness variations of the overall structure. Despite those limitations, biopolymers are a suitable model system to study bundles of semiflexible polymers since they are readily experimentally accessible and can be investigated via a multitude of experimental methods such as fluorescence microscopy [13,14,15], optical tweezers [16,17], atomic force microscopy [18,19,20,21], light scattering [22,23], electron microscopy [14,24,25,26], and X-ray diffraction [27,28]. Throughout this review, we mainly focus on the protein actin as one of the most abundant biological components but also emphasize findings for other biopolymers such as microtubules to illustrate mutual, rather general concepts.

Studies on single bundles require sophisticated experiments to uncover their physical principles to gain a deeper understanding of their crucial cellular functions. In biological systems, for instance, bundles are mainly formed by crosslinking proteins, which add further parameters to the WLC model and introduce their own inherent mechanical contributions. First mathematical descriptions incorporate these material properties for distinct boundary conditions of the bundle within the frame of the worm-like bundle model (WLB) [29,30].

To determine contributions of crosslinkers, in vitro experiments of single bundles formed without any crosslinkers are inevitable. Commonly employed bundling mechanisms are summarized and arising effects without crosslinking components are highlighted and explained within according theoretical approaches. Subsequently, the large impact of crosslinkers is reviewed from an experimental as well as theoretical perspective to illustrate their influences to mechanical as well as dynamical properties of bundled structures. For the interested reader, we also recommend a related review by Benetatos and Jho [31], which focuses on recent theoretical and computational advances from a different perspective.

BUNDLE FORMATION

Actin filaments are negatively charged along their surface. To achieve bundling of these filaments, attractions are needed to overcome arising electrostatic repulsions. An overview of three potential bundling mechanisms is given in Figure 1displaying two fundamental physical principles—depletion forces and counterion condensation—which counteract the repulsive forces. The third mechanism employs additional, crosslinking components connecting two or more filaments as found in biological systems. In cells, these crosslinking proteins usually undergo binding and unbinding events with different on- and off- rates for differing crosslinkers [2,32,33].

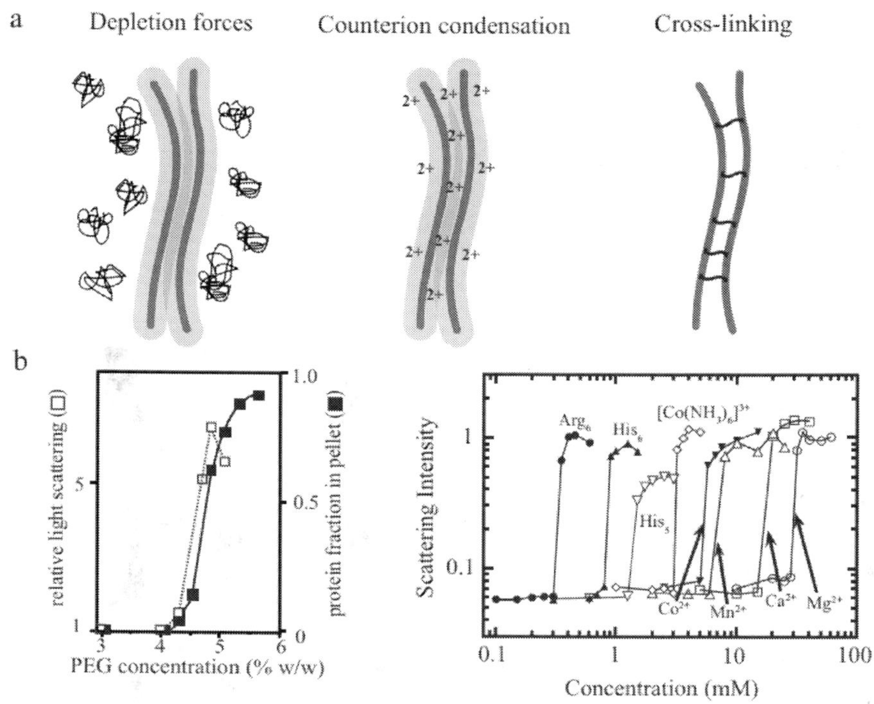

Figure 1. **(a)** Shown are three bundling mechanisms for actin filaments used for experiments described within this review. Actin filaments can be either bundled by fundamental physical effects such as depletion forces (e.g., induced by

polyethylene glycol (PEG)) or counterion condensation but also by additional crosslinking proteins; (**b**) light scattering experiments revealed specific threshold concentrations and displayed rather sharp transitions from isotropically distributed filaments to ordered bundles when depletion forces or counterion condensation were employed in a concentration dependent manner. The graphs are adapted from Suzuki et al. as well as from J.X. Tang and P.A. Janmey [22,23].

The addition of crosslinking proteins significantly alters a bundle's properties. Their features contribute to the mechanical behavior of the overall structure as described within the WLB theory. To investigate bundles without additional contributions, it is necessary to arrange actin filaments without accessory proteins. A first possibility is bundle formation via depletion forces. Due to molecular crowding, filaments experience an attracting force leading to an overlap of their excluded volumes. Within this process, the free energy of the system is decreased and entropy increased [34,35,36]. A second approach employs high concentrations of positive, divalent ions in solution to cause bundling via counterion condensation, where divalent counterions can compensate the partial surface charges of two filaments establishing an ion-based crosslinker [23].

The interplay of the rich cellular protein pool and fundamental physical principles renders in vivo studies of actin bundles difficult. To determine their mechanical properties and to distinguish between contributions of involved components, in vitro investigations are inevitable. The potential to form bundles without crosslinkers allows for investigation of their mechanical properties decoupled from contributions of additional components.

Depletion Interaction

In a crowded environment, suspended colloidal particles experience a force, which tends to group them into clusters. This phenomenon of induced depletion forces was described originally as an arrangement of beads in a polymeric solution [34,35], but can be readily extended to suspended, rod-like particles [17,36,37,38,39,40,41]. Attractive forces between these particles are generated solely by the presence of free, non-interacting polymers (e.g., polyethylene glycol (PEG)) in the surrounding solution. In these environments, a small volume directly surrounding the colloids can be described as an excluded volume. This volume is hardly penetrated by polymers of the solution (Figure 2a). If a polymer enters this region, its degrees of motion will be confined due to the vicinity to the colloidal particle. A confinement of motion would lead to an entropy decrease and is thus unfavorable [34,35].

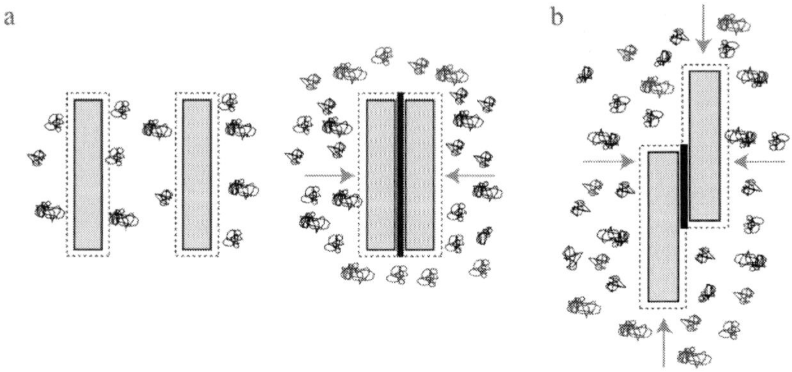

Figure 2: (**a**) High concentrations of non-interacting polymers—so-called crowding agents—in solution are causing attractive forces between colloidal, rod-like particles [35,36] when exceeding a threshold concentration Co. If excluded volumes of these suspended particles overlap, the total volume available for the non-interacting polymers increases. The newly available volume for polymers in solutions leads to an increase in entropy for the whole system; and (**b**) depletion forces arrange these rods with vertical as well as horizontal components (as indicated by the red arrows). If the rods are deflected along their axes, a constant restoring force arises, driving the system back to the equilibrium state with a maximized overlap (**black**) of excluded volumes. This figure is inspired by S. Asakura and F. Ooosawa, M. Hosek and J.X. Tang, and M. Kinoshita [35,36,37].

By increasing the polymer concentration, a grouping effect of the suspended colloids appears, resulting in an overlap of these excluded volumes. Accordingly, the total volume accessible for non-interacting polymers is increased leading to an entropy increase in the system (Figure 2a). Bundling, for instance, of actin filaments due to the presence of an inert polymer has been described previously [36]. The authors describe that the inert polymer PEG induces a bundling phenomenon in actin filament solutions when its concentration exceeds a critical onset value Co (Figure 1b). Over a limited range of PEG's molecular weight and ionic strength, Co can be expressed as a function of these two variables. The process is reversible, but hysteresis is also observed in the dissolution of bundles, with ionic strength having a large influence. Additional actin filaments are able to join previously formed bundles, but PEG polymers are not incorporated into the actin structures [36]. In this picture, actin takes the role of the colloidal, rod-like particle, because PEG is much smaller and acts as the depleting agent. Bundling of rod-like particles does not only involve vertical clustering (Figure 2a) but also an axial arrangement (Figure 2b). Thus, if the ends of two filaments are shifted along the axis at the onset of the bundling process, the filaments slide along each other until the maximal overlap of their excluded volumes is reached. In turn, these filaments can be deflected from their equilibrium position by externally induced axial sliding since they are not crosslinked. Theoretical approaches have included these possibilities, revealing that two axially deflected rods in a crowded environment tend to restore the energetic minimum by a sliding motion. Exerted restoring forces have found to be of linear shape since the energy gain per unit length is constant [37,38,39]. This means that each step towards the equilibrated state contributes with the same amount of free energy when assuming a constant inter-filament spacing. Recently, these predictions have been supported by experimental findings. A system comprised of microtubules (relatively stiff biopolymers) and the depletant PEG revealed a constant restoring force when microtubules were attracted along their axes. Addition of a third rod showed that forces arise due to a pairwise additivity of the depletion interaction [40]. However, if depletion forces become very large due to high molecular crowding, the inter-filament spacing becomes smaller and smaller. In this case, electrostatic repulsion might not be sufficient to properly separate the filaments anymore and they experience frictional forces, which can effectively stall the sliding motion [41].

In general, molecular crowding effects represent a fundamental physical interaction which cannot be switched off even in active cellular systems. The cytoplasm itself is a densely crowded environment filled with macromolecules (up to 40%) [42]. Thus, sliding motions between filaments can be induced solely by depletion forces, a process which does not rely on active transport. In this regard, the depletion induced sliding or contraction of filaments can be considered a complementary or even competing process with respect to the active, dissipative cellular processes.

Counterion Condensation

Commonly studied biopolymers are polyelectrolytes, i.e., they carry a surface net charge. Cytoskeletal polymers like actin, microtubules, and intermediate filaments are strong polyelectrolytes, whereas extracellular proteins like collagen and fibrin exhibit low surface charges [43]. Bundling in the presence of (mostly multivalent) counterions has been described for various polymer-counterion combinations [43] and theoretical models describing the origin of attractive forces between like-charged molecules mostly refer to polyelectrolyte rods [34,44,45,46,47,48].

Although actin filaments are semiflexible, this approximation leads to valid results since the effective persistence length of a charged polymer in solution is increased due to monomer–monomer repulsion [49,50,51]. In contrast to the description of bundles formed by chemical crosslinkers or depletion interactions, forces mediated by charge fluctuations cannot be added pairwise [47,48]. However, we note that there has been some debate about this topic [52]. Calculations usually involve the "Manning criterion", which states that mobile ions in solution condensate on oppositely charged rods when $\xi|z|>1$ [53]. In this formulation, ξ is the Manning parameter giving the ratio of characteristic electrostatic repulsion to thermal energy [45,53] and z is the ion valency. Consequentially, the charge per unit length on the (previously highly charged polyelectrolyte) rods decreases dramatically and electrostatic repulsion between them disappears for sufficiently small distances [45].

Ray and Manning have explored the complex interplay between free energy contributions of two polyelectrolytes with condensed counterions [44]. In the frame of their model, they considered the free energy of electrostatic repulsion between charge sites on the same rod, the free energy associated with repulsion between the two rods, and the energy contribution from transferring a counterion from the solution into the condensed state. Note that in the condensed state, ions are not fixed in their position, but can move freely within the condensation region. Generally, the origin of attractions is not the increased number of condensed counterions when the distance between rods becomes smaller [44]. Instead, the size of the condensation volume is the determining quantity. If the rods approach each other (mathematically from infinity), the condensation volume is initially slightly decreased yielding repulsive interactions at large distances due to a decrease in the translational entropy of the condensed counterions. By decreasing the rods' distance further, the condensation volumes interpenetrate, which increases the translational entropy of the condensed counterions, thus leading to an attractive interaction. When compacting the rods even closer, the condensation volume shrinks below a critical threshold resulting in a repulsion for short inter-rod distances.

Another analytical description, supported by Brownian-dynamics simulations, was given for two stiff polyecletrolyte rods where all counterions are condensed [45]. Attractive interactions were found to appear at higher counterion valency. Ha and Liu assumed that not all counterions should be condensed but that free and condensed counterions are freely exchangeable [46,47]. The calculation of the interplay between two like-charged rods in a solution of counterions resulted in an interaction consisting of two parts: a repulsive term since not all counterions are in a condensed state and an attractive term due to charge fluctuations along their contours [46]. As a consequence, the number of condensed ions and the charge fluctuations between the rods increase upon decreasing the inter-rod distance. Taking the non-pairwise additivity of counterion-mediated interactions into account, the theory can be extended to describe bundles featuring more than two polyelectrolyte rods. However, in this extension the emanating attraction does not only depend on the Manning parameter ξ, but also on the separation between condensed ions on one rod [45,47].

Even though some of the aforementioned studies do not only involve analytical calculations but also computer simulations capturing the effects of counterion condensation [44,45,48], we would like to point to a number of publications approaching this topic via simulations [54,55,56,57].

Crosslinkers

Crosslinkers can be classified into physical and chemical crosslinkers. Physical crosslinkers such as van der Waals forces, depletion interactions, or certain crosslinking molecules are reversible and can be treated as an effective pairwise potential under specific conditions [58]. Chemical crosslinkers, on the other hand, are irreversible and drive the system into a frozen, disordered state. Approaches applying gelation theory to these anisotropic networks of macromolecules in combination with permanent crosslinkers [59,60,61,62] modeled the bundled state as a nematic gel and have been extensively reviewed by Benetatos and Jho [31].

Bundle formation in the presence of crosslinkers has been treated as phase transitions from isotropic polymer solutions to bundled structures [63,64]. A first approach treats polymers within solution as rigid rods and the formation of bundles occurs when the free energy of an isotropic network and the bundle phase becomes equal [63]. It comprises four contributions: (1) the increased configurational entropy of crosslinkers distributed along the filaments relative to the isotropic network; (2) the excluded volume repulsion of the rods; (3) the translational entropy of the bundles; and (4) the reduction in rotational entropy of the bundles relative to the isotropic phase. The competition of these energies and namely the entropy gain of crosslinkers in bundles compared to crosslinkers in isotropic networks favor bundle formation at low temperatures.

Within a second approach, polymers are treated as charged rods in the presence of inter-rod linkers [64]. Those linkers can be either chemical crosslinkers or counterions and are modeled as effective potentials. The energy potentials reigning the transition behavior from isotropic to bundled phase are the short-range attractive potential mediated by the linker and the long-range electrostatic repulsion between the charged rods. Each interaction is accounted for small and large angles between filaments. By adopting the generalized Onsager theory, the free energy per unit volume consists of (1) the translational and rotational entropy of the rods; (2) the interaction between two rods and (3) the contribution of free linkers as well as (4) an energy contribution of linkers adsorbed to isolated rods. Using this formulation of the free energy, the formation of a bundle phase can be deduced. In general, electrostatic repulsion favors large angles between filaments, but in bundles steric repulsions are minimized and linker binding is maximized (as opposed to a nematic phase where inter-filament repulsion is still strong).

In another study, a critical crosslinker concentration has been reported to be responsible for bundling and unbundling processes [58,65]. Above this critical concentration, filaments form either a single, large bundle or a number of sub-bundles, whereas below all filaments are unbound.

This transition occurs from the competition of (1) the bending energy of single filaments; (2) the intra-filament crosslinker interaction energy; (3) the repulsion potential between single filaments; and (4) the attraction mediated by crosslinkers. Since the most dominant contributions are the energy gain by bundling due to the attractive potential and the loss of conformational entropy of the filaments, bundling occurs more easily if the filaments are stiffer.

The special case of actin bundles formed by crosslinking molecules such as fascin or espin was treated in a coarse-grained model [27,66]. It describes the thermodynamic transition from parallel unbound actin filaments to overtwisted bundles, where actin filaments are twisted with a –28/13 symmetry [15,27]. In the native configuration, actin filaments exhibit a helicity of –13/6: 6 rotations per 13 monomer repetitions [15]. Thus, parallel actin filaments are incommensurate for crosslinkers which preferably bind monomers on adjacent filaments minimally separated from each other. Overtwisting the filaments improves the binding geometry, but is accompanied with the cost of torsional energy. This competes with the binding energy of the crosslinkers, being the sum of the energy gain of perfectly aligned crosslinkers between monomers and the energy penalty of distorting the bonds. The transition from unbound, natively twisted filaments to fully bound, overtwisted filaments was found to be strongly sensitive on the crosslinker stiffness. Stiff crosslinkers cause a thermodynamic phase transition of second-order, whereas soft crosslinkers cause a smooth increase of bound crosslinkers and overtwist with increasing chemical potential. Under thermal fluctuations of filament orientation and crosslinker position, the thermodynamic distinction between the two bundling regimes is preserved [27]. Note that this model only accounts for bundles of infinite length and width [66], and the relationship between twist in bundles and finite bundle size is discussed in Section 6.

A recent description of bundle formation is based on parallel, pre-aligned polymers [67]. Two interaction terms determine the phase behavior: a short-range attraction that can be counterion-induced, hydrophobic or mediated by crosslinks or hydrogen bonds and a comparatively long-range repulsion such as electrostatic interaction between like-charged polyelectrolytes. The lateral density perpendicular to the nematically aligned polymers was studied for various interaction strengths and ranges as well as the tensions and lengths of the polymers. Depending on the interplay of the competing interactions, a macroscopic phase separation with infinite bundle size or a microphase separation with finite bundles occur. As reviewed by Benetatos and Jho [31], a similar mechanism leads to bundles in brushes of grafted polymers where the attachment points can move freely on a surface [68]. Due to the grafting, a repulsive interaction is not needed for likewise bundle formation.

DYNAMICS IN UNCROSSLINKED BUNDLES—CONTRACTIONS WITHOUT MOLECULAR MOTORS

The initial step to understand the properties of bundles of semiflexible polymers is to investigate their properties emerging without any additional components or inner-bundle interactions. Studying effects which can be attributed to the semiflexible polymers themselves then allows to evaluate effects which are driven by additional factors such as crosslinkers or counterion condensation. Arranging filaments without these additional contributions can be achieved by crowding effects inducing depletion forces (Subsection 2.1). An example of how these bundles can be even used as a reference systems for crosslinked bundles can be found in Section 5. However, already depletion force induced bundles show a surprisingly rich repertoire of dynamical behaviors. Depletion forces cannot only arrange semiflexible polymers laterally into bundled structures but can also induce an axial sliding in a biologically relevant force regime. In the two filament case, theoretical approaches have shown that axially deflected filaments in a crowded environment slide against each other to achieve a maximized overlap of their excluded volumes (Figure 2b). The arising energy potential of this process has been reported to yield a constant force driving filaments to the equilibrium position [37,38,39]. These theoretical predictions have been experimentally verified with relatively stiff microtubules, which were grouped by the depletion agent PEG [40,69]. Using an optical tweezers setup, Hilitski et al. report that the attraction force along the axis of two microtubules is indeed constant.

Upon addition of a third filament, they further revealed that arising forces are pairwise additive [40]. However, if induced depletion forces become very large due to a high molecular crowding, the inter-filament spacing becomes small. At a certain threshold concentration, electrostatic repulsion of the like-wise charged filaments might not be sufficient to properly separate them anymore. Consequently, they experience frictional forces, which can effectively stall the sliding motion [41]. However, scaling up pairwise depletion force induced filament contractions to a multi-filament scale reveals completely different dynamics [17]. To probe these dynamics, bundles of actin filaments were formed by the depletant methyl cellulose and resulting effects were only induced by attractive filament–filament interactions. For contact-free manipulations via optical tweezers, actin filaments were enriched with biotin triggering the attachment of polystyrene beads via biotin-streptavidin bonds. Actin bundles were deflected from their equilibrium position by pulling forces (Figure 3a), yielding strains that exceed normal elastic deformations (up to 175% of the initial contour length). Due to actin's rigidity, these elongations can neither be attributed to thermal fluctuations of single filaments nor to stretching of the filament backbone. Thus, the stretching process effectively pulled filaments apart. This sliding decreased the overlap of excluded volumes of filaments and, accordingly, the freely available space for polymers within the solution. After the deflecting laser was switched off, bundles started to contract, aiming to restore a maximized overlap again (Figure 3a,b) [17]. Note that crosslinking proteins would have frozen this sliding motion and thus contractions are only induced by entropic arguments.

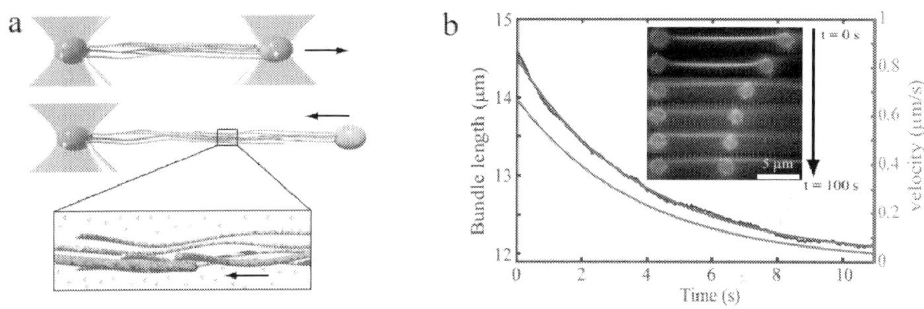

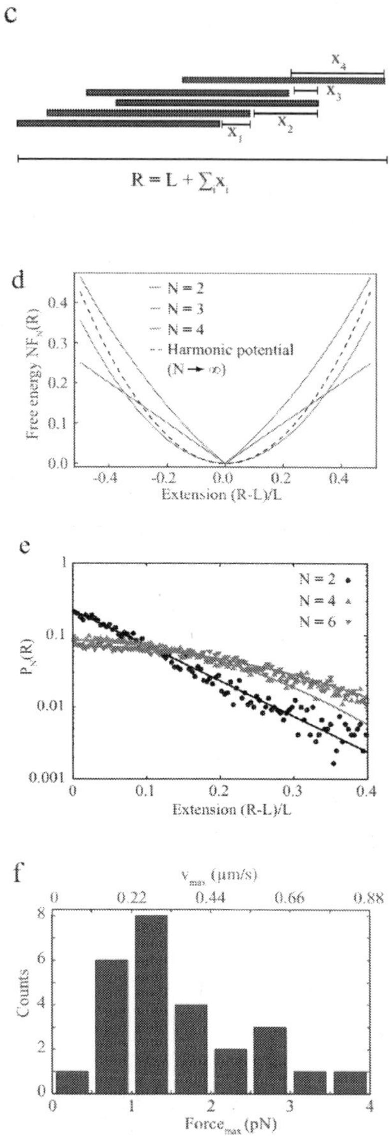

Figure 3: (**a**) Bundles formed by depletion forces can be deflected from their energetic minimum by inducing pulling forces via optical tweezers. After switching off the external force, bundles contract to restore the maximal overlap of excluded volumes; (**b**) recorded bundle lengths over time during contraction processes (inset) can be well described by an exponential decay function. This fitting function can be differentiated allowing an evaluation of the contraction velocity (**red** graph); (**c**) the exponential decay can be described within a

mathematical model in an idealized 2D-scenario, where forces are applied at the first (i=1) and at the last filament (i=N), with Lbeing the uniform filament length, R the length of the stretched bundle, and xi the displacement between filaments; (**d**) plotting the free energies FN(R) vs. the extension R−L shows a strong dependency on the filament number N. A two-filament bundle (N=2) has a linear energy landscape [37,38,39,40], but with only a few filaments (N=4) the asymptotic, harmonic form (dashed) is nearly reached [17]; (**e**) the transition from a linear to a harmonic potential shape was further confirmed by simulations. A two-filament bundle follows an exponential probability distribution as given by the Boltzmann weight. Bundles formed by six filaments yield a Gaussian probability distribution corresponding to a harmonic potential; and (**f**) measured maximal forces compare well to forces exerted by single myosin motors [2]. The figure is adapted from Schnauß et al. [17].

In contrast to the two-filament case where contractions proceed with a constant velocity [37,38,39,40,41], multi-filament bundles display fundamentally different dynamics with an exponentially decreasing velocity (Figure 3b). These dynamics translate into an exponential force decay, which, in turn, implies a harmonic energy potential. Within a mathematical model, these surprising dynamics are explained as an emergent phenomenon of rod-like colloids (actin filaments) when taking pairwise interactions to a multi-filament scale. Bundles were modeled as two-dimensional arrangements of laterally stacked pairs of rigid filaments (Figure 3c), which employs the recent experimental finding that forces are pairwise additive [40]. This approach yields an arising harmonic potential in an overdamped (highly viscous) environment (Figure 3d) when the system is composed of multiple filaments. The harmonic form of the potential and the according exponential force decay are further verified by simulations, which are independent of the mathematical model (Figure 3e) [17].

In general, molecular crowding effects represent fundamental physical interactions, which cannot be switched off even in active systems such as cells. This argument is supported by the fact that the cytoplasm itself is a densely filled environment [42]. Additionally, these bundle contractions already appear at macromolecular contents which are well below the crowding in cells emphasizing the potential biological relevance of this contractile process. Kinetics and maximal forces (Figure 3f) are in a regime of active processes based on single molecular motors, but are independent of the conversion of chemical energy into mechanical work [2].

MECHANICAL PROPERTIES OF BUNDLES

Worm-Like Chain Model

Mechanical properties of semiflexible polymers can neither be understood in the classical physical picture of flexible polymers nor rigid rods. Due to their non-vanishing backbone stiffness, they remain in an outstretched configuration while still showing strong thermal fluctuations [2]. Within the frame of the WLC model, the polymer is described as a differentiable curve $\vec{r}(s)$ and statistical properties are determined by the effective free energy [70]

$$H = \frac{\kappa}{2} \int_0^L ds \left(\frac{\partial^2 \vec{r}(s)}{\partial s^2} \right)^2 , \quad (1)$$

with κ being the bending stiffness and L the contour length of the polymer. The tangent vector $\vec{t}(s) = \partial \vec{r}(s)/\partial s$ at the arc length s of $\vec{r}(s)$ is used to calculate material properties of the polymer. Fluctuations at a certain temperature allow for evaluation of the stiffness of the polymer by correlating the tangent vectors along the backbone. The arising correlation function decays exponentially with a decay constant called the persistence length L_p:

$$\langle \vec{t}_0 \cdot \vec{t}_s \rangle = e^{-\frac{|s|}{L_p}} .$$

Thus, L_p is a quantitative measure for polymer stiffness and if $L_p \cong L$, the polymer is considered semiflexible. However, L_p depicts thermal fluctuations showing a temperature dependency $L_p \sim 1/T$ and cannot be considered a material constant in general. Introducing the bending stiffness, κ, eliminates the temperature dependency by multiplying the thermal energy to the persistence length: $\kappa = k_B T L_p$ [1].

The WLC can be incorporated into theoretical approaches of assemblies of filaments into networks. However, to describe filament bundles, the theory has to be extended to depict the according behaviors sufficiently.

Worm-Like Bundle Model

The WLB is an extension of the WLC, which is able to describe the behavior of a bundle of worm-like chains. In nature, these filament bundles are usually held together by crosslinking proteins adding their own mechanical properties to the overall structure. Coherent with the WLC, the intrinsic material parameter for the WLB is the bending stiffness κ_b of the whole bundle. This bending stiffness, however, is state-dependent and arises from the interplay of the individual filament stiffness and the relative sliding motions between filaments within the bundle [29,30,71]. This interplay leads to a sensitivity of κ_b to time and length scales on which a bundle is probed.

Within the frame of this theory, N filaments of unit length L and bending stiffness κ_f are building up a crosslinked bundle. Neighboring filaments are discretely and irreversibly crosslinked with a mean axial spacing δ. Additionally, crosslinkers are compliant with a shear stiffness k_x and modeled to be inextensible and transverse to the main axis while filaments are fixed at a distance b (see Figure 4a). To describe the behavior of the whole bundle, three Hamiltonians are needed [29,30,71]:

$$H_{\text{WLB}} = H_{\text{bend}} + H_{\text{stretch}} + H_{\text{shear}}. \tag{2}$$

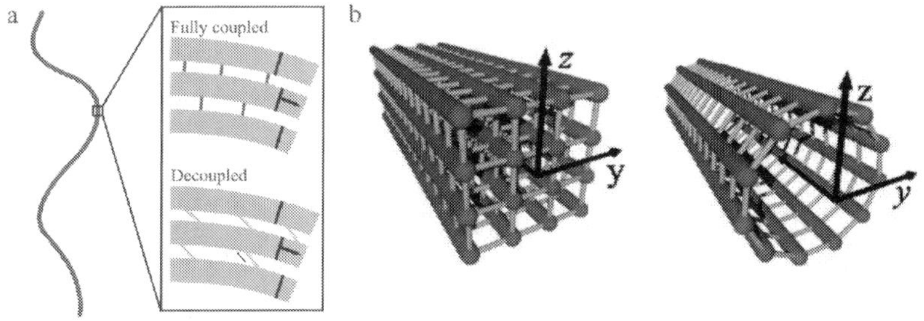

Figure 4: The worm-like bundle (WLB) model. (**a**) A bundle is deformed by in-plane bending, which leads to bending and stretching forces on the internal filaments (**light red**) and shearing of the crosslinkers (**blue**). These three contributions yield the overall Hamiltonian describing the behavior of the bundle. Depending on the mechanical properties of the crosslinker, two limiting cases arise: the fully coupled regime (**dark blue**—crosslinkers resist shear deforma-

tions) and the decoupled regime (**light blue**—crosslinkers deform under shear deformation). The illustration is inspired by Bathe et al. [30]; and (**b**) various bundle geometries can be considered, which yields different scaling predictions for differing bundle cross sections. This figure is adapted from Heussinger et al. [71].

These contributions correspond to weak bending of single filaments (H_{bend}), filament stretching ($H_{stretch}$), and crosslinker shear (H_{shear})

$$H_{\text{bend}} = \frac{N\kappa_f}{2}\int_0^L \mathrm{d}s \left(\frac{\partial^2 \vec{r}_\perp}{\partial s^2}\right)^2,$$

$$H_{\text{stretch}} = Mk_s\delta \int_0^L \mathrm{d}s \sum_{i=-M}^{M-1}\left(\frac{\partial u_i}{\partial s}\right)^2,$$

$$H_{\text{shear}} = \frac{Mk_\times}{\delta}\int_0^L \mathrm{d}s \sum_{i=-M}^{M-1}\left(\Delta u_i + b\frac{\partial \vec{r}_\perp}{\partial s}\right)^2, \tag{3}$$

with M describing the according bundle layer, u_i the stretching deformation of filament i, k_s the filament stretching stiffness on the scale of the crosslinker spacing δ and $\Delta u_i = u_i - u_{i-1}$ [29].

Due to the N-dependency of the first term, non-trivial properties arise. A coupling parameter $\alpha = \frac{k_\times L^2}{k_s \delta^2}$ can be introduced by similarity transformations allowing descriptions of the relative stiffness of a bundle's stretching and shearing modes. Interactions of these parameters result in a new Hamiltonian

$$\tilde{H} = M\int_{\tilde{s}} \left[2M\tilde{\kappa}_f \tilde{r}_\perp''^2 + \sum_{i=-M}^{M-1} \tilde{u}_i'^2 + \alpha \sum_{i=-M}^{M-1}\left(\Delta u_i + \tilde{r}_\perp'\right)^2\right], \tag{4}$$

describing, for instance, how crosslinker concentration affects the overall stiffness [29]. These interactions render a bundle a tunable structure.

Using the coupling parameter α, two limiting cases arise due to differing κ_b caused by different crosslinker properties. If $\alpha \to \infty$, crosslinkers resist shear and filaments appear as connected structures. Under deformation, the bundle's response is dominated by filament stretching [29]. These so-called fully coupled bundles display a quadratic dependency of κ_b on N. If $\alpha \to 0$, filaments bend independently under deformation since crosslinkers do not resist shearing forces. These so-called decoupled bundles display a linear dependency of κ_b on N.

In the intermediate regime, a transition from the fully coupled to the decoupled state results in a change of the scaling behavior of κ_b from N^2 to N. Thus, the model describes two competing mechanisms with increasing N, crosslinker shearing and filament stretching. This cooperative effect adds new dimensions or parameters to bundled structures allowing the bundle's response to be tuned against external deformation (see Section 5). Additionally, a combination of various crosslinkers allows to tune these properties even further. The emergence of different coupling regimes has been predicted for stereocilial bundles [72].

We note that the persistence length of bundles of semiflexible filaments has been studied in the WLB [29,30,71] but also for microtubules [71,73] as bundles of protofilaments. An important finding of the WLB is that the effective bending stiffness is not constant as for the WLC, but mode-dependent. Thus, depending on how the bundle is probed, different bending stiffnesses and, accordingly, different persistence lengths may be measured. Modeling a microtubule as an array of cylindrically arranged, fully coupled protofilaments yields a stiffness of $\kappa_b \sim N^3$ [71]. This seems to contradict the aforementioned scaling $\kappa_b \sim N^2$ for a fully coupled bundle, but the two results have been calculated for different bundle cross sections. A homogeneous (square lattice) bundle cross section to resemble actin bundle architecture yields N^2 and protofilaments arranged on the surface of a cylinder like in microtubules yields N^3 (see Figure 4b). The railway-track model for bundles of worm-like chains connected by harmonic springs also proposed an arclength-dependent persistence length or a renormalized bending stiffness [74].

Alternatively, microtubules can be modeled as hollow cylindrical beams that can be treated with linear elasticity theory [73]. With highly compliant inter-protofilament bonds, microtubules are able to absorb shear stresses by shifting neighboring protofilaments relatively to each other, which is in contradiction to the WLB extension for microtubules [71]. Thermal fluctuations are dominated by the high compliance of the bonds between protofilaments. Consequentially, a relationship between bundle length, i.e., microtubule length, and persistence length emerges, which agrees well with experimental results [73,75].

In the WLB, filaments are explicitly modeled as extensible polymers connected by inextensible crosslinkers with a certain shear modulus. In contrast, both the railway-track model [74] and a tension-induced bundling model [76] incorporate inextensible filaments bound together by harmonic springs. In the latter, a longitudinal tension aligns the filaments in parallel and thus induces crosslinker binding. With increasing strength and number of crosslinkers, the bundle is found to have an increased effective stiffness [76]. For infinitely strong crosslinkers, the alignment effect is even higher since transverse filament fluctuations are suppressed. Within a bundle, the extensibility of filaments and their various moduli (bending, shear, torsion) as well as the extensibility of crosslinkers and their shear modulus are needed to minimize the free energy under mechanical load and during bundle formation (see also Subsection 5.2). The mechanical response of a bundle of inextensible filaments and inextensible, infinitely stiff crosslinkers would resemble that of a stiff beam.

TRANSIENT CROSSLINKERS

Crosslinking proteins add an additional parameter to the properties defining the behavior of a bundle. In nature, bundles are mainly arranged by these additional components and thus understanding their role is of central importance to understand cellular processes. In general, these crosslinkers can be considered transient, i.e., they constantly bind and unbind from filaments. Experimental and theoretical approaches have revealed a rich repertoire of a bundle's response against external stimuli.

Tunable Mechanical Responses of Transiently Crosslinked Actin Bundles

α-actinin, for instance, is a prominent example of actin crosslinking proteins forming networks as well as bundles. It has two actin binding sites and is arranged as an anti-parallel dimer [77]. Furthermore, α-actinin is considered a dynamic, transient crosslinker able to constantly bind and unbind from actin filaments. This allows rearrangements within actin structures as well as active or passive responses to external forces [16]. In unbound states, these crosslinkers can diffuse and subsequently bind at other positions [78].

This transient binding is manifested in different responses against external forces when compared to permanently crosslinked bundles [13]. Thus, varying binding affinities and crosslinker dimensions are further parameters impacting a bundle's mechanical properties. One of the first studies on these effects determined the bending stiffness of bundles enriched with differing crosslinker proteins such as fascin, plastin, or α-actinin [13]. By evaluating thermal fluctuations recorded with epi-fluorescence microscopy (Figure 5), the dependency of bundle size, involved concentrations, and crosslinker type to the overall bending stiffness were systematically investigated. Depending on the crosslinker type, distinct bending stiffness regimes were observed that differed by orders of magnitudes [13]. These results were a first indication of the large impact of actin accessory proteins, which has been widely underestimated before.

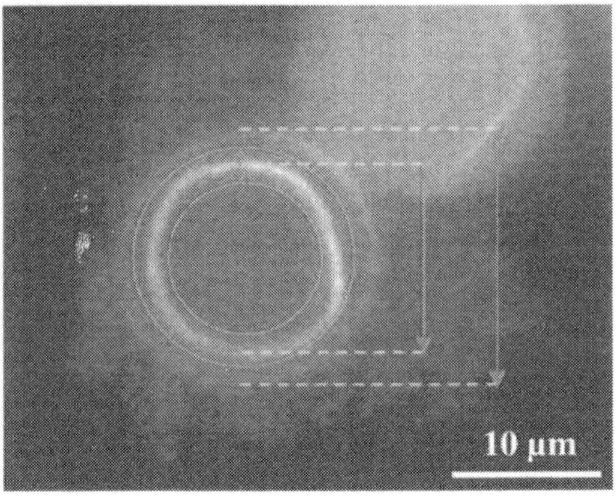

Figure 5: Analysis of thermal fluctuations of the actin bundle backbone (white circular shape) revealed different bending stiffnesses depending on the type of the crosslinkers. The two red circles mark the region the bundle can occupy in the time course of its thermal fluctuations. The picture is taken from Claessens et al. [13].

These in vitro experiments already draw a complex picture of possible bundle responses against external stimuli. As illustrated by Strehle et al., transient crosslinking of bundled actin filaments causes time-dependent mechanical responses due to dynamic rearrangements of crosslinking proteins [16]. In this study, α-actinin crosslinked bundles were deformed by large bendings induced by optical tweezers. Bundles displayed elastic as well as plastic responses depending on the time scale of the deformations (Figure 6). Short deformation times of 5 s yielded a fully elastic response for crosslinked as well as uncrosslinked bundles (formed by depletion forces), which restored their original positions after stress release. Tests showed the consistency of elastic responses when deforming a bundle multiple times on short time scales ensuring that the bundles were not damaged during deformation processes [16]. However, dynamics of elastic relaxations have been found to strongly depend on the bundle structure. Visible inhomogeneities within the bundle caused an additional relaxation time resulting in a distinct relaxation profile [16].

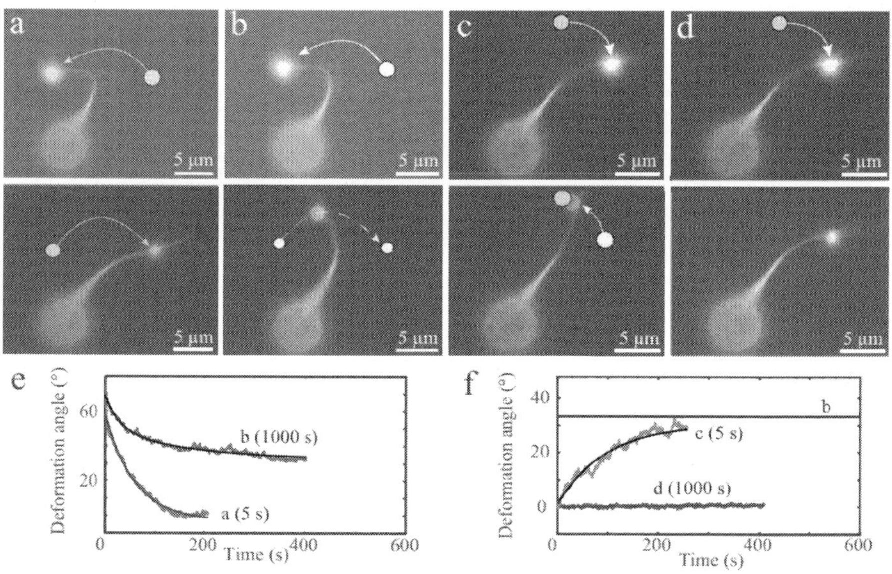

Figure 6: Shown are deformation experiments and according graphs (**e,f**) of an α-actinin crosslinked actin bundle and its time dependent responses to external deformations. (**a**) the bundle was deformed on a short time scale by moving the trapped bead along a predefined arc with a subsequent holding time of 5 s in the deformed configuration. For short deformations, a crosslinked bundle restored its original configuration after stress release; (**b**) deformations on a long time scale with a deformation time of 1000 s lead to dynamic rearrangements of crosslinkers causing a plastic deformation. The bundle relaxed only partially to a new equilibrium position; (**c**) the bundle was deformed to its former zero position with a holding time of 5 s. After stress release, the bundle returned

to its new equilibrium position; and (**d**) this procedure was repeated with a deformation time of 1000 s, which restored the initial equilibrium position of the bundle by dynamic crosslinker rearrangements. Thus, the effect is plastically reversible with a rearranged internal structure. This figure is adapted from Strehle et al. [16].

For long deformation times of 1000 s, the two types of bundles responded differently. While uncrosslinked bundles restored their initial position similar to 5 s deformations, crosslinked bundles only relaxed partially and remained in a bent shape [16]. The plastic response can be attributed to transient crosslinkers, which can unbind, diffuse, and rebind. This structural rearrangement supported the new bundle conformation induced by external forces and could be reversed by another plastic deformation [16].

These experiments illustrate how crosslinking components can change mechanical properties of single bundles. Transient binding even facilitates dynamical rearrangements, enabling time-dependent responses against external stimuli. These binding and unbinding events can even induce further dynamic effects such as contractions as recently illustrated in a system of two microtubules crosslinked by components of the Ase1/PRC1/Map65 family [69,79,80]. We speculate that this crosslinker induced sliding can be extended to the bundle level.

Unbundling/Rebundling of Crosslinked Bundles

While there are a number of studies investigating the formation of crosslinked bundles, only few discuss reversible crosslinkers [58,64,65,81] and unbundling [58,65,82,83,84].

The driving mechanism for "thermal unbundling" in bundles with a sufficiently low crosslinker density are lateral fluctuations of the filaments transverse to the bundle axis [58,65]. For a given crosslinker above a critical temperature, a discontinuous unbundling transition occurs since the gain in conformational entropy of the filaments wins over the energy gain from binding due to the attractive potential mediated by the crosslinker.

Heussinger and coworkers theoretically studied unbundling driven by an external force deforming the bundle [82,83,84]. Their work was inspired by the experiments described above showing the transience of actin crosslinkers and according plastic deformation of actin bundles if the external force was applied sufficiently long [16]. In their analysis, they extended the established worm-like bundle model to account for transient crosslinkers and discussed the effect of crosslinker stiffness on unbundling. With the model applying to highly crosslinked bundels, lateral fluctuations of the filaments were neglected.

In the first step [82], the crosslinker shearing energy that also occurs in the WLB model [29,30] was modified by multiplication with an occupation parameter n. This variable can have the values 0 and 1, corresponding to the unbound and bound state of a crosslinker, respectively. To characterize the unbundling under bundle deformation, the average crosslinker occupation $\langle n \rangle$ was calculated. Without a deformation, crosslinkers were found to unbind (corresponding to decreasing $\langle n \rangle$) upon decreasing their binding affinity or increasing their stiffness. With increasing bundle deformation, the average crosslinker occupation decreases, eventually leading to an unbundling transition. This effect is attributed to an increasing mismatch between binding sites of neighboring filaments under load. Another parameter affecting the occupation $\langle n \rangle$ is the stiffness of the crosslinker. For stiff crosslinkers, $\langle n \rangle$ exhibits a discontinuous drop with increasing bending deformation indicating a cooperative unbinding when filaments can no longer be stretched out by the strong crosslinkers. Bundles with soft crosslinkers exhibit a smooth decrease in $\langle n \rangle$ with increasing bending. In those bundles, crosslinkers are claimed to unbind one after another.

A more detailed analysis including extensive Monte Carlo simulations was performed by Vink and Heussinger [83]. Again, the two competing energies result from the crosslinker shear with the occupation parameter and from stretching the filaments due to bending. Accordingly, filaments are more prone to stretching if bundled by stiff crosslinkers. In the case of more than two filaments bundled by stiff crosslinkers, a series of first-order transitions with increasing deformation is predicted, resulting in several sub-bundles.

In a third step, the possibility of crosslinker remodeling was considered [84], which can lead to plastic deformations of a bundle under external load [16]. The shearing energy was further modified to allow for rebinding of a crosslinker which remained attached to one filament. The second binding domain of the crosslinker can either rebind to its original binding site directly opposite on the neighboring filament or to the next right neighboring site. The latter rebinding case induces a certain mismatch between the two filaments, considered a defect in the bundle. By comparing bundles of inextensible and extensible filaments, it was found that filament stretching delayed the deformation of defects.

The effect of tensile stress on a bundle of two weakly bending WLCs connected by reversible crosslinkers has also been investigated [81]. As an extension of the tension-induced bundle model presented by von der Heydt et al. [76] and discussed in Subsection 2.3, a binary variable (comparable to the occupation parameter as explained above) was introduced to account for bound and unbound crosslinkers. Consequently, the average fraction of bound crosslinkers was analyzed for a range of tensile forces. The result displays a discontinuous phase transition from weakly bound polymers to strongly bound polymers with increasing tension. At low forces, the filaments are still capable of strong transverse fluctuations, while at higher tension suppressing these fluctuations comes not at a high entropy cost and the filaments can be bound together more tightly. The sudden crossover to a tight bundle might be interpreted as a force-stiffening mechanism [81].

FINITE BUNDLE SIZE

Theories considering the bundle formation process as a phase transition from a solution of free polymers to a bundled phase predict infinite bundles [31,63,64]. However, experiments have shown that bundles exhibit a certain range of diameters [14,25]. The origin of finite bundle sizes has been attributed to a number of effects [85,86,87,88,89,90,91].

Especially for counterion induced bundles, the formation of an infinitely large bundle, spanning the whole sample, is thought to be prevented by an energy barrier impeding addition of more polymers to an already existing bundle [85,86]. This energy barrier could originate from an angle-dependent repulsion between charged bundles and polyelectrolyte rods [85]: if a rod and a bundle meet each other in a non-parallel way, the electrostatic repulsion between them is too high for further bundle growth. Thus, with an increasing bundle size, the energy barrier should grow as well [85]. Another explanation for an energy barrier hindering bundle growth are steric interactions inside counterion induced bundles [86]. Since ions have a finite size, steric interactions between them might prevent dense packing of the rods and the neutralization of charges necessary to add more rods. Additionally, during bundle growth the rods might encounter frustration also inhibiting complete charge neutralization [86].

As mentioned in Subsection 2.3, finite bundles occur in a microphase separation in a system with short-range attraction and longer-range repulsion between parallel pre-aligned polymers [67]. As long as the attractive forces are not too strong and the polymers have finite tension, the areal density of the polymers is modulated with a characteristic wavelength corresponding to bundle size [67].

For neutral, achiral filaments in a bundle, structural defects have been discussed as a potential principle leading to a limited bundle size (Figure 7a) [87]. A defect inside the bundle decreases the adhesion energy and increases the elastic energy of bending and twisting the filaments, which contradicts the bundle growth. Consequentially, a minimum energy configuration arises which corresponds to a finite bundle size.

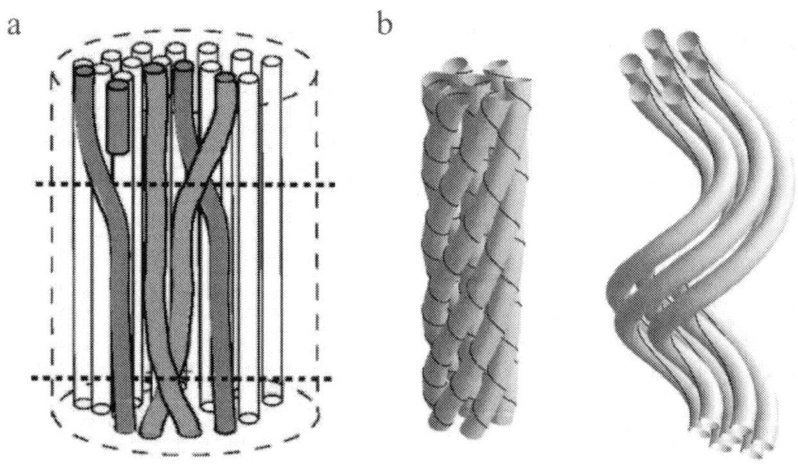

Figure 7: (**a**) Structural defects of the inner-bundle geometry impede the energy minimization during the bundle formation process, thus limiting the bundle size [87]; and (**b**) the molecular structure of the filaments impact the overall bundle geometry. Chiral filaments twist themselves around the bundle axis (**left**) while helical filaments cause a writhe of the bundle axis (**right**) [89]. Figures are adapted from N.S. Gov and G.M. Grason [87,89].

Several studies refer to chirality and twist in bundles as the main restriction factor of bundle growth [88,89,90,91]. Twist in bundles has been observed in a number of experiments [15,18,27]. Even one of the first models describing the mechanics of a bundle of two worm-like chains predicted twist in three dimensions [74]. Grason and coworkers have investigated the energy contributions for hexagonally ordered bundles of chiral and helical filaments and found that both geometrical constrictions limit bundle growth [88,89]. These results concur with the overtwisting bundle formation process [27,66], which is summarized in Subsection 2.3. Chiral filaments have to twist around the bundle axis and the energy cost of adding as well as twisting new filaments limits the bundle diameter. Helical filaments force the bundle axis itself to twist and outer filaments added to the bundle have to unbend from their intrinsic geometry (Figure 7b). The competition between the energetic cost of unbending a helical filament and the cohesive energy of the bundle leads to a finite bundle size. Whereas helical filaments are modeled with an intrinsically twisted backbone, chirality of single filaments is accounted for by a distinct crosslinker distribution on the filaments. Crosslinker sites that are helically distributed on the filaments have also been considered in an extension of the worm-like bundle model [91]. For sufficiently stiff crosslinkers, filaments inside the bundle are found to twist superhelically. The competition between the bending energy of the filaments and crosslinker induced twist determines the magnitude of twist and the equilibrium bundle width.

Monte Carlo simulations investigating the chirality in bundles also confirmed that the energetic cost of twisting filaments grows faster than the gain in adhesion energy by adding a new filament layer to a bundle [90]. Further simulations examining finite lateral bundle size have been reviewed extensively by Benetatos and Jho [31].

BUNDLES IN BIOLOGICAL SYSTEMS

Generally, polymers can be produced synthetically but also appear in a wide variety in nature. Well known biopolymers are DNA and proteins, which are crucial elements for life. In particular, semi-flexible polymers are the basis of biological compounds since they stabilize systems already at a low volume fraction. Additionally, they remain in an outstretched configuration due to their stiffness (enabling active transport) but still show Brownian fluctuations (enabling stochastic processes). In nature, semiflexible filaments such as actin are often arranged into bundled structures. One of the main cellular functions of actin bundles is the structural support of the cytoskeleton realized by so-called stress fibers. While they provide mechanical stability due to their large bending stiffness (Figure 8), they also fulfill a variety of active functions. Astonishingly, these seemingly contradictory tasks are realized by the same key component. Furthermore, filaments within a bundle are either aligned parallel or anti-parallel, where the latter enables contractile forces by myosin motors. Additional factors such as formin can drastically alter polymerization kinetics of the underlying filaments [92,93] forming an even more complex picture of actin bundle dynamics.

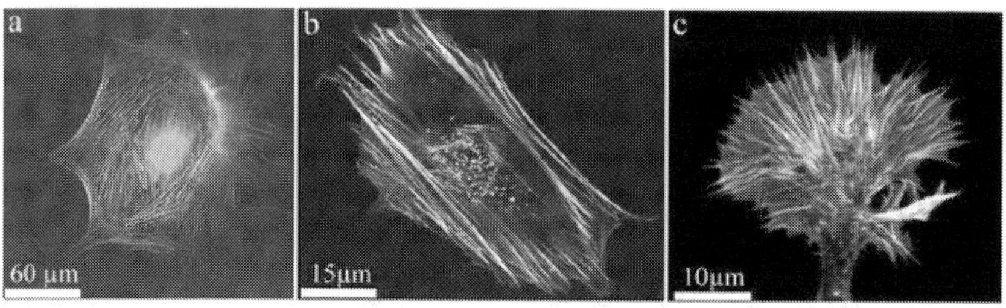

Figure 8: Actin bundles build up stress fibers within the cytoskeleton giving a cell structural stability. (**a**,**b**) Stress fibers entirely span throughout a stationary Balb3T3 fibroblast ((**a**) myosin labeled **green** and actin **red**; (**b**) actin **red**); (**c**) actin bundles stabilize a NG108 neuronal growth cone and according filopodia. Images were taken by Thomas Fuhs and Daniel Koch and the figure is adapted from Huber et al. 2013 [2].

In cellular systems, actin bundles are found, for instance, in brush border microvilli where they support epithelial plasma membrane protrusions. These finger-like, highly regular entities are comprised of 20 to 25 crosslinked actin filaments per bundle and increase the cell surface area to enhance material exchange with the environment, e.g., in the small intestines [32,33,94]. Furthermore, these protrusions are involved in cell adhesion and mechanotransduction. Closely related are so-called stereocilia, which are also membrane protrusions of epithelial cells stabilized by actin bundles. They are typically longer than microvilli and underlying actin bundles are up to 10μm long and contain up to 900 actin filaments [94,95,96,97]. Strictly speaking, stereocilia are no real cilia since they do not contain microtubules but are solely stabilized and driven by actin bundles. These bundles are the key structure in hair cell stereocilia, which are found in arrays in the inner ear at the moveable part of mechanoreceptors [98]. They facilitate the transduction of mechanical stimuli (e.g., sound waves) into electrical signals enabling the basic principle of hearing [99].

In general, if cells need to dynamically respond to external stimuli, crosslinked actin structures are rather static and impede dynamic processes. However, dynamical, transient crosslinkers allow to dissolve actin bundles within minutes facilitating cellular dynamics such as migration. The ability to dissolve as well as to retain bundle integrity is balanced by prestress induced by molecular motors. Myosin activity counteracts bundle disintegration by applying a steady force that enables a bundle to maintain its integrity while it is also capable of rapid structural rearrangements via dynamic crosslinking [100,101].

An important biological example of the dynamic, active behavior of actin bundles are filopodia of neuronal growth cones, which are small cytoplasmic protrusions (Figure 8c). Growth cones use these filopodia to determine the motion and growth of the entire neuron and thus the development of the nerve system. To create intact neuronal structures during embryogenesis appropriately, growth cones have to analyze their environment carefully to grow along the right paths. The involved filopodia are among the most crowded cellular entities and are filled up to 40% with macromolecules [42]. Comprising bundles are formed by crosslinkers and actuate outgrowth and retraction. Fascin and fimbrin are the prominent crosslinker examples and are known as relatively stable crosslinkers, which initially oppose these dynamics [102]. However, their binding affinity can be altered by phosphorylation yielding a drastically reduced actin binding affinity [103,104]. This mechanism itself can be regulated by interactions of the cell with its environment, which potentially acts as a trigger to initiate bundle dynamics. Thus, filopodia and their underlying bundle structure can explore their environment and responses are activated by external signals.

We note that these actin bundles stretch from the tip of the growth cone to the base of the filopodia. Interestingly, they are highly contractile entities despite they are not directly contracted by active myosin motors, which are located at the interior actin cortex [105]. This conceptual idea is supported by the natural outgrowth of actin filaments within filopodia effected by filament polymerization at the tip. This leads to highly polar, parallel filaments in the bundle rendering active myosin contractions implausible.

CONCLUSIONS

This review gives a comprehensive overview of experimental and theoretical approaches describing the diverse behavior of single bundles of semiflexible polymers. These structures can be arranged through a variety of different effects allowing to test the influence of specific parameters. However, bulk measurements such as shear rheology cannot be applied since single bundles are rather anisotropic structures within solutions. Their investigations have to be performed at the mesoscopic level with rather low throughput experiments. A major limitation is that the number of filaments within a bundle is hardly detectable, which impedes direct correlations to existing theories. Additionally, the internal arrangement can be barely explored since single filaments cannot be easily visualized or according imaging techniques influence the bundle itself leading to discordant results. Due to internal structures and additional components such as crosslinkers, a complex interplay emerges leading to a variety of dynamical phenomena. Mechanical properties of these bundled structures cannot be comprehensively described within the frame of the WLC, which led to the extension known as the WLB model [29,30,71].

Generally, semiflexible polymers have the tendency to arrange in bundles, a fact which is widely employed in nature. In small entities such as cells, bundles are potentially even more important than networks since they provide mechanical stability at a much lower volume fraction. They further allow to inhomogeneously distribute mechanical properties throughout the cell body, which is a rigorous advantage for in vivo systems. Additionally, cells are very crowded entities, which inherently trigger bundle formation even without additional components such as crosslinkers or active molecular motors. Complementing cells, the extracellular matrix is also comprised of bundles, which are built up by collagen fibers [19,106,107,108]. They stabilize tissue and also affect cellular behavior, which emphasizes the biological importance of these structures even further [109,110].

We like to note that single bundles or WLB can be even further arranged into higher ordered structures such as asters. These formations of semiflexible bundles have been reported for actin [111,112,113,114,115] and microtubule [116,117,118] systems. Recently, these structures have been also documented for experiments employing DNA nanotubes [119] as a new model system for semiflexible polymers [120,121,122]. Due to structural programming, these tubes can be precisely varied allowing to tune their persistence length [122]. We speculate that this new model system will lead to additional insights for single bundles allowing, for instance, to test the dependency of the reported phenomena on the persistence lengths of the underlying polymers.

ACKNOWLEDGMENTS

We gratefully acknowledge funding by the Deutsche Forschungsgemeinschaft (DFG) (1116/14-1) and the Leipzig School of Natural Sciences "BuildMoNa" (GSC 185). Tina Händler acknowledges funding by the European Social Fund (ESF—100077106).

AUTHOR CONTRIBUTIONS

Jörg Schnauß, Tina Händler, Josef A. Käs conceived the article, extensively researched the literature in this field and contributed to writing process.

Abbreviations

The following abbreviations are used in this manuscript:

WLC	Worm-like chain
WLB	Worm-like bundle
PEG	Polyethylene glycol
κ	Bending stiffness
Lp	Persistence length

REFERENCES

1. Doi, M.; Edwards, S.F. The Theory of Polymer Dynamics (International Series of Monographs on Physics); Clarendon Press: Oxford, UK, 1988; Volume 73.

2. Huber, F.; Schnauß, J.; Rönicke, S.; Rauch, P.; Müller, K.; Fütterer, C.; Käs, J. Emergent complexity of the cytoskeleton: From single filaments to tissue. Adv. Phys. 2013, 62, 1–112.
3. MacKintosh, F.; Käs, J.; Janmey, P. Elasticity of Semiflexible Biopolymer Networks. Phys. Rev. Lett. 1995, 75, 4425–4428.
4. Morse, D.C. Viscoelasticity of Concentrated Isotropic Solutions of Semiflexible Polymers. 2. Linear Response. Macromolecules 1998, 31, 7044–7067.
5. Morse, D. Tube diameter in tightly entangled solutions of semiflexible polymers. Phys. Rev. E 2001, 63.
6. Gardel, M.L.; Valentine, M.T.; Crocker, J.C.; Bausch, A.R.; Weitz, D.A. Microrheology of Entangled F-Actin Solutions. Phys. Rev. Lett. 2003, 91.
7. Fletcher, D.A.; Mullins, R.D. Cell mechanics and the cytoskeleton. Nature 2010, 463, 485–492.
8. Broedersz, C.; MacKintosh, F. Modeling semiflexible polymer networks. Rev. Mod. Phys. 2014, 86, 995–1036.
9. Golde, T.; Schuldt, C.; Schnauß, J.; Strehle, D.; Glaser, M.; Käs, J. Fluorescent beads disintegrate actin networks. Phys. Rev. E 2013, 88.
10. Lieleg, O.; Claessens, M.M.A.E.; Heussinger, C.; Frey, E.; Bausch, A.R. Mechanics of Bundled Semiflexible Polymer Networks. Phys. Rev. Lett. 2007, 99.
11. Müller, K.W.; Bruinsma, R.F.; Lieleg, O.; Bausch, A.R.; Wall, W.A.; Levine, A.J. Rheology of Semiflexible Bundle Networks with Transient Linkers. Phys. Rev. Lett. 2014, 112.
12. Broedersz, C.P.; Depken, M.; Yao, N.Y.; Pollak, M.R.; Weitz, D.A.; MacKintosh, F.C. Cross-Link-Governed Dynamics of Biopolymer Networks. Phys. Rev. Lett. 2010, 105.
13. Claessens, M.; Bathe, M.; Frey, E.; Bausch, A.R. Actin-binding proteins sensitively meditate F-actin bundle stiffness. Nat. Mater. 2006, 5, 748–753.
14. Lai, G.H.; Coridan, R.; Zribi, O.V.; Golestanian, R.; Wong, G.C.L. Evolution of Growth Modes for Polyelectrolyte Bundles. Phys. Rev. Lett. 2007, 98.

15. Claessens, M.M.A.E.; Semmrich, C.; Ramos, L.; Bausch, A.R. Helical twist controls the thickness of F-actin bundles. Proc. Natl. Acad. Sci. USA 2008, 105, 8819–8822.

16. Strehle, D.; Schnauß, J.; Heussinger, C.; Alvarado, J.; Bathe, M.; Käs, J.; Gentry, B. Transiently crosslinked F-actin bundles. Eur. Biophys. J. 2011, 40, 93–101.

17. Schnauß, J.; Golde, T.; Schuldt, C.; Schmidt, B.S.; Glaser, M.; Strehle, D.; Händler, T.; Heussinger, C.; Käs, J.A. Transition from a Linear to a Harmonic Potential in Collective Dynamics of a Multifilament Actin Bundle. Phys. Rev. Lett. 2016, 116.

18. Shao, Z.; Shi, D.; Somlyo, A.V. Cryoatomic force microscopy of filamentous actin. Biophys. J. 2000, 78, 950–958.

19. Yang, L.; van der Werf, K.O.; Koopman, B.F.; Subramaniam, V.; Bennink, M.L.; Dijkstra, P.J.; Feijen, J. Micromechanical bending of single collagen fibrils using atomic force microscopy. J. Biomed. Mater. Res. Part A 2007, 82, 160–168.

20. Ikawa, T.; Hoshino, F.; Watanabe, O.; Li, Y.; Pincus, P.; Safinya, C.R. Molecular Scale Imaging of F-Actin Assemblies Immobilized on a Photopolymer Surface. Phys. Rev. Lett. 2007, 98.

21. Gilmore, J.L.; Kumeta, M.; Takeyasu, K. AFM Investigation of the Organization of Actin Bundles Formed by Actin-Binding Proteins. J. Surf. Eng. Mater. Adv. Technol. 2013, 3, 13–19.

22. Suzuki, A.; Yamazaki, M.; Ito, T. Polymorphism of F-actin assembly. 1. A quantitative phase diagram of F-actin. Biochemistry 1996, 35, 5238–5244.

23. Tang, J.X.; Janmey, P.A. The Polyelectrolyte Nature of F-actin and the Mechanism of Actin Bundle Formation. J. Biol. Chem. 1996, 271, 8556–8563.

24. Schmid, M.F.; Agris, J.M.; Jakana, J.; Matsudaira, P.; Chiu, W. Three-dimensional structure of a single filament in the Limulus acrosomal bundle: Scruin binds to homologous helix-loop-beta motifs in actin. J. Cell Biol. 1994, 124, 341–350.

25. Haviv, L.; Gov, N.; Ideses, Y.; Bernheim-Groswasser, A. Thickness distribution of actin bundles in vitro. Eur. Biophys. J. 2007, 37, 447–454.

26. Nolting, J.F.; Möbius, W.; Köster, S. Mechanics of Individual Keratin Bundles in Living Cells. Biophys. J. 2014, 107, 2693–2699.
27. Shin, H.; Drew, K.R.P.; Bartles, J.R.; Wong, G.C.L.; Grason, G.M. Cooperativity and Frustration in Protein-Mediated Parallel Actin Bundles. Phys. Rev. Lett. 2009, 103.
28. Hémonnot, C.Y.J.; Reinhardt, J.; Saldanha, O.; Patommel, J.; Graceffa, R.; Weinhausen, B.; Burghammer, M.; Schroer, C.G.; Köster, S. X-rays Reveal the Internal Structure of Keratin Bundles in Whole Cells. ACS Nano 2016, 10, 3553–3561.
29. Heussinger, C.; Bathe, M.; Frey, E. Statistical mechanics of semiflexible bundles of wormlike polymer chains. Phys. Rev. Lett. 2007, 99.
30. Bathe, M.; Heussinger, C.; Claessens, M.M.A.E.; Bausch, A.R.; Frey, E. Cytoskeletal bundle mechanics. Biophys. J.2008, 94, 2955–2964.
31. Benetatos, P.; Jho, Y. Bundling in semiflexible polymers: A theoretical overview. Adv. Colloid Interface Sci. 2016, 232, 114–126.
32. Heintzelman, M.B.; Mooseker, M.S. Assembly of the intestinal brush border cytoskeleton. Curr. Top. Dev. Biol. 1992, 26, 93–122.
33. Bartles, J.R.; Zheng, L.; Li, A.; Wierda, A.; Chen, B. Small espin: A third actin-bundling protein and potential forked protein ortholog in brush border microvilli. J. Cell Biol. 1998, 143, 107–119.
34. Asakura, S.; Oosawa, F. On Interaction between Two Bodies Immersed in a Solution of Macromolecules. J. Chem. Phys. 1954, 22, 1255–1256.
35. Asakura, S.; Oosawa, F. Interaction between particles suspended in solutions of macromolecules. J. Polym. Sci. 1958, 33, 183–192.
36. Hosek, M.; Tang, J.X. Polymer-induced bundling of F actin and the depletion force. Phys. Rev. E 2004, 69.
37. Kinoshita, M. Interaction between big bodies with high asphericity immersed in small spheres. Chem. Phys. Lett. 2004, 387, 47–53.
38. Galanis, J.; Nossal, R.; Harries, D. Depletion forces drive polymer-like self-assembly in vibrofluidized granular materials. Soft Matter 2010, 6, 1026–1034.

39. Li, W.; Ma, H.R. Depletion interactions between two spherocylinders. Eur. Phys. J. E Soft Matter 2005, 16, 225–231.
40. Hilitski, F.; Ward, A.R.; Cajamarca, L.; Hagan, M.F.; Grason, G.M.; Dogic, Z. Measuring Cohesion between Macromolecular Filaments One Pair at a Time: Depletion-Induced Microtubule Bundling. Phys. Rev. Lett. 2015, 114.
41. Ward, A.; Hilitski, F.; Schwenger, W.; Welch, D.; Lau, A.W.C.; Vitelli, V.; Mahadevan, L.; Dogic, Z. Solid friction between soft filaments. Nat. Mater. 2015, 14, 583–588.
42. Ellis, R. Macromolecular crowding: Obvious but underappreciated. Trends Biochem. Sci. 2001, 26, 597–604.
43. Janmey, P.A.; Slochower, D.R.; Wang, Y.H.; Wen, Q.; Cebers, A. Polyelectrolyte properties of filamentous biopolymers and their consequences in biological fluids. Soft Matter 2014, 10, 1439–1449.
44. Ray, J.; Manning, G.S. An attractive force between two rodlike polyions mediated by the sharing of condensed counterions. Langmuir 1994, 10, 2450–2461.
45. Grønbech-Jensen, N.; Mashl, R.J.; Bruinsma, R.F.; Gelbart, W.M. Counterion-Induced Attraction between Rigid Polyelectrolytes. Phys. Rev. Lett. 1997, 78, 2477–2480.
46. Ha, B.Y.; Liu, A.J. Counterion-Mediated Attraction between Two Like-Charged Rods. Phys. Rev. Lett. 1997, 79, 1289–1292.
47. Ha, B.Y.; Liu, A.J. Effect of Non-Pairwise-Additive Interactions on Bundles of Rodlike Polyelectrolytes. Phys. Rev. Lett.1998, 81, 1011–1014.
48. Podgornik, R.; Parsegian, V.A. Charge-Fluctuation Forces between Rodlike Polyelectrolytes: Pairwise Summability Reexamined. Phys. Rev. Lett. 1998, 80, 1560–1563.
49. Barrat, J.L.; Joanny, J.F. Persistence Length of Polyelectrolyte Chains. EPL (Europhys. Lett.) 1993, 24, 333.
50. Ha, B.Y.; Thirumalai, D. Electrostatic Persistence Length of a Polyelectrolyte Chain. Macromolecules 1995, 28, 577–581.
51. Dobrynin, A.V. Electrostatic Persistence Length of Semiflexible and Flexible Polyelectrolytes. Macromolecules 2005, 38, 9304–9314.
52. Shklovskii, B.I. Wigner Crystal Model of Counterion Induced Bundle

Formation of Rodlike Polyelectrolytes. Phys. Rev. Lett. 1999, 82, 3268–3271.

53. Manning, G.S. Limiting Laws and Counterion Condensation in Polyelectrolyte Solutions. III. An Analysis Based on the Mayer Ionic Solution Theory. J. Chem. Phys. 1969, 51, 3249–3252.
54. Stevens, M.J. Bundle Binding in Polyelectrolyte Solutions. Phys. Rev. Lett. 1999, 82, 101–104.
55. Lee, K.C.; Borukhov, I.; Gelbart, W.M.; Liu, A.J.; Stevens, M.J. Effect of Mono- and Multivalent Salts on Angle-Dependent Attractions between Charged Rods. Phys. Rev. Lett. 2004, 93.
56. Fazli, H.; Mohammadinejad, S.; Golestanian, R. Salt-induced aggregation of stiff polyelectrolytes. J. Phys. Condens. Matter 2009, 21.
57. Mohammadinejad, S.; Golestanian, R.; Fazli, H. Chiral structure of F-actin bundle formed by multivalent counterions. Soft Matter 2012, 8, 3649–3656.
58. Kierfeld, J.; Lipowsky, R. Unbundling and desorption of semiflexible polymers. EPL (Europhys. Lett.) 2003, 62, 285–291.
59. Benetatos, P.; Zippelius, A. Anisotropic Random Networks of Semiflexible Polymers. Phys. Rev. Lett. 2007, 99.
60. Kiemes, M.; Benetatos, P.; Zippelius, A. Orientational order and glassy states in networks of semiflexible polymers. Phys. Rev. E 2011, 83.
61. Ulrich, S.; Zippelius, A.; Benetatos, P. Random networks of cross-linked directed polymers. Phys. Rev. E 2010, 81.
62. Benetatos, P.; Ulrich, S.; Zippelius, A. Force-extension relation of cross-linked anisotropic polymer networks. New J. Phys. 2012, 14.
63. Zilman, A.G.; Safran, S.A. Role of cross-links in bundle formation, phase separation and gelation of long filaments. EPL (Europhys. Lett.) 2003, 63, 139–145.
64. Borukhov, I.; Bruinsma, R.F.; Gelbart, W.M.; Liu, A.J. Structural polymorphism of the cytoskeleton: A model of linker-assisted filament aggregation. Proc. Natl. Acad. Sci. USA 2005, 102, 3673–3678.

65. Kierfeld, J.; Kühne, T.; Lipowsky, R. Discontinuous Unbinding Transitions of Filament Bundles. Phys. Rev. Lett. 2005, 95.

66. Shin, H.; Grason, G.M. Structural reorganization of parallel actin bundles by crosslinking proteins: Incommensurate states of twist. Phys. Rev. E 2010, 82.

67. Dutta, S.; Benetatos, P.; Jho, Y.S. Bundle formation in parallel aligned polymers with competing interactions. EPL (Europhys. Lett.) 2016, 114.

68. Benetatos, P.; Terentjev, E.M.; Zippelius, A. Bundling in brushes of directed and semiflexible polymers. Phys. Rev. E2013, 88.

69. Braun, M.; Lansky, Z.; Hilitski, F.; Dogic, Z.; Diez, S. Entropic forces drive contraction of cytoskeletal networks. Bioessays 2016, 38, 474–481.

70. Kroy, K.; Frey, E. Force-Extension Relation and Plateau Modulus for Wormlike Chains. Phys. Rev. Lett. 1996, 77, 306–309.

71. Heussinger, C.; Schüller, F.; Frey, E. Statics and dynamics of the wormlike bundle model. Phys. Rev. E 2010, 81.

72. Howard, J.; Ashmore, J.F. Stiffness of sensory hair bundles in the sacculus of the frog. Hear. Res. 1986, 23, 93–104.

73. Pampaloni, F.; Lattanzi, G.; Jonáš, A.; Surrey, T.; Frey, E.; Florin, E.L. Thermal fluctuations of grafted microtubules provide evidence of a length-dependent persistence length. Proc. Natl. Acad. Sci. USA 2006, 103, 10248–10253.

74. Everaers, R.; Bundschuh, R.; Kremer, K. Fluctuations and Stiffness of Double-Stranded Polymers: Railway-Track Model. EPL (Europhys. Lett.) 1995, 29, 263–268.

75. Taute, K.M.; Pampaloni, F.; Frey, E.; Florin, E.L. Microtubule Dynamics Depart from the Wormlike Chain Model. Phys. Rev. Lett. 2008, 100.

76. Von der Heydt, A.; Wilkin, D.; Benetatos, P.; Zippelius, A. Elasticity of cross-linked semiflexible biopolymers under tension. Phys. Rev. E 2013, 88.

77. Ylänne, J.; Scheffzek, K.; Young, P.; Saraste, M. Crystal Structure of the alpha-Actinin Rod Reveals an Extensive Torsional Twist. Structure 2001, 9, 597–604.

78. Aratyn, Y.; Schaus, T.; Taylor, E.; Borisy, G. Intrinsic dynamic behavior of fascin in filopodia. Mol. Biol. Cell 2007, 18, 3928–3940.
79. Lansky, Z.; Braun, M.; Lüdecke, A.; Schlierf, M.; ten Wolde, P.R.; Janson, M.E.; Diez, S. Diffusible Crosslinkers Generate Directed Forces in Microtubule Networks. Cell 2015, 160, 1159–1168.
80. Schnauß, J.; Glaser, M.; Schuldt, C.; Golde, T.; Händler, T.; Schmidt, B.U.S.; Diez, S.; Käs, J.A. Motor-free force generation in biological systems. Diffus. Fundam. 2015, 23, 1–15.
81. Benetatos, P.; von der Heydt, A.; Zippelius, A. Tension-induced binding of semiflexible biopolymers. New J. Phys.2014, 16.
82. Heussinger, C. Cooperative crosslink (un)binding in slowly driven bundles of semiflexible filaments. Phys. Rev. E2011, 83.
83. Vink, R.L.C.; Heussinger, C. Cross-linked biopolymer bundles: Cross-link reversibility leads to cooperative binding/unbinding phenomena. J. Chem. Phys. 2012, 136.
84. Sadhukhan, P.; Schumann, O.; Heussinger, C. Elasto-plastic response of reversibly crosslinked biopolymer bundles. Eur. Phys. J. E 2014, 37, 1–9.
85. Ha, B.Y.; Liu, A.J. Kinetics of bundle growth in DNA condensation. EPL (Europhys. Lett.) 1999, 46, 624–630.
86. Henle, M.L.; Pincus, P.A. Equilibrium bundle size of rodlike polyelectrolytes with counterion-induced attractive interactions. Phys. Rev. E 2005, 71.
87. Gov, N.S. Packing defects and the width of biopolymer bundles. Phys. Rev. E 2008, 78.
88. Grason, G.M.; Bruinsma, R.F. Chirality and Equilibrium Biopolymer Bundles. Phys. Rev. Lett. 2007, 99.
89. Grason, G.M. Braided bundles and compact coils: The structure and thermodynamics of hexagonally packed chiral filament assemblies. Phys. Rev. E 2009, 79.
90. Yang, Y.; Meyer, R.B.; Hagan, M.F. Self-Limited Self-Assembly of Chiral Filaments. Phys. Rev. Lett. 2010, 104.
91. Heussinger, C.; Grason, G.M. Theory of crosslinked bundles of helical filaments: Intrinsic torques in self-limiting biopolymer assemblies. J. Chem. Phys. 2011, 135.

92. Romero, S.; Le Clainche, C.; Didry, D.; Egile, C.; Pantaloni, D.; Carlier, M.F. Formin is a processive motor that requires profilin to accelerate actin assembly and associated ATP hydrolysis. Cell 2004, 119, 419–429.
93. Kovar, D.R.; Pollard, T.D. Insertional assembly of actin filament barbed ends in association with formins produces piconewton forces. Proc. Natl. Acad. Sci. USA 2004, 101, 14725–14730.
94. Krause, W.J. Krause's Essential Human Histology for Medical Students; Universal Publishers: Boca Raton, FL, USA, 2005.
95. Tilney, M.S.; Tilney, L.G.; Stephens, R.E.; Merte, C.; Drenckhahn, D.; Cotanche, D.A.; Bretscher, A. Preliminary biochemical characterization of the stereocilia and cuticular plate of hair cells of the chick cochlea. J. Cell Biol. 1989, 109, 1711–1723.
96. Tilney, L.G.; Tilney, M.S.; DeRosier, D.J. Actin filaments, stereocilia, and hair cells: How cells count and measure. Ann. Rev. Cell Dev. Biol. 1992, 8, 257–274.
97. Bartles, J.R. Parallel actin bundles and their multiple actin-bundling proteins. Curr. Opin. Cell Biol. 2000, 12, 72–78.
98. Pickles, J.O. An Introduction to the Physiology of Hearing; Emerald Group Publishing: Bingley, UK, 2012.
99. Roberts, W.M.; Howard, J.; Hudspeth, A.J. Hair cells: Transduction, tuning, and transmission in the inner ear. Ann. Rev. Cell Dev. Biol. 1988, 4, 63–92.
100. Xu, J.; Wirtz, D.; Pollard, T.D. Dynamic cross-linking by alpha-actinin determines the mechanical properties of actin filament networks. J. Biol. Chem. 1998, 273, 9570–9576.
101. Fabry, B.; Maksym, G.; Butler, J.; Glogauer, M.; Navajas, D.; Fredberg, J. Scaling the Microrheology of Living Cells. Phys. Rev. Lett. 2001, 87.
102. Vignjevic, D.; Kojima, S.; Aratyn, Y.; Danciu, O.; Svitkina, T.; Borisy, G.G. Role of fascin in filopodial protrusion. J. Cell Biol. 2006, 174, 863–875.
103. Yamakita, Y.; Ono, S.; Matsumura, F.; Yamashiro, S. Phosphorylation of human fascin inhibits its actin binding and bundling activities. J. Biol. Chem. 1996, 271, 12632–12638.
104. Ono, S.; Yamakita, Y.; Yamashiro, S.; Matsudaira, P.T.; Gnarra, J.R.;

Obinata, T.; Matsumura, F. Identification of an actin binding region and a protein kinase C phosphorylation site on human fascin. J. Biol. Chem. 1997, 272, 2527–2533.

105. Romero, S.; Quatela, A.; Bornschlogl, T.; Guadagnini, S.; Bassereau, P.; Tran van Nhieu, G. Filopodium retraction is controlled by adhesion to its tip. J. Cell Sci. 2012, 125, 4999–5004.

106. Gentleman, E.; Lay, A.N.; Dickerson, D.A.; Nauman, E.A.; Livesay, G.A.; Dee, K.C. Mechanical characterization of collagen fibers and scaffolds for tissue engineering. Biomaterials 2003, 24, 3805–3813.

107. Yang, L.; Fitié, C.F.C.; van der Werf, K.O.; Bennink, M.L.; Dijkstra, P.J.; Feijen, J. Mechanical properties of single electrospun collagen type I fibers. Biomaterials 2008, 29, 955–962.

108. Dutov, P.; Antipova, O.; Varma, S.; Orgel, J.P.R.O.; Schieber, J.D. Measurement of Elastic Modulus of Collagen Type I Single Fiber. PLoS ONE 2016, 11, e0145711.

109. Ma, X.; Schickel, M.E.; Stevenson, M.; Sarang-Sieminski, A.; Gooch, K.; Ghadiali, S.; Hart, R. Fibers in the Extracellular Matrix Enable Long-Range Stress Transmission between Cells. Biophys. J. 2013, 104, 1410–1418.

110. Mohammadi, H.; Janmey, P.A.; McCulloch, C.A. Lateral boundary mechanosensing by adherent cells in a collagen gel system. Biomaterials 2014, 35, 1138–1149.

111. Backouche, F.; Haviv, L.; Groswasser, D.; Bernheim-Groswasser, A. Active gels: Dynamics of patterning and self-organization. Phys. Biol. 2006, 3, 264–273.

112. Smith, D.; Ziebert, F.; Humphrey, D.; Duggan, C.; Steinbeck, M.; Zimmermann, W.; Käs, J. Molecular motor-induced instabilities and cross linkers determine biopolymer organization. Biophys. J. 2007, 93, 4445–4452.

113. Huber, F.; Strehle, D.; Käs, J. Counterion-induced formation of regular actin bundle networks. Soft Matter 2012, 8, 931–936.

114. Huber, F.; Strehle, D.; Schnauß, J.; Käs, J. Formation of regularly spaced networks as a general feature of actin bundle condensation by entropic forces. New J. Phys. 2015, 17, 043029.

115. Köster, D.V.; Husain, K.; Iljazi, E.; Bhat, A.; Bieling, P.; Mullins, R.D.; Rao, M.; Mayor, S. Actomyosin dynamics drive local membrane component organization in an in vitro active composite layer. Proc. Natl. Acad. Sci. USA 2016, 113, E1645–E1654.
116. Nédélec, F.J.; Surrey, T.; Maggs, A.C.; Leibler, S. Self-organization of microtubules and motors. Nature 1997, 389, 305–308.
117. Surrey, T.; Nedelec, F.; Leibler, S.; Karsenti, E. Physical properties determining self-organization of motors and microtubules. Science 2001, 292, 1167–1171.
118. Laan, L.; Pavin, N.; Husson, J.; Romet-Lemonne, G.; van Duijn, M.; López, M.P.; Vale, R.D.; Jülicher, F.; Reck-Peterson, S.L.; Dogterom, M. Cortical Dynein Controls Microtubule Dynamics to Generate Pulling Forces that Position Microtubule Asters. Cell 2012, 148, 502–514.
119. Yin, P.; Hariadi, R.F.; Sahu, S.; Choi, H.M.T.; Park, S.H.; LaBean, T.H.; Reif, J.H. Programming DNA Tube Circumferences. Science 2008, 321, 824–826.
120. Glaser, M.; Schnauß, J.; Tschirner, T.; Schmidt, B.U.S.; Moebius-Winkler, M.; Käs, J.A.; Smith, D.M. Self-assembly of hierarchically ordered structures in DNA nanotube systems. New J. Phys. 2016, 18.
121. Williams, M.A.K. Hey hey hey hey, it was the DNA. New J. Phys. 2016, 18.
122. Schiffels, D.; Liedl, T.; Fygenson, D.K. Nanoscale Structure and Microscale Stiffness of DNA Nanotubes. ACS Nano2013, 7, 6700–6710.

SECTION II:

SYNTHETIC POLYMERS BY RADICAL POLYMERIZATION

CHAPTER 6

ORIGINS AND DEVELOPMENT OF INITIATION OF FREE RADICAL POLYMERIZATION PROCESSES

Dietrich Braun

Deutsches Kunststoff-Institut, Schlossgartenstr. 6, 64289 Darmstadt, Germany

ABSTRACT

At present worldwide about 45% of the manufactured plastic materials and 40% of synthetic rubber are obtained by free radical polymerization processes. The first free radically synthesized polymers were produced between 1910 and 1930 by initiation with peroxy compounds. In the 1940s the polymerization by redox processes was found independently and simultaneously at IG Farben in Germany and ICI in Great Britain. In the 1950s the systematic investigation of azo compounds as free radical initiators followed. Compounds with labile C–C-bonds were investigated as initiators only in the period from the end of the 1960s until the early 1980s. At about the same time, iniferters with cleavable S–S-bonds were studied in detail. Both these initiator classes can be designated as predecessors for "living" or controlled free radical polymerizations with nitroxyl-mediated polymerizations, reversible addition fragmentation chain transfer processes (RAFT), and atom transfer radical polymerizations (ATRP).

INTRODUCTION

Since the 1980s about 50% of the total production of synthetic polymers worldwide used as plastics were obtained by free radical polymerization processes. Due to the increasing importance of metal organic initiation systems for plastics and synthetic rubber, the participation of radical polymerization is slightly decreasing, but nevertheless nowadays it still amounts annually to nearly 45% or about 100 million tons of plastic materials and 40% or 4.6 million tons of synthetic rubber.

In technical polymerization processes, peroxy compounds play since about 100 years the most important role beside about 60 years old redox systems and the azo initiators. Since about 30 years, the modern possibilities to control free radical polymerization reactions especially by suppressing termination and transfer processes attract increasing attention and technical interest, even if their share in the total polymer production is still rather small.

In most textbooks and monographs on free radical polymerization, only very few if at all is said on the history of this part of macromolecular chemistry. Therefore, the following review deals mainly with the development of initiation of free radical polymerization processes and the most important initiator classes. Due to the limitation to history of the initiating systems, the more recent developments in theory and details of the mechanisms and the kinetics of free radical polymerization processes will not be discussed.

THE TERMS "POLYMERIZATION" AND "FREE RADICALS"

In 1832 the Swedish chemist Jöns Jacob Berzelius (1779–1848) recognized the existence of compounds with the same proportionate composition but different numbers of constituent atoms. He called this phenomenon "polymeric" (in German "Polymerie") (from the Greek πολυμερής, "consisting of many parts"). This term soon came in general use. So a chemical dictionary of the 1860s offered the following, very modern sounding definition:

"Bodies are said to be polymeric when they have the same percentage composition, but different molecular weights; the olefins C^nH^{2n} for example."

Berthelot (1827–1907) called in a lecture at 27th of April 1863 in Paris the conversion reaction of certain compounds such as styrene to their polymerides "polymeric transformation" [1] (later commonly referred to as "polymerization").

Berthelot postulated that all substances, which can add hydrogen, chlorine or water, are able to polymerize and predicted that "incomplete" (in the present nomenclature "unsaturated") compounds should also be polymerizable. He discussed three possibilities to initiate such reactions.

(i)by direct application of heat; but he suspected also, that the evolution of heat due to reactions between the "monomer" molecules and the initiating species was the fundamental reason of polymerization.(ii)by a simultaneous chemical reaction, what nowadays could be described as the formation of initiating species.(iii)due to the influence of a "nascent" state according to Joseph Priestly (1733–1804), who applied this concept to an enhanced reactivity of hydrogen at the time of its release, which we now could call a reactive intermediate.

With these ideas, Berthelot can be considered the first "polymer chemist." He was an unusually creative scientist and published many papers on various fields of chemistry but also served in two French governments as Minister of Foreign Affairs (1895) and Minister of Education (1896/97).

The term "radical" was used since the middle of the 19th century for atoms or groups of several atoms, which up to day in many cases are considered constituents of chemical compounds.

As far back to the early days of the modern valence theory, organic chemists discussed the possible existence of free radical species and attempted to isolate them or at least their direct reaction products. In the 1850s, Charles Adolphe Wurtz proposed that the reaction of sodium with alkyl iodides (RI) produced compounds with empirical formulas corresponding to alkyl radicals ($R^{\bullet}$):

$$RI + Na \longrightarrow NaI + R^{\bullet}$$

$$2R^{\bullet} \longrightarrow R - R$$

But in his experiments the obtained gases did not contain the free radicals but their dimers, as was deduced from their vapour densities.

In the following years with its success in explaining many of the known aspects of organic chemistry in terms of tetravalent carbon, little effort was made to isolate organic free radicals until the turn of the century. In 1900, Gomberg reported the existence of triphenylmethyl as the first stable radical, formally produced by the dissociation of "hexaphenylethane," which was prepared by the reaction of triphenylmethyl chloride either with silver or zinc dust:

$$2(C_6H_5)_3C - Cl \longrightarrow 2(C_6H_5)_3C^{\bullet} \longrightarrow \text{Dimer}$$

Since this invention, the existence of free radicals soon became a part of organic chemistry. However, it lasted almost about forty years before the importance of free radicals as species with unpaired electrons and as short-living reaction intermediates began to be fully accepted.

EARLY HISTORY OF THE THEORY OF FREE RADICAL POLYMERIZATION

In 1920 Staudinger published his first paper "Über Polymerisation" [2] where he described the formation of long chain molecules by repeated addition of monomers and created the term "Makromolekül." At this time the mechanism of such reactions was uncertain, but Staudinger proposed the participation of species with trivalent carbon atoms, which we now would call radicals. He also postulated [3], that the chain ends contain "free valences," which are unstable and grow until any at this time unknown kind of inactivation proceeds.

In a paper of 1929, Staudinger et al. [4] concluded that the polymerization is going on by an active state of the chains until the process is terminated by ring formation as at this time nothing was known about the chain length and the nature of the chain ends. In 1931 Staudinger and Kohlschütter [5] explained the polymerization of acrylic acid by an activation of the monomer molecule and a very fast addition of further monomer molecules in a chain reaction.

In 1935 Schulz suggested the activation of monomers for polymerization processes by "catalysts" [6]. In 1937 Flory described the kinetics of the vinyl polymerization as a chain reaction with participation of free radicals and postulated that two active (growing) chains can be terminated only by bimolecular combination or disproportionation reactions [7]. This paper of Flory can be considered the first complete formulation of the free radical polymerization, where the overall rate R_w in dependence on concentrations of the monomer [M] and the initiator [I] can be described with the overall rate coefficient K by the "square-root law": $R_w = K \cdot [M] \cdot [I]^{1/2}$.

POLYMERIZATION PROCESSES BEFORE HERMANN STAUDINGER

Long before the initiation of polymerization processes by "catalysts" or initiators became known, some polymerizations initiated thermally or by light were described in the early literature. In all these cases it is not clear to which extent the presence of molecular oxygen took part in these reactions. Especially the so-called thermal polymerization of styrene and of some other vinyl monomers may at least have been partially initiated by primarily formed peroxides, as was suggested by Lautenschläger [8] already in 1913.

Probably the earliest description of a polymerization process was given by Simon [9] in 1839. He isolated styrene from storax, a liquid product from a plane tree (liquidamber orientalis) and observed that the liquid during storing became more and more viscous and finally a glass-like solid. He called the obtained resin "styroloxid," which a few years later by Blyth and Hofmann [10] was named "metastyrol," as it contained no oxygen. Interesting is also the remark of Hofmann and Blyth in 1845: "It is not unthinkable that metastyrol could be used for optical applications."

Blyth and Hofmann also observed that this metastyol was formed when styrene was exposed to sunlight, while it remained unchanged in the dark. According to Morawetz, [11] this is probably the first report on a light-induced polymerization. Another early example is the observation of Baumann in 1872, who described the transformation of vinyl chloride and vinyl bromide by sunlight to "isomeric bodies" [12].

Stobbe and Posnjak reported the first kinetic investigation of a vinyl polymerization in 1909 [13]. They studied the conversion of styrene to "metastyrol" by following the viscosity after heating at 200°C and found that freshly distilled styrene polymerized only after an induction period of seven hours, while no induction period was observed if the polymerization was started after a two weeks aging at room temperature before polymerization. This result was later confirmed by measurements of the refractive index to follow the reaction [14]. Stobbe and Posnjak concluded that in the aged styrene "a polymerization nucleus must have been formed which leads to the acceleration of the polymerization." 1913 Lautenschläger suggested [8] in his much later [15] published dissertation that the observations of Stobbe and Posnjak should be explicable by the formation of small amounts of peroxides, which acted as polymerization catalysts.

MOST IMPORTANT INITIATOR CLASSES FOR FREE RADICAL POLYMERIZATION REACTIONS

Peroxy Compounds

Besides the mentioned suggestion of Lautenschläger [8] on the action of peroxides in the styrene polymerization, it seems that Fritz Klatte at about 1912 for the first time used peroxy compounds to initiate the vinyl polymerization [16]. In 1913 in a patent "the preparation of horn substitute, artificial fibers, lacquers, and so forth. from poly(vinyl chloride)" was claimed. In a number of following inventions Klatte described the preparation of several monomers by the addition of inorganic acids like hydrogen chloride and a number of organic acids like chloroacetic acid to acetylene using mercury catalysts and their polymerization by heat or light "with and without additives." In one of his patents [17], he mentioned the remarkable acceleration of the polymerization by so-called catalysts like peroxides, ozonides, or oxygen forming products like perborates or percarbonates. But he also realized that the formation of peroxides from the used monomers can initiate polymerization processes similar to that with benzoyl peroxide. He also found that copper or copper compounds can inhibit the polymerization.

But it lasted many years until the generation of free radicals from peroxides was generally accepted as a method for initiation of polymerizations. Even in a review of 1937, [18] this was considered controversial, and in many papers rather vague formulations like "activation of monomers" (e.g., Schulz in 1937 [19]) were used. But—as already mentioned—in the same year Flory had no doubt that free radicals at the end of the growing polymer chains were responsible for the chain propagation and he gave the first description of the mechanism and the kinetics of the free radical polymerization of vinyl compounds as a chain reaction.

In the 1940s Price and Tate [20], Nozaki and Bartlett [21], Kern and Kämmerer [22] and others, and some years later Bevington et al. [23, 24] reported that initiator fragments from peroxides were attached to each polymer chain. For these investigations, peroxides with chemically (e.g., with bromine atoms), spectroscopically (e.g., with azo groups or IR sensitive structural units) or ^{14}C-labeled peroxides were used, which are also early examples for end group analyses for molecular weight determination of vinyl polymers. Kern summarized these results in 1947/48 [25] and formulated the primary step of free radical polymerizations by peroxides with substituents X:

$$X-Ph-CO-O-O-CO-Ph-X \longrightarrow 2X-Ph-CO-O^{\bullet}$$

$$X-Ph-CO-O^{\bullet}+CH_2=CHR \longrightarrow X-Ph-CO-O-CH_2CHR^{\bullet}$$

Besides the so-obtained ester end groups especially at higher polymerization temperatures the benzoxy radicals can decarboxylate to X-Ph • -radicals which contrary to the ester groups are not cleavable by acids or alkali. This can be used to determine the degree of decarboxylation during the initiation step.

Redox Systems

Long before the explanation of the free radical polymerization by its various steps like initiation, propagation, and termination, the retardation of the polymerization of unsaturated compounds and the acceleration of polymerization by molecular oxygen were observed. These phenomena were explained by the formation of polymeric peroxides and their thermal cleavage to free radicals. The first idea was that such polymeric peroxides decompose under formation of "active" oxygen, which could initiate the polymerization, whereas molecular oxygen would inhibit the polymerization. So in the beginning of these investigations, it was uncertain why oxygen under some conditions accelerates the polymerization and under others acts as an inhibitor. Later this effect was understood as a competition between autoxidation and inhibition [26].

From experiments to suppress the inhibition of polymerization processes by molecular oxygen resulted one of the technically most important method of initiation of polymerization, originally called redox catalysis. Of large importance in this respect was the observation of Winnacker and Patat that the reproducibility of the emulsion polymerization of chloroprene in presence of oxygen in alkaline medium can be improved by adding reduction agents like sulfites or hyposulfites, which remarkably increased the polymerisation rate [27].

On the basis of these observations in the German laboratories of IG Farben in Höchst and Leverkusen it was soon realized that this effect was caused by the reaction of reducing and oxidating reagents, that is, by a redox process. Kern finally explained the primary step of such redox polymerizations by the formation of free radicals from peroxides and ferrous ions [28]:

$$Y-CO-O-O-CO-Y+Fe^{2+} \longrightarrow Y-CO-O^{\bullet}+Y-CO-O^{-}+Fe^{3+}$$

$$Y-CO-O^{\bullet}+CH_2=CHR \longrightarrow Y-CO-O-CH_2CHR^{\bullet}$$

Later on it was observed that this type of reaction could also be extended to combinations of many nonperoxidic oxidating agents with organic reducing compounds.

As a result of the Yalta agreement between the Allied Powers, after the Second World War Germany was not permitted to continue these research activities, which were of high interest for the production of synthetic rubber. Therefore the cold rubber technology on the basis of redox systems was developed in the United States, where in 1954 about 70% of the total synthetic rubber was produced by this method.

In September 1945 the Faraday Society organized a "General Discussion on Oxidation" with a section on electron transfer reactions.

At this meeting Bacon reported on extensive studies of redox-initiated polymerizations that had been carried out since October 1940 in the laboratories of Imperial Chemical Industries [29, 30]. From Evans' laboratory followed an account of the study of vinyl polymerization in aqueous solution by hydrogen peroxide and ferrous ions [31]:

$$H_2O_2 + Fe^{2+} \longrightarrow HO^\bullet + Fe^{3+} + HO^-$$

$$HO^\bullet + CH_2 = CHR \longrightarrow HO - CH_2 - CHR^\bullet \longrightarrow \text{Polymer}$$

Similarly, polymerization was induced by the reaction of a reducing agent to a persulfat containing solution. So the redox polymerization has two independent roots, as due to the war none of the participating groups was aware of the work carried out in the other country.

Azo Compounds

The use of triphenyl azobenzene by Schulz in 1939 drew first attention to azo compounds as polymerization initiators [32]:

$$(C_6H_5)_3C - N = N - C_6H_5$$

Gomberg obtained this azo compound in 1897 [33]. The first and up to now mostly used aliphatic azo initiator azoisobutyronitrile (AIBN) was discovered by Thiele and Heuser in 1896 [34]. The observation that AIBN yields initiating radicals at the same rate in a variety of solvents by several authors [35, 36] made AIBN the initiator of choice in kinetic studies of free radical polymerizations. But only in the early 1950s the systematic investigation of this initiator class started, and now in the Polymer Handbook about 300 suitable aliphatic azo initiators are summarized. Most of them are easily obtained by adding HCN to ketones followed by reaction with hydrazine und oxidation with, for example, bromine.

AIBN 

One of the problems of this type of azo compounds is their content of cyano groups, which may cause toxic side products during their thermal decomposition.

Therefore later several nitrile free azo compounds were synthesized and used as initiators, for example, azo esters [37–39]:

$$C_6H_5R^1R^2C - N = N - CR^2R^1C_6H_5$$

With R^1 = alkyl, R^2 = OOC-alkyl

Some other azo initiators with functional groups like carboxy create reactive end groups on the produced polymer chains for further modification, for example, for the synthesis of block copolymers.

Till now, there is not enough information available about the theoretical relationships between chemical structure and decomposition behaviour but nevertheless, nowadays a broad variety of azo initiators for a large range of polymerization temperatures is available.

Initiators with Labile C–C-Bonds

Since many years it is known that the strength of C–C-bonds and the stability of the radicals formed by homolytic thermal cleavage of such bonds depend mainly on the kind and size of the substituents on these carbon atoms. For example, tetraphenylsuccinic acid dinitrile (X = CN in the general formula below) is cleaved at 100°C in solution into diphenylmethyl cyanomethyl radicals. Even as the degree of dissociation is small, these radicals are sufficiently reactive to initiate the polymerization of styrene, as was described by Schulz and Wittig already in 1939 [40].

These investigations did not find scientific or practical interest, probably due to the much easier available peroxides. Much later, in 1969, another example for the initiation of free radical polymerizations by C–C-cleavage was found by chance by Braun and Becker [41, 42]. They observed, that benzopinacol (X = OH) by thermal treatment yields radicals that are able to initiate the polymerization of vinyl monomers, but not the initially produced diphenyl hydroxy methyl radicals are the initiating species; these radicals transfer hydrogen atoms to the monomer under initiation of a growing chain and formation of benzophenone (Scheme 1). The authors called that process transfer initiation.

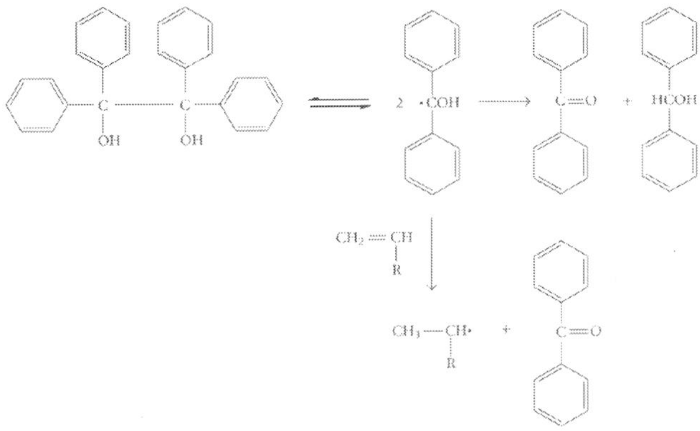

Scheme 1: Thermal decomposition of benzopinacol and initiation of vinyl polymerization by the formed diphenyl hydroxy methyl radicals.

Continuing these observations, in 1981 Bledzki and Braun [43] and Braun [44] investigated this until that time not noted initiator class of tetraaryl-substituted alkanes with various constituents X (see Scheme 2) in more detail.

$$\begin{array}{ccc}(C_6H_5)_2C & — & C(C_6H_5)_2 \\ | & & | \\ X & & X\end{array}$$

$$X = OH, OCH_3, O - C_6H_5, CN, CH_3, O - Si(CH_3)_3$$

Scheme 2: Tetraalkyl-substituted alkanes with constituents X.

They found that by thermal cleavage of tetraaryl ethanes with X = CN, OR, but also with halogen or alkyl groups substituted diaryl methyl radicals cannot only initiate the vinyl polymerization but also cause rather strong primary radical termination.

The by primary radical termination resulting and due to the large and—depending on the structure of the last monomer unit—thermally labile end groups can undergo reversible cleavage at the chain ends. Therefore the so obtained polymers consist of "dormant" macromolecules that can reinitiate the polymerization reaction and are in this respect a predecessor of the later, called living or controlled free radical polymerization, as shown in the example with tetraphenyl succinic acid dinitrile in Scheme 3.

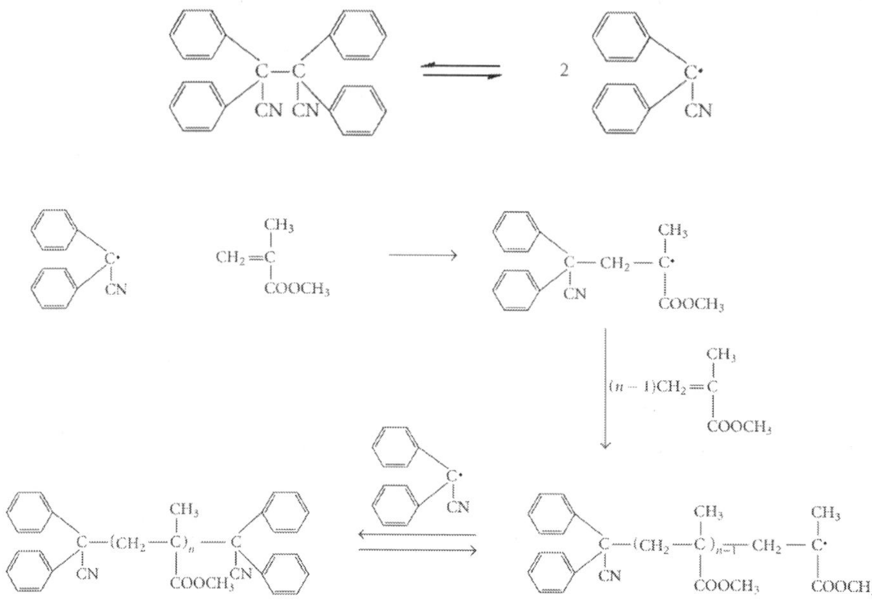

Scheme 3: Polymerization of methyl methacrylate with succinic acid dinitrile.

Due to this "stepwise" addition of monomer molecules the degree of polymerization increases with increasing conversion, as is shown in Figure 1 for the bulk polymerization of methyl methacrylate with 1.2-diphenoxy-1.1.2.2-tetraphenyl ethane.

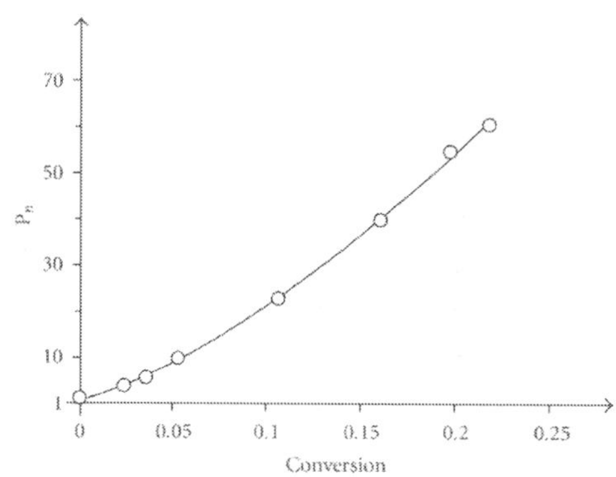

Figure 1: Bulk polymerization of methyl methacrylate with 1.2-diphenoxy-1.1.2.2-tetraphenylethane (30 mmol.l^{-1}) at 70°C); (data from [43]).

At about the same time Otsu and Yoshida [45] investigated initiators with labile –S–S–, –CO–S–, –CS–S–, and –O–CS–S– bonds like diaryl disulfides, dibenzoyl disulfide, tetraalkyldithiuram disulfides, and (bisalkylxanthogen) disulfides, for example,

$$\mathrm{Ph-S-S-Ph} \qquad \mathrm{Ph-CO-S-S-CO-Ph}$$

$$\mathrm{(alkyl)_2N\text{-}CS\text{-}S\text{-}S\text{-}CS\text{-}N(alkyl)_2} \qquad \mathrm{alkyl\text{-}O\text{-}CS\text{-}S\text{-}S\text{-}CS\text{-}O\text{-}alkyl}$$

Such compounds may be decomposed thermally or photochemically to initiating radicals, but the chain transfer to the initiator is the most important mechanism for the termination of growing polymer chains. In most cases the obtained endgroups are thermally stable, but photochemically labile, which can be described as a photoinitiated living polymerization.

General scheme of the polymerization of monomer M by disulfides with substituents R:

Initiation by heat or light:	$\mathrm{R-S-S-R \longrightarrow 2R-S^{\bullet}}$
Propagation	$\mathrm{R-S^{\bullet}} + n\mathrm{M \longrightarrow Pol^{\bullet}}$
Chain transfer	$\mathrm{R-S-S-R + Pol^{\bullet} \longrightarrow Pol{-}SR + R-S^{\bullet}}$
Primary radical termination/re-initiation	$\mathrm{Pol^{\bullet} + R-S^{\bullet} \longleftrightarrow Pol{-}SR}$

Otsu named such initiators that at the same time are starters, chain transfer agents, and terminators "iniferters (initiator, transfer agent, chain terminator) or initers (initiator, chain terminator), in analogy to the similar terminology used by Kennedy [46] for cationic systems.

Controlled Free Radical Polymerization

The usual free radical polymerization is determined by chain termination and chain transfer processes which limit the length of growing chains and are the reason for normally rather broad molecular weight distributions. Therefore, many researchers looked for radical processes, where these irreversible terminating steps are avoided or at least strongly restricted, similar to the so called anionic "living polymerization," which term was introduced in 1956 by Szarwc et al. [47, 48] for the anionic polymerization of styrene in aprotic solvents. The task was to minimize the rate of reactions between the radical species themselves (termination and chain transfer) and to convert reactive growing species reversibly into dormant species.

One of the first examples are the polymerizations that were initiated with highly substituted alkanes (see above). Here the initiating radicals are also chain-terminating agents. The relatively weak bond between the last monomer unit and the terminating radical opens the chance of reactivating the chains, which insofar can be named "dormant." With tetraaryl alkanes due to the relatively frequent terminating steps the obtainable degrees of polymerization are rather small and the molecular weight distributions are not very narrow.

Therefore, the controlling effect in this type of "living" polymerization is rather limited, but these initiators fulfil the principle requirements for controlling the radical polymerization.

(i) The radicals formed by initiator decomposition must be relatively long living and they can initiate the polymerization with rather slow rates.

(ii) Primary radical termination and/or chain transfer to the initiator should be the only or at least main mechanism for interruption of the chain growth.

(iii)The bond of the initiator fragment to the chain end must be thermally or photochemically labile.

The following Scheme 4 shows the principle of living free radical polymerization processes with I-X as the initiating species in a dynamic equilibrium with a relatively small number of growing free radicals and a large number of deactivated "dorming" species [49]. In such reactions irreversible chain transfer and chain termination are not completely impossible but pushed back to a high degree.

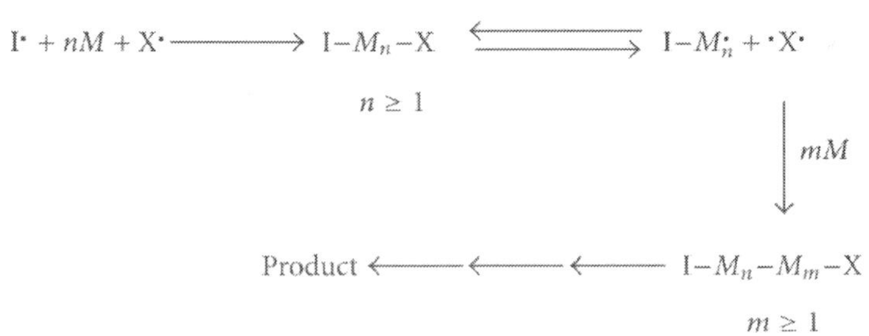

Scheme 4: Living free radical polymerization with I-X as the initiating species.

The actual investigations on living free radical polymerizations are characterized by a lot of systems where the growing chain adds more or less stable free radicals that terminate the chains by forming of relatively weak bonds that can be cleaved again thermally or photochemically.

An interesting reaction of this type is the nitroxyl-mediated polymerization, invented by Solomon and Rizzardo in the middle of the 1980s [50]. The reaction is initiated by classical radical initiators like peroxides or azo compounds. The termination of the growing chains by nitroxyls like 2.2.6.6-tetramethylpiperidine-1-oxyl (TEMPO) yields C–ON-bonds at the chain ends which are reversibly cleavable with the rate constants k_a and k_d above 100°C (k_p: rate constant for chain propagation):

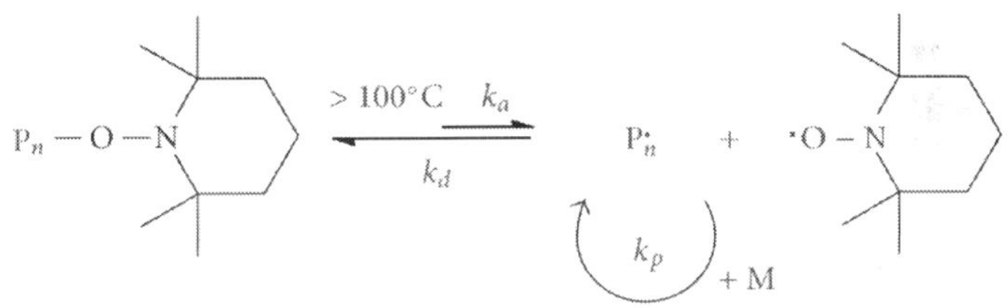

But in principle, this method is connected—not to the same extent—with similar problems like the polymerization by substituted tetraarylalkanes (see above).

Another possibility for controlling the free radical polymerization by "Reversible Addition Fragmentation Chain Transfer" (RAFT) was invented by Krstina et al. in 1995 [51]. In this case the polymerization is initiated with small concentrations of usual initiators. The growing chain (P_n•) reacts with substituted dithioesters by a "radical transesterification" to an equilibrium between active and deactivated species:

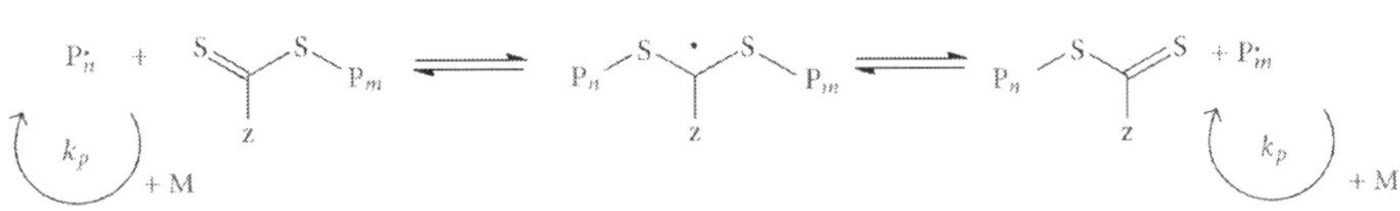

But most interest during the recent years got the "Atom Transfer Radical Polymerization" (ATRP), described by Matyjaszewski et al. [52, 53] as a catalytic process, where the proportion of radicals (or the polymerization rate) can be easily controlled by the amount and the activity of the catalyst. It is a repetitive addition of a monomer to growing radicals generated from "dormant" alkyl halides (R-X) by a reversible redox process catalyzed by transition metal compounds, such as cuprous halides complexed, for example, by two 2,2'-bipyridine (bipy) molecules:

$$P_n{-}X + Cu(I)/2bipy \rightleftarrows P_n^{\bullet} + X - Cu(II)/2bipy$$

In the initiation step radicals are formed from alkyl halides R-X by a reversible redox process with a transition metal halide like cuprous chloride:

$$R - X + Cu(I)X \longleftrightarrow R^{\bullet} + Cu(II)X_2$$

Due to a dynamic equilibrium between dorming (P_n -X) and active ($P_n^{\bullet}$) polymer chains the concentration of active species is very small. The radicals add monomer molecules to growing polymer chains that were reversibly terminated by halide addition from the Cu(II) halide:

$$P_n^{\bullet} + Cu(II)X_2 \longleftrightarrow P_nX + Cu(I)X$$

The process is terminated as in the conventional radical polymerization by combination or disproportionation of two active growing chains, but due to the small concentration of active polymer chains only a small degree of these chains is irreversibly terminated. The molecular weight of the polymer can be determined by the ratio of used monomer to initiator concentration: $P_n = \triangle\ [M]/[I]_0$. The so obtainable polydispersity M_w/M_n is between 1.0 and 1.5. With suitable catalysts (metal salt and ligand) and initators in the ideal case the molecular weight increases as typical for living polymerizations linearly with the conversion (Figure 2).

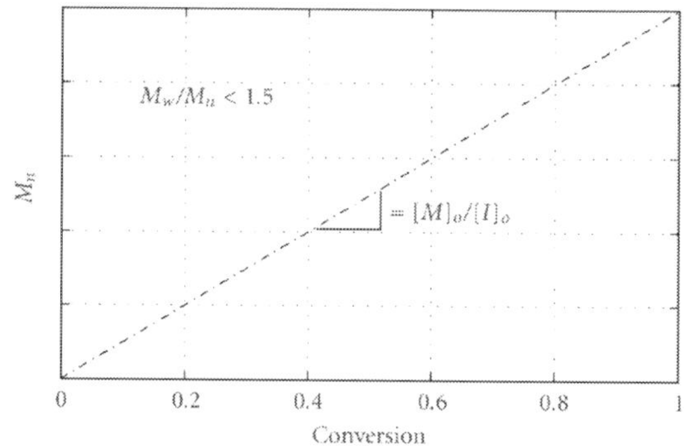

Figure 2: Relationship between molecular weight and conversion in the ideal free radical living polymerization.

The task of the ligand is to dissolve the metal salt in organic solvents and to control the redox potential of the metal with respect to reactivity [54]. For this purpose a large number of ligands were investigated, as the examples of nitrogen containing ligands for the ATPR with copper catalyst show (Scheme 5).

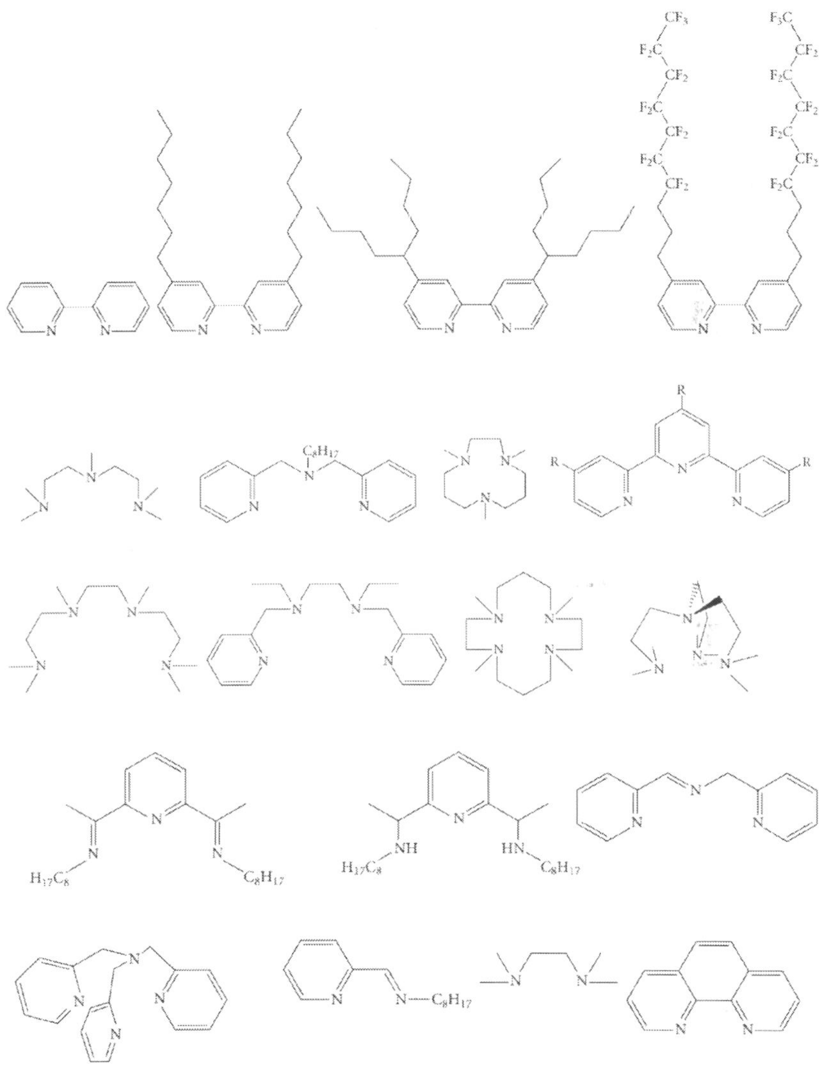

Scheme 5: Nitrogen ligands for the ATPR with copper catalysts.

By using alkyl halides with functional groups as initiators it is possible to create functional end groups at the polymer chains, which can be used to synthesize block, star, or graft copolymers in subsequent reactions. Some for such purposes suited initiators can be seen in Scheme 6.

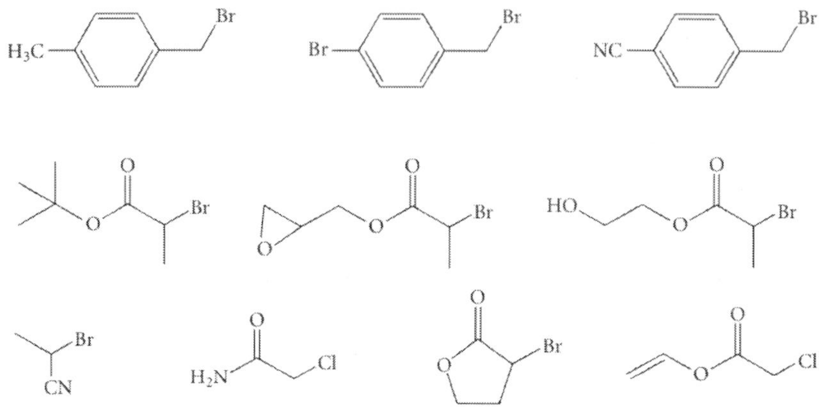

Scheme 6: Alkyl halides as initiators for ATPR.

The investigation of free radical polymerizations that are completely or at least to a large extend without any termination steps is presently one of the most actual fields of free radical polymerization, even if the discussion on the nomenclature of such processes ("living" or "controlled") is by no means finished [55, 56].

This report on the history of initiation of polymerization processes by the presently most important classes of free radicals demonstrates that these reactions concern by no means a "ripe" field of polymer science. In spite of its chain reaction character with extremely fast steps it is now possible to control and to tailor the free radical polymerization by methods which were unthinkable in its beginning more than 80 years ago. It can be expected that in the near future the further development of the controlled free radical polymerization will find increasing practical application.

ACKNOWLEDGMENT

The author thanks the Dr. Illing Stiftung für Makromolekulare und Technische Chemie for the generous support of his research on free radical polymerization.

REFERENCES

1. M. Berthelot, Lecons de Chimie Professées en 1864 et 1865, Société de Paris, Paris, France, 1866.
2. H. Staudinger, "Über Polymerisation," Berichte der Deutschen Chemischen Gesellschaft, vol. 53, no. 6, pp. 1073–1085, 1920.
3. H. Staudinger, "Die Chemie der hochmolekularen organischen Stoffe im Sinne der Kekuléschen Strukturlehre," Berichte der Deutschen Chemischen Gesellschaft, vol. 59, pp. 3019–3043, 1926.
4. H. Staudinger, M. Brunner, K. Frey, P. Garbsch, R. Signer, and S. Wehrli, "Über das Polystyrol, ein Modell des Kautschuks," Berichte der Deutschen Chemischen Gesellschaft, vol. 62, no. 1, pp. 241–263, 1929.
5. H. Staudinger and H. W. Kohlschütter, Berichte der Deutschen Chemischen Gesellschaft, vol. 64, p. 2019, 1931.
6. G. V. Schulz, "On the relationship between the rate of reaction and composition of the reaction products in macro polymerisation processes," Zeitschrift für Physikalische Chemie, vol. 30, p. 379, 1935.
7. P. J. Flory, "The mechanism of vinyl polymerizations," Journal of the American Chemical Society, vol. 59, no. 2, pp. 241–253, 1937.
8. L. Lautenschläger, , Ph.D. thesis, University of Karlsruhe, Karlsruhe, Germany, 1913.
9. E. Simon, Justus Liebigs Annalen der Chemie, vol. 31, p. 267, 1839.
10. J. Blyth and A. W. Hofmann, "Ueber das Styrol und einige seiner Zersetzungsprodukte," Justus Liebigs Annalen der Chemie, vol. 53, no. 3, pp. 289–329, 1845.
11. H. Morawetz, Polymers: The Origins and Growth of a Science, John Wiley & Sons, New York, NY, USA, 1985.
12. E. Baumann, "Ueber einige Vinylverbindungen," Justus Liebigs Annalen der Chemie, vol. 163, no. 3, pp. 308–322, 1872.
13. H. Stobbe and G. Posnjak, Justus Liebigs Annalen der Chemie, vol. 371, no. 3, p. 259, 1909.
14. H. Stobbe, "Polymerisation des Styrols," Justus Liebigs Annalen der Chemie, vol. 409, no. 1, pp. 1–13, 1915.

15. H. Staudinger and L. Lautenschläger, "Über Polymerisation und Autoxydation," Justus Liebigs Annalen der Chemie, vol. 488, no. 1, pp. 1–8, 1931.

16. M. Kaufman, The History of PVC, chapter 2, Maclaren and Sons, Glasgow, Scotland, 1969.

17. DRP, 281 688, Griesheim-Elektron, 1913.

18. D. H. Hey and W. A. Waters, "Some organic reactions involving the occurrence of free radicals in solution," Chemical Reviews, vol. 21, no. 1, pp. 169–208, 1937.

19. G. V. Schulz, "On the relationship between the rate of reaction and composition of the reaction products in macro polymerisation processes," Zeitschrift für Physikalische Chemie, vol. 30, p. 379, 1935.

20. C. C. Price and B. E. Tate, "The polymerization of styrene in the presence of 3,4,5-tribromobenzoyl peroxide," Journal of the American Chemical Society, vol. 65, no. 4, pp. 517–520, 1943.

21. K. Nozaki and P. D. Bartlett, "The kinetics of decomposition of benzoyl peroxide in solvents. I," Journal of the American Chemical Society, vol. 68, no. 9, pp. 1686–1692, 1946.

22. W. Kern and H. Kämmerer, "Die chemische Molekulargewichtsbestimmung von Polystyrolen, II," Journal für Praktische Chemie, vol. 161, no. 11-12, pp. 289–292, 1943.

23. J. C. Bevington, H. W. Melville, and R. P. Taylor, "The termination reaction in radical polymerizations. Polymerizations of methyl methacrylate and styrene at 25∘," Journal of Polymer Science, vol. 12, no. 1, pp. 449–459, 1954.

24. J. C. Bevington, H. W. Melville, and R. P. Taylor, "The termination reaction in radical polymerizations. II. Polymerizations of styrene at 60∘ and of methyl methacrylate at 0 and 60∘, and the copolymerization of these monomers at 60∘," Journal of Polymer Science, vol. 14, no. 77, pp. 463–476, 1954. ·

25. W. Kern, "Über den Primärakt der mit Hilfe von Peroxyden ausgelösten Polymerisation ungesättigter Verbindungen," Die Makromolekulare Chemie, vol. 1, no. 3, pp. 229–248, 1948.

26. W. Kern, "Über den Einfluß des molekularen Sauerstoffes auf die

Polymerisation ungesättigter Verbindungen," Die Makromolekulare Chemie, vol. 1, no. 3, pp. 199–208, 1948.

27. W. Kern, "Die Katalyse der Polymerisation ungesättigter Verbindungen mit Hilfe von Redoxsystemen," Die Makromolekulare Chemie, vol. 1, no. 3, pp. 209–228, 1948.

28. W. Kern, "Die Metallredoxkatalyse der Polymerisation ungesättigter Verbindungen," Die Makromolekulare Chemie, vol. 1, no. 3, pp. 249–268, 1948.

29. R. G. R. Bacon, "Reduction activation. A new polymerisation technique," Transactions of the Faraday Society, vol. 42, pp. 140–155, 1946.

30. R. G. R. Bacon, "The initiation of polymerisation processes by redox catalysts," Quarterly Reviews, vol. 9, no. 3, pp. 287–310, 1955.

31. J. H. Baxendale, M. G. Evans, and C. S. Park, "The mechanism and kinetics of the initiation of polymerisation by systems containing hydrogen peroxide," Transactions of the Faraday Society, vol. 42, pp. 155–169, 1946.

32. G. V. Schulz, "Anregung von Kettenpolymerisationen durch freie Radikale: II. Entwicklung der Radikale durch Zersetzung einer Azoverbindung," Die Naturwissenschaften, vol. 27, no. 39, pp. 659–660, 1939.

33. M. Gomberg, "Tetraphenylmethan," Berichte der Deutschen Chemischen Gesellschaft, vol. 30, no. 2, pp. 2043–2047, 1897.

34. J. Thiele and K. Heuser, "Ueber Hydrazinderivate der Isobuttersäure," Justus Liebigs Annalen der Chemie, vol. 290, no. 1, pp. 1–43, 1896.

35. F. M. Lewis and M. S. Matheson, "Decomposition of aliphatic azo compounds," Journal of the American Chemical Society, vol. 71, no. 2, pp. 747–748, 1949.

36. C. G. Overberger, M. T. O›Shaughnessy, and H. Shalit, "The preparation of some aliphatic azo nitriles and their decomposition in solution," Journal of the American Chemical Society, vol. 71, no. 8, pp. 2661–2666, 1949.

37. D. Braun and R. Jakobi, "Synthesis and decomposition of azo-initiators, I," Monatshefte für Chemie, vol. 113, no. 12, pp. 1403–1414, 1982.

38. D. Braun and R. Jakobi, "Synthese und Zerfall von Azoinitiatoren, II," Monatshefte für Chemie, vol. 114, no. 1, pp. 37–45, 1983.
39. D. Braun and R. Jakobi, "Neue Initiatoren zur Polymerisationsauslösung," Die Angewandte Makromolekulare Chemie, vol. 105, no. 1, pp. 217–230, 1982.
40. G. V. Schulz and G. Wittig, "Anregung von Kettenpolymerisationen durch freie Radikale," Naturwissenschaften, vol. 27, no. 22, pp. 387–388, 1939.
41. D. Braun and K. H. Becker, "Aromatische Pinakole als Polymerisationsinitiatoren," Die Angewandte Makromolekulare Chemie, vol. 6, no. 1, pp. 186–189, 1969.
42. D. Braun and K. H. Becker, "Kinetik der Polymerisationsauslösung mit aromatischen Pinakolen," Die Makromolekulare Chemie, vol. 147, no. 1, pp. 91–99, 1971.
43. A. Bledzki and D. Braun, "Polymerisationsauslösung mit substituierten Ethanen, 1. Polymerisation von Methylmethacrylat mit 1,1,2,2-Tetraphenyl-1,2-diphenoxyethan," Makromolekulare Chemie, vol. 182, no. 4, pp. 1047–1056, 1981.
44. D. Braun, "Initiation of free radical polymerization by thermal cleavage of carbon-carbon bonds," Macromolecular Symposia, vol. 111, pp. 63–71, 1996.
45. T. Otsu and M. Yoshida, "Role of initiator-transfer agent-terminator (iniferter) in radical polymerizations: polymer design by organic disulfides as iniferters," Die Makromolekulare Chemie, vol. 3, no. 2, pp. 127–132, 1982.
46. J. P. Kennedy, "Functional polymers by cationic techniques," Journal of Macromolecular Science, vol. 13, no. 5, pp. 695–714, 1979.
47. M. Szwarc and A. Rembaum, "Polymerization of methyl methacrylate initiated by an electron transfer to the monomer," Journal of Polymer Science, vol. 22, no. 100, pp. 189–191, 1956.
48. M. Szwarc, M. Levy, and R. Milkovich, "Polymerization initiated by electron transfer to monomer. A new method of formation of block polymers," Journal of the American Chemical Society, vol. 78, no. 11, pp. 2656–2657, 1956.

49. M. K. Georges, R. P. N. Veregin, G. K. Hamer, and P. M. Kazmaier, Macromolecular Symposia, vol. 88, p. 89, 1994.
50. D. H. Solomon, E. Rizzardo, and P. Cacioli, "Polymerization process and polymers produced thereby," US patent no. 4581429, 1986.
51. J. Krstina, G. Moad, E. Rizzardo, C. L. Winzor, C. T. Berge, and M. Fryd, "Narrow polydispersity block copolymers by free-radical polymerization in the presence of macromonomers," Macromolecules, vol. 28, no. 15, pp. 5381–5385, 1995.
52. J.-S. Wang and K. Matyjaszewski, "Controlled "living" radical polymerization. Atom transfer radical polymerization in the presence of transition-metal complexes," Journal of the American Chemical Society, vol. 117, no. 20, pp. 5614–5615, 1995.
53. K. Matyjaszewski, "Mechanistic and synthetic aspects of atom transfer radical polymerization," Journal of Macromolecular Science, vol. 34, no. 10, pp. 1785–1801, 1997.
54. G. Kickelbick and K. Matyjaszewski, "4,4′,4″-tris(5-nonyl)-2,2′:6′,2″-terpyridine as ligand in atom transfer radical polymerization (ATRP)," Macromolecular Rapid Communications, vol. 20, no. 6, pp. 341–346, 1999.
55. B. Ivan, "Macromolecular nomenclature note no. 19: terminology and classification of quasiliving polymerizations and ideal living polymerizations on the basis of the logic of elementary polymerization reactions, and comments on using the term "controlled"," Macromolecular Chemistry and Physics, vol. 201, no. 17, pp. 2621–2628, 2000.
56. T. R. Darling, T. P. Davis, M. Fryd et al., "Living polymerization: rationale for uniform terminology," Journal of Polymer Science, vol. 38, no. 10, pp. 1706–1752, 2000.

CHAPTER

7

AQUEOUS FREE-RADICAL POLYMERIZATION OF NON-IONIZED AND FULLY IONIZED METHACRYLIC ACID

Eric Jean Fischer, Giuseppe Storti and Danilo Cuccato

Institute for Chemical and Bioengineering, ETH Zurich, Vladimir-Prelog-Weg 1, 8093 Zurich, Switzerland

ABSTRACT

Water-soluble, carboxylic acid monomers are known to exhibit peculiar kinetics when polymerized in aqueous solution. Namely, their free-radical polymerization rate is affected by several parameters such as monomer concentration, ionic strength, and pH. Focusing on methacrylic acid (MAA), even though this monomer has been largely addressed, a systematic investigation of the effects of the above-mentioned parameters on its polymerization rate is missing, in particular in the fully ionized case. In this work, the kinetics of non-ionized and fully ionized MAA are characterized by in-situ nuclear magnetic resonance (NMR). Such accurate monitoring of the reaction rate enables the identification of relevant but substantially different effects of the monomer and electrolyte concentration on polymerization rate in the two ionization cases. For non-ionized MAA, the development of a kinetic model based on literature rate coefficients allows us to nicely simulate the experimental data of conversion versus time at a high monomer concentration. For fully ionized MAA, a novel propagation rate law accounting for the electrostatic interactions is proposed: the corresponding model is capable of predicting reasonably well the electrolyte concentration effect on polymerization rate. Nevertheless, further kinetic information in a wider range of monomer concentrations would be welcome to increase the reliability of the model predictions.

Keywords: methacrylic acid; free radical polymerization; modeling; propagation; termination; electrostatic interactions; electrostatic screening; kinetics; NMR

INTRODUCTION

Water soluble polymers are very attractive materials which have applications in many different fields: they are particularly employed in the manufacturing of pharmaceuticals and cosmetics, in enhanced oil separation and water purification processes, and as additives for thickening, flocculation, coating, etc. [1,2,3]. The peculiar properties of such polymers come from their building blocks, namely water soluble vinyl monomers exhibiting polarizable, ionizable, or charged moieties. Typical examples of such monomers are acrylic and methacrylic acid as well as their esters, acrylamides and vinyl amides, and ammonium salts. The presence of charges and polarized groups in such monomers makes the corresponding polymers suitable for establishing electrostatic interactions between the polymer chains as well as between the polymer and its environment. As a consequence, the solution and surface properties of the polymer as well as the viscoelastic behavior of the material are affected by such interactions, determining the peculiar features of water-soluble polymers which make them so interesting and versatile [4,5].

In addition to the effect on the material properties, these interactions involving monomer and polymer moieties have an impact on the reaction kinetics during aqueous radical polymerization, which is the usual method of synthesis of water-soluble polymers. The presence of charges or dipoles can induce interactions of various nature between the reacting species, and they can in turn substantially affect the observed reaction order of some relevant kinetic steps with respect to the reactants [6]. Monomer-monomer association [7], hindrance of the internal motions of reaction complexes due to intermolecular interactions [8], and electrostatic forces producing diffusion limitations [9,10,11] are some of the relevant phenomena which may affect the kinetic behavior of water-soluble, ionized, or ionizable monomers, especially with respect to propagation reactions.

A general effect of reduction of the propagation rate coefficient upon increasing monomer concentration has been detailed by studies on various water-soluble monomers, and it has been explained in terms of "fluidization" of the transition state structure due to its interactions with the surrounding species, in particular hydrogen bonding involving water molecules [8,12,13]. The electrostatic effects on the propagation kinetics of ionized or ionizable monomers have been discussed in terms of a reduced diffusion of the monomer to the radical site of an active chain when both are similarly charged, as a result of the repulsion forces. In this view, the sensitivity of the propagation kinetics to the initial monomer concentration is explained by the phenomenon of electrostatic screening. Increasing the concentration of electrolyte in a solution containing charged species reduces the strength of the electrostatic interactions among the charges by screening of the repulsive forces. In the case of polymerization of ionized monomers, the monomer itself acts as an electrolyte in solution: accordingly, increasing the concentration of ionized monomer enhances its propagation kinetics [6,11,14]. The most recent studies on the effect of charge interactions on the kinetics of water-soluble, ionized, or ionizable monomers have been largely focused on copolymer composition: in this way, it is possible to isolate the contribution of propagation reactions and specifically address the sensitivity of propagation reactivity ratios upon changes in ionic strength, monomer concentration and ionization, pH, etc. [10,11,15,16]. On the other hand, the investigation of such effects on the polymerization rate is of great interest due to its implications on the final polymer properties (e.g., molar masses).

In this context, methacrylic compounds offer the advantage of not involving secondary reactions, i.e., backbiting, which can strongly affect the polymerization rate and make it difficult to focus on the effect of charge interactions only [17,18,19]. The kinetics of methacrylic acid (MAA) has been the subject of several studies by pulsed-laser polymerization (PLP), which have been focused especially on the estimation of propagation [12,20] and termination [21,22] rate coefficients. In addition, some attempts of characterizing and modeling the polymerization kinetics of MAA have been carried out recently, in some cases together with the analysis of the molar masses and of the effect of chain transfer reactions [23,24,25]. Nevertheless, the majority of the existing studies are focused on non-ionized MAA and limited to medium-large monomer concentrations. Experimental and modeling studies on fully ionized MAA are missing in the literature, with the notable exception of PLP analysis [20]. Moreover, specific studies on the influence of non-monomeric electrolyte addition on the polymerization kinetics cannot be found.

For all these reasons, in this work we focus on MAA with the aim of elucidating the effects of monomer concentration and ionization on the rate of polymerization. Namely, the time evolution of conversion is characterized by in-situ NMR for both non-ionized and fully ionized acids, with focus on the impact of the monomer concentration and ionic strength on the kinetic behavior. Experimental reactions are carried out in the medium-low monomer concentration range (i.e., 1 to 10 wt % of initial monomer). In combination with fundamental kinetic modeling, substantial differences between the polymerization behaviors of the two ionization forms of MAA are revealed. Finally, the addition of NaCl to the fully ionized system is applied to shed light on the effect of the electrostatic interactions on the reaction kinetics.

MATERIALS AND METHODS

The polymerization reactions are carried out in deuterium oxide (D_2O, D, 99%, 99.5% chemical purity, Cambridge Isotope Laboratories, Inc., Tewksbury, MA, USA) using NMR tubes (5 mm NMR tube, Type 5UP (Ultra Precision), 178 mm, ARMAR AG, 5312 Döttingen, Switzerland) as reactors, which are directly inserted in an operating NMR spectrometer (UltraShield 500 MHz/54 mm magnet system, Bruker Inc., Billerica, MA, USA). For each experiment, the monomer MAA (99.5%, extra pure, stabilized with ca. 250 ppm 4-methoxyphenol (MEHQ), Acros Organics, 2440 Geel, Belgium) is dissolved in D_2O. If the reaction is to be carried out at full monomer ionization, i.e., at the initial degree of acid dissociation α=1, a solution of NaOH (pellets, analytical reagent grade, Fisher Scientific Ltd., Loughborough, LE11 5RG, UK) is added. If the ionic strength is to be altered, pure NaCl (EMSURE, ACS, ISO, Reag. Ph Eur, for analysis, Merck KGaA, 64271 Darmstadt, Germany) is added. As the last step, a solution of the radical initiator 2,2′-azobis(2-methylpropionamidine) dihydrochloride (V-50, 98%, Acros Organics, 2440 Geel, Belgium) in D_2O is added. All the reagents are used as received, and the V-50 solution is renewed once a week to avoid any degradation. The reaction mixture is degassed with N_2 for 5 min at room temperature, as cooling with ice water leads to partial monomer precipitation. About 0.6 mL of the reaction mixture is transferred into the NMR tube, which is purged with N_2 during and after the transfer in order to avoid any contamination with oxygen. Once closed, the NMR tube is immediately put into the magnet at room temperature, where the reaction is monitored by in-situ NMR. It takes about 10 min to reach a constant reaction temperature of 50 °C. During this relatively short time of temperature adjustment, the inhibitor and possible impurities associated to the monomer are consumed, as no polymer formation is detected by NMR. The monomer conversion is measured as a function of the polymerization time by a series of ^{1}H NMR acquisitions. Since the relaxation time T1 of MAA is ca. 5 s, the instrument is set to wait ca. 25 s between each scan. As each data point consists of an average of 4 scans and considering the acquisition time, a data point is generated every 2.5 min. This resolution is enough to follow the evolution of the reaction. The conversion is evaluated as

$$\chi(t) = 1 - \frac{A(t)}{A_0} \qquad (1)$$

where A(t) is the sum of the areas of the two hydrogen atoms at the double-bond carbon (sp^2 hybridized) and A_0 is the area of the same hydrogens at time zero. More details about the in-situ NMR procedure adopted are given elsewhere [11].

In the experiments, the initial monomer concentration is varied from low to medium values (1 to 10 wt %). In order to mimic the ionic strength of higher monomer concentration, sodium chloride (NaCl) is added. For example, to obtain the ionic strength of the 10 wt % MAA reference (1.16 mol/kg), the reaction mixture contains 1 wt % MAA and

6.1 wt % NaCl. MAA concentrations in D_2O higher than 10 wt % are avoided due to solubility limitations at room temperature, which hinders an accurate transfer of the mixture into the NMR tube. The initial V-50 concentration is kept constant at a level high enough to ensure adequate kinetic rates, i.e., 0.02 wt % for the reactions at α=0 and 0.10 wt % for the reactions at α=1(which are considerably slower). In each reaction, a certain inhibition time is observed, which is due to the presence of MEHQ in the monomer. During this time, polymer formation is instead detected by NMR, contrary to what was observed throughout the temperature adjustment. The rate of polymer formation is very slow at the beginning of the reaction and then increases with time as the inhibitor is consumed. As a consequence, the derivative of the conversion-time curve reaches its maximum value only with some delay, when the inhibitor is completely depleted. Therefore, the time axis of the conversion versus time plots is adjusted to remove this effect: a delay time is considered for each experiment so that the tangent to the conversion-time curve at its maximum slope runs through the origin. It is worth noting that the observed inhibition time increases with monomer content (e.g., from ca. 20 min at 1 wt % MAA to ca. 70 min at 10 wt % MAA), in agreement with a larger inhibitor content.

RESULTS

Non-Ionized System

Reactions are carried out at 50 °C with 1, 5, and 10 wt % of initial monomer content ($w_{M,0}$, weight fraction) and a constant amount of V-50 (0.02 wt %). With "non-ionized MAA" we mean the situation of natural dissociation of the monomer in the absence of any added base. The conversion versus time plots are shown in Figure 1. At 10 wt % of initial monomer, the slope of the conversion-time curve has reached its maximum value (i.e., the inhibitor is completely consumed) and remains constant until ca. 40% of conversion. After about 60 min, an increase of the slope is noticeable, which is a sign of diffusion limitations to termination (gel effect). This is fully consistent with the experimental results of Buback et al. [23], who ran their reactions at larger initiator concentrations (5 mmol/L, equivalent to ca. 0.12 wt %). At lower monomer concentrations, the initial polymerization rate is faster than at 10 wt %, in agreement with the increase in the propagation rate coefficient as a function of $w_{M,0}$, revealed by PLP studies [12]. It is also worth noting that at 5 wt % and 1 wt % of MAA, the autoacceleration of the reaction due to gel effect is less relevant: this behavior is consistent with the lower viscosity of the reacted solution observed in the NMR tube in the low monomer concentration cases compared to the situation at 10 wt % of initial MAA. Furthermore, it appears that at high conversion, the polymerization rate is slowing down: such an effect is particularly pronounced at 1 wt % of MAA, where full conversion is reached only hours later. The reproducibility of the obtained conversion versus time curves has been tested for the reaction at 1 wt % and 10 wt % of initial monomer. For the repetition of the reaction with 10 wt % MAA, the difference in the slope above 40% conversion is a bit larger. Since this discrepancy (of similar size) at the same reaction conditions has already been reported in the literature [23], we consider this accuracy to be sufficient.

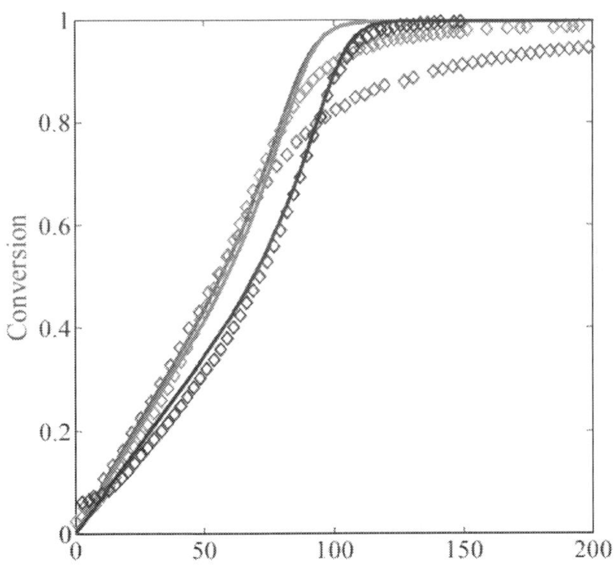

Figure 1. Monomer conversion versus time profiles of the radical batch polymerization of non-ionized methacrylic acid (MAA) in aqueous solution at 50 °C, 0.02 wt % of 2,2′-azobis(2-methylpropionamidine) dihydrochloride (V-50) initiator, and initial monomer concentration equal to 1 wt % (red), 5 wt % (green), and 10 wt % (blue): comparison between experimental data by in-situ nuclear magnetic resonance (NMR) (diamonds) and simulated curves (lines).

A kinetic model suitable for evaluating the reaction rate and number average molecular weight of the active chains has been developed. The most conventional free-radical kinetic scheme is considered, with all the reactions listed in Table 1. The initiator I_2 decomposes to form two radicals $I^\cdot$ with rate coefficient k_d. Assuming that all terminations are accounted for by an initiator efficiency factor f, and applying the quasi-steady-state assumption for the radical fragments $I^\cdot$, the rate of propagation of $I^\cdot$ is equal to the rate of initiation (i.e., the first term in the monomer balance equation in Table 2). The radicals of any chain length $R_n^\cdot$ undergo propagation with a monomer unit M (rate coefficient k_p), and termination by combination (k_{tc}) and by disproportionation (k_{td}). Chain transfer to the monomer is also considered (k_{ctm}). The population balance equations (PBEs) are given in Table 2, where the definitions of the zero- and first-order moments of the chain length distribution of the active chains (λ_0 and λ_1, respectively) and of their number average molecular weight (M_n^R) are as follows:

$$\lambda_0 = \sum_{n=1}^{\infty} R_n^\cdot \tag{2}$$

$$\lambda_1 = \sum_{n=1}^{\infty} nR_n^\cdot \tag{3}$$

$$M_n^R = MW_{MAA}\frac{\lambda_1}{\lambda_0} \tag{4}$$

where MW_{MAA} is the molar mass of MAA, thus the average chain length is defined in g mol^{-1}. The reaction rate coefficients are defined in Table 3, where χ is the conversion and P and T are set to 1.013 bar and 323 K, respectively. In the definition of the composite model for the termination rate coefficient, a critical average chain length $M_{n,C}^R = 68\ MW_{MAA}$ is considered, according to Wittenberg et al. [25]. Since we do not include the population balance equations for the dead chains, we cannot compute the weight-average molecular weight required to calculate the viscosity parameter C_η as detailed by Wittenberg et al. [25]; nevertheless, we obtained a satisfactory overlap of our model with the k_t versus conversion profiles at 30 wt % of MAA reported in the reference by setting C_η to 9.

In the model equations, a constant density of 1.1 g mL^{-1} is assumed for D_2O solutions.

Table 1: Reaction scheme used in the modeling of non-ionized methacrylic acid (MAA) polymerization

Reaction	Scheme
Initiator decomposition	$I_2 \xrightarrow{k_d} 2I^{\cdot}$
Propagation of initiator fragment	$I^{\cdot} + M \xrightarrow{k_p} R_1^{\cdot}$
Propagation of active chain	$R_n^{\cdot} + M \xrightarrow{k_p} R_{n+1}^{\cdot}$
Chain transfer to monomer	$R_n^{\cdot} + M \xrightarrow{k_{ctm}} P_n + R_1^{\cdot}$
Termination by combination	$R_n^{\cdot} + R_m^{\cdot} \xrightarrow{k_{tc}} P_{n+m}$
Termination by disproportionation	$R_n^{\cdot} + R_m^{\cdot} \xrightarrow{k_{td}} P_n + P_m$

Table 2: Population balance equations on the involved species I_2, M and the zero and first order moments of the active chain distribution, λ_0 and λ_1

Reaction
$\frac{dI_2}{dt} = -k_d I_2$
$\frac{dM}{dt} = -2fk_d I_2 - k_p M \sum_{m=1}^{\infty} R_m^{\cdot} - k_{ctm} M \sum_{m=1}^{\infty} R_m^{\cdot}$
$\frac{d\lambda_0}{dt} = 2fk_d I_2 - 2(k_{td} + k_{tc})\lambda_0^2$
$\frac{d\lambda_1}{dt} = 2fk_d I_2 + k_p M\lambda_0 - k_{ctm} M(\lambda_1 - \lambda_0) - 2(k_{td} + k_{tc})\lambda_1\lambda_0$

Table 3: Reaction rate coefficients used in the modeling of non-ionized MAA polymerization

Reaction
Initiation [23,25]
$k_d/s^{-1} = 2.24 \cdot 10^{15} exp\left(-\frac{1.52 \cdot 10^4}{T}\right)$ $f = 0.8$
Propagation [21]
$k_p/(L\ mol^{-1}s^{-1}) = 4.1 \cdot 10^6 exp\left(-\frac{1.88 \cdot 10^3}{T}\right) \left\{0.08 + 0.92 exp\left[\frac{-5.3 w_{M,0}(1-\chi)}{1-\chi w_{M,0}}\right]\right\} exp\left\{\frac{p}{T}\left[0.096 + \frac{0.11 w_{M,0}(1-\chi)}{1-\chi w_{M,0}}\right]\right\}$
Termination [22,23,25]

$k_t^{1,1}/(L\,mol^{-1}s^{-1}) = 2.29{\cdot}10^{12}exp\left(-\frac{2.64{\cdot}10^3}{T}\right)$
$k_t^{CLD}/(L\,mol^{-1}s^{-1}) = \begin{cases} k_t^{1,1}(M_n^R)^{-0.61} & @\, M_n^R \leq M_{n,c}^R \\ k_t^{1,1}(M_{n,c}^R)^{-0.444}(M_n^R)^{-0.166} & @\, M_n^R > M_{n,c}^R \end{cases}$
$k_t/(L\,mol^{-1}s^{-1}) = k_t^{CLD}\left[0.96 + 0.04exp(\chi C_\eta)\right]^{-1} + C_{RD}w_{M,0}(1-\chi)k_p$
$k_{tc}/(L\,mol^{-1}s^{-1}) = 0.2k_t$
$k_{td}/(L\,mol^{-1}s^{-1}) = 0.8k_t$
Chain transfer to monomer [23]
$k_{ctm}/(L\,mol^{-1}s^{-1}) = 5.37{\cdot}10^{-5}k_p$

Once all parameter values are set, the model is used to predict the kinetic profiles for our experiments at 1 wt %, 5 wt %, and 10 wt % of MAA. The resulting curves are shown in Figure 1: a good agreement between the experimental data and the model predictions is found, with the initial slope perfectly reproduced at all the concentrations. At 10 wt % of initial monomer, both the simulated curve and the experimental data show an increase in polymerization rate due to the gel effect; moreover, they are nicely superimposed. On the other hand, at 1 wt % and 5 wt % the model predicts a slight increase in reaction rate, which is not observed experimentally. This discrepancy could reflect an increase in termination or a decrease in the propagation rate not accounted for by the model.

To better elucidate the possible reason for this disagreement, the experimental data have been processed as follows: (i) the local slope $\Delta\chi/\Delta t$ is calculated at each time step; (ii) the product between the propagation rate coefficient and the actual radical concentration, λ0, is evaluated as:

$$k_p\lambda_0 = \frac{\Delta\chi/\Delta t}{1-\chi} \tag{5}$$

(iii) using the k_p value defined as in Table 3, the radical concentration is then obtained; and (iv) the apparent rate coefficient of termination is finally estimated as:

$$k_t^{exp} = \frac{fk_dI_2}{{\lambda_0}^2} \tag{6}$$

These resulting k_t^{exp} values are depicted in Figure 2: the increase in k_t reflecting the slowdown mentioned above becomes relevant at residual monomer concentrations below 0.5 wt %, i.e., at large polymer contents. Generally, a decrease (instead of an increase) in termination rate towards higher conversion is expected, thus the observed deceleration in the conversion versus time profiles should be imputed to a decrease in the propagation rate coefficient. Since k_t is chain-length dependent (cf. Table 3), there is also the possibility that at high monomer conversion, the termination rate coefficient increases due to reduced monomer concentration leading to the production of shorter chains. However, from Figure 2 such an increase in the literature k_t is observed only at very large conversion (i.e., above 95%): before that point, the termination rate coefficient decreases monotonously with conversion.

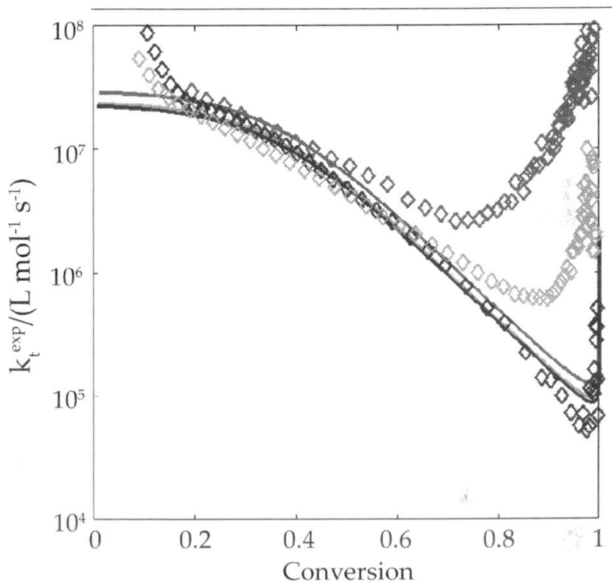

Figure 2: Termination rate coefficient as a function of the monomer conversion calculated from the propagation rate coefficient reported in Table 3 and the experimental data of conversion versus time of non-ionized MAA polymerization (diamonds) at an initial monomer concentration equal to 1 wt % (red), 5 wt % (green), and 10 wt % (blue): comparison with the simulated kt using the model equations (lines).

Therefore, a similar procedure as above has been applied to calculate the propagation rate coefficient deduced from experiments, this time by using the termination rate coefficient as in Table 3. Since k_t is function of k_p, the procedure is not as straightforward: at each (t,χ) step, the set of PBEs in Table 2 is solved for different values of k_p while minimizing the error between the experimental and calculated conversion. Figure 3 shows the resulting k_p^{exp} curves as a function of conversion: all the curves exhibit a strong decrease at high conversion, more or less when the residual MAA concentration falls below 0.5 wt %. A similar behavior has been previously reported for the polymerization of acrylic acid (AA): a decrease of the apparent k_p at concentrations below 3 wt % has been noticed and imputed to differences in the local and overall monomer concentrations in solution, which may be a result of preferential solvation [26]. This decrease explains the decreasing reaction rate at high conversion shown in Figure 1. Apart from this discrepancy, the k_p predicted by the model equations provides a reasonable description of k_p^{exp} as shown by Figure 3. Therefore, we can conclude that the developed model is suitable for reproducing the experimental behavior of the system at medium-high monomer concentration and to qualitatively capture the trends of the propagation and termination rate coefficients as a function of conversion. The residual weakness in predicting the polymerization rate at lower monomer concentration is imputed to inaccurate k_p evaluation at high conversion.

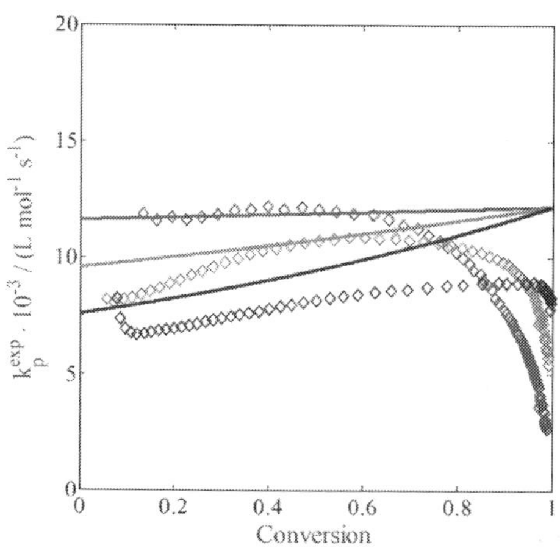

Figure 3: Propagation rate coefficient as a function of the monomer conversion calculated from the termination rate coefficient reported in Table 3 and the experimental data of conversion versus time of non-ionized MAA polymerization (diamonds) at an initial monomer concentration equal to 1 wt % (red), 5 wt % (green), and 10 wt % (blue): comparison with the simulated kp using the model equations (lines).

Fully Ionized System

Three different sets of experiments are carried out for the polymerization of fully ionized MAA at 50 °C and 0.10 wt % V-50: the first set at different values of initial monomer concentration; the second set at different values of initial monomer concentration while keeping the ionic strength constant at an ionic strength equivalent to a monomer concentration of 10 wt %; and the third set at different values of ionic strength while keeping $w_{M,0}$ constant at 5 wt %. In the second and third cases, the ionic strength is adjusted by the addition of NaCl. An overview over all these polymerization reactions is provided in Table 4. A remarkably good reproducibility of the experiments in the fully ionized case has been observed for the reactions at 1 wt % and 10 wt % initial MAA.

Table 4: List of all polymerization reactions of fully ionized MAA (alpha = 1) in aqueous solution at 50 °C and 0.1 wt % of 2,2′-azobis(2-methylpropionamidine) dihydrochloride (V-50) initiator

Reaction No.	Initial Monomer Concentration $w_{M,0}$ (wt %)	Ionic Strength Corresponding to (wt %)	Ionic Strength (kg mol^{-1})
1	1	1	0.12
2	2.5	2.5	0.29
3	5	5	0.58
4	10	10	1.16
5	1	5	0.58
6	1	10	1.16
7	5	10	1.16
8	5	15	1.74
9	5	20	2.32
10	5	30	3.48

Let us consider the experiments at different monomer concentrations. When comparing the resulting conversion versus time curves shown in Figure 4 to those reported above for the non-ionized case (Figure 1), the initial reaction rates are much slower at full MAA ionization: this feature can be explained by the electrostatic repulsion between the charges of the ionized reactants in solution. An increase in the polymerization rate is observed when moving from 1 wt % of MAA to the higher concentrations, whereas between 2.5 wt % and 10 wt % the kinetics appears to be unaffected by the monomer concentration. The former effect corresponds to an increase in the propagation kinetics due to an increase in the solution ionic strength, in agreement with the so-called phenomenon of electrostatic screening: namely, higher initial concentration of the monomer corresponds to higher concentration of the electrolyte in the system [6,11]. The latter effect is more surprising, and reveals either a saturation of the electrostatic screening already at 2.5 wt % of MAA or a competition between electrostatic and non-electrostatic effects of monomer concentration on propagation kinetics. Another counterbalancing effect to the electrostatic-driven increase in k_p could be represented by the known dependence of termination kinetics of fully ionized MAA on monomer concentration [27].

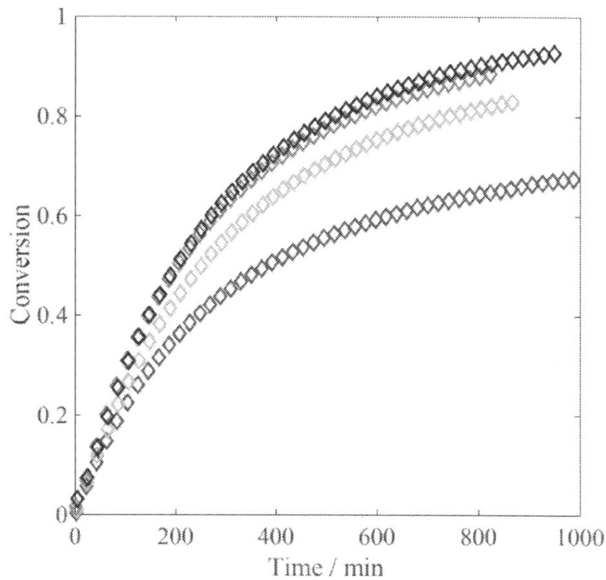

Figure 4. Monomer conversion versus time profiles of the radical batch polymerization of fully ionized MAA in aqueous solution at 50 °C, 0.1 wt % of V-50 initiator, and initial monomer concentration equal to 1 wt % (red), 2.5 wt % (yellow), 5 wt % (green), and 10 wt % (blue).

The impact of monomer concentration on the reacting system at constant ionic strength (equivalent to 10 wt % MAA) and 0.10 wt % of initiator is shown in Figure 5. The idea behind this set of experiments is to reproduce the non-electrostatic influence of the monomer concentration, keeping the ionic strength constant. As shown in the figure, the initial reaction rate does not change with the monomer concentration. Only at a conversion higher than 50% does the curve at low monomer concentration seem to become slower than those at higher monomer concentration, where the propagation rate seems to be fully independent of the non-electrostatic (i.e., intrinsic) reactivity.

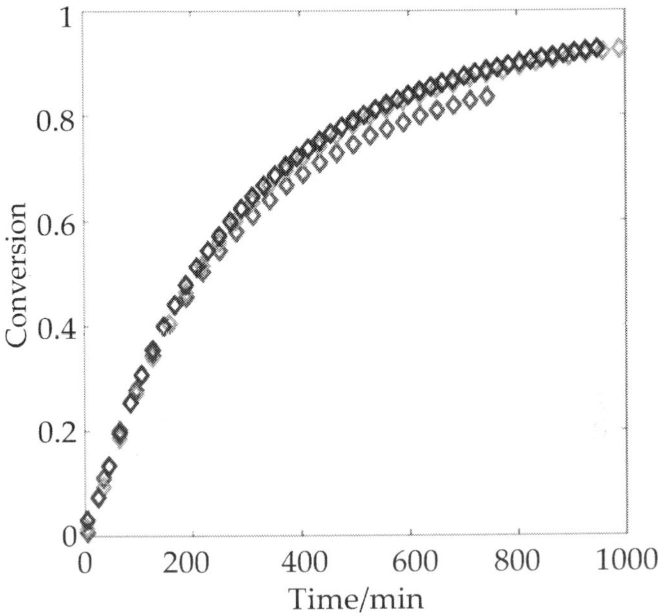

Figure 5: Monomer conversion versus time profiles of the radical batch polymerization of fully ionized MAA in aqueous solution at 50 °C, 0.1 wt % of V-50 initiator and constant ionic strength equivalent to 10 wt % MAA by addition of NaCl. Variation of monomer concentration: 1 wt % (red), 5 wt % (green), and 10 wt % (blue, no NaCl addition).

The impact of NaCl addition on the reacting system at constant monomer concentration is shown in Figure 6. The aim is to reproduce the ionic strength of higher monomer concentrations while keeping constant the initial monomer concentration, in order to separate the effect of the charges from the mixed effects associated to an increase in monomer content (i.e., electrostatic and non-electrostatic ones). As shown in the figure, the initial reaction rate increases continuously with increasing amounts of added salt. Additionally, there is no visible effect of auto acceleration on the rate of polymerization in this case.

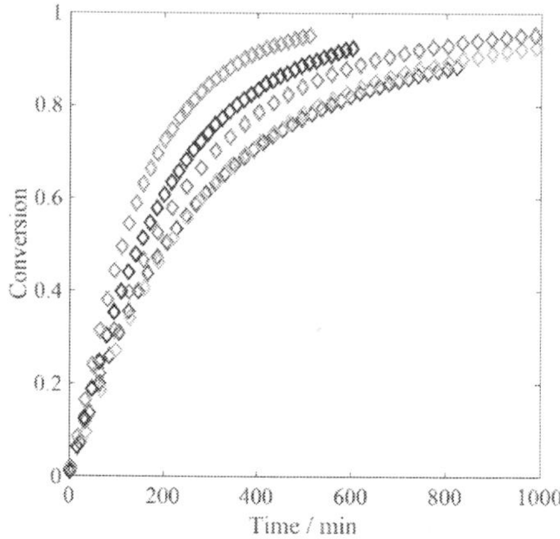

Figure 6: Monomer conversion versus time profiles of the radical batch polymerization of fully ionized MAA in aqueous solution at 50 °C, 0.1 wt % of V-50 initiator, initial monomer concentration equal to 5 wt %, without addition of salt (red), salt addition to obtain an ionic strength equivalent to a monomer concentration of 10 wt % (yellow), 15 wt % (green), 20 wt % (blue), and 30 wt % (purple).

Since the propagation reaction seems to be affected by the electrolyte concentration in the entire explored range (i.e., up to 30 wt % of equivalent MAA), this same reaction is expected to never become fully kinetically-controlled: namely, the effects of electrostatic interactions and electrostatic screening on the propagation kinetics are likely to play a role at all explored values of initial monomer concentration, and some kind of "electrostatic saturation" at higher monomer content (i.e., above 2.5 wt % of MAA) cannot be assumed.

Aiming to develop a kinetic model for the polymerization of fully ionized MAA, the reaction scheme and population balances in Table 1 and Table 2 are again considered, whereas the validity of the reaction rate coefficients in Table 3 is critically discussed hereinafter.

The propagation rate coefficient proposed by Beuermann et al. [21] has been improved by Lacik et al. to take into account the effect of acid ionization [20]. However, as this equation has been proposed for the specific range of initial monomer concentration 5 wt %$\leq w_{M,0} \leq$40 wt %, it cannot be used for our experiments at 1 wt % and 2.5 wt % of initial monomer content, where the calculated kp would be negative. Furthermore, this equation cannot be used to simulate the effect of ionic strength on the propagation kinetics coming from NaCl addition. For these reasons, we propose a rate law of propagation of fully ionized MAA composed of an intrinsic kinetics term and a diffusion term, the latter being driven by electrostatic interactions and thus a function of the electrolyte concentration, similar to the equation developed in a previous work [11]:

$$k_p = \left(\frac{1}{k_{p,i}^0 \exp(-Bw_{M,0})} + \frac{1}{k_{D,0} C_E^{\beta}} \right)^{-1} \tag{7}$$

where $k_{p,i}^0$, $k_{D,0}$, β, and B are fitting parameters, while C_E is the electrolyte concentration in mol kg^{-1}, which is a function of $w_{M,0}$:

$$C_E = 2\left(\frac{w_{S,0}}{MW_{NaCl}} + \frac{w_{M,0}\alpha}{MW_{MAA}} \right) \tag{8}$$

where $w_{S,0}$ and MW_{NaCl} are the initial weight fraction and molar mass of NaCl, respectively.

The first part of Equation (7) introduces the non-electrostatic dependence on the monomer concentration, similar to what has been done by Beuermann et al. for non-ionized MAA [21]. However, it should be noted that such dependence can be in principle different for the fully ionized case: for this reason, the parameter B is estimated as well. For simplicity, the conversion-dependent term of the exponent $(1-\chi)/(1-w_{M,0\chi})$ has been omitted. The parameter $k_{p,i}^0$ identifies the intrinsic kinetics rate coefficient of fully ionized MAA, thus without any electrostatic interaction effect. The electrostatic and non-electrostatic dependence of k_p on $w_{M,0}$ for the fully ionized case, defining respectively an increase and a decrease of the rate coefficient as the initial monomer concentration increases, are expected to compensate each other at higher monomer concentrations, thus reproducing the experimental behavior observed in Figure 4.

In order to verify if the proposed rate law for k_p is reasonable, the propagation rate coefficient is evaluated as a function of conversion using the same procedure described for the non-ionized case, namely using the experimental data reported in Figure 4 and Figure 5. As for k_t, the rate coefficient from a recent publication by Kattner et al. [27] has been adopted: the authors proposed an expression for the termination rate coefficient at monomer concentrations of 5 and 10 wt % as shown in Table 5.

Table 5: Termination rate coefficients for the fully ionized MAA [27]. For 10 wt %, $M_{n,C}^{R}$ is equal to 80 MW_{MAA}

Ionic Strength Equivalent to $w_{M,0}$/wt %	$k_t^{1,1}/\left(\text{L mol}^{-1}\text{s}^{-1}\right)$	$k_t/\left(\text{L mol}^{-1}\text{s}^{-1}\right)$
5	$1.97\cdot10^{8}\exp\left(-\frac{999}{T}\right)$	$k_t^{1,1}M_n^{R\,-0.59}$
10	$7.24\cdot10^{8}\exp\left(-\frac{1049}{T}\right)$	$\begin{cases} k_t^{1,1}M_n^{R\,-0.59} & @\,M_n^R \le M_{n,C}^R \\ k_t^{1,1}{M_{n,C}^{R}}^{-0.41}M_n^{R\,-0.18} & @\,M_n^R > M_{n,C}^R \end{cases}$

Similar to the propagation rate, it can be assumed that the termination reaction is also made of two contributions, one related to an intrinsic kinetics and the other driven by electrostatic interactions. Due to the limited amount of available data, we need to assume that only one of these two contributions majorly affects the termination rate. Since bimolecular termination of growing radicals is a reaction with a low activation energy barrier (mostly controlled by diffusion), we assume that the change in ionic strength upon change in monomer concentration is principally responsible for the k_t variation as a function of $w_{M,0}$. This is considered more realistic than an effect of monomer concentration on the fluidity of the transition state, as previously considered in the non-electrostatic effect of monomer concentration on the propagation kinetics of other water soluble compounds [8]. Therefore, we will focus on those experiments with an ionic strength (obtained by a certain amount of monomer concentration or addition of NaCl) equivalent to a monomer concentration of 5 wt % or 10 wt % when considering the kt expressions reported in Table 5.

Equivalent to the procedure for the non-ionized case (cf. Figure 3), the experimental propagation rates are shown in Figure 7 and Figure 8 for ionic strengths equal to 5 wt % and 10 wt %, respectively. The propagation rate coefficient of 5 wt % MAA without addition of NaCl can be compared with the experimental results of Lacik et al. [20]: at our reaction temperature (50 °C), the Lacik equation predicts a propagation rate of about 770 L mol^{-1} s^{-1}, whereas our experimentally deduced propagation rate is about five times smaller. At monomer concentrations of 5 wt % and 10 wt %, the propagation rate increases at low conversion, which is probably due to the competition between propagation and radical scavenging by the inhibitor. Above 50% conversion, a general decrease is visible. Apart from these minor variations with conversion, the results reveal that the reaction rate increases with ionic strength. In fact, for any monomer concentration, the propagation rate at an ionic strength of 10 wt % monomer concentration is 2–3 times faster than at 5 wt %. At constant ionic strength, a reduction of monomer concentration leads to a slight increase of the propagation rate coefficient. However, this effect is less relevant than the change in ionic strength. The observed competition between the influence of monomer concentration and ionic strength confirms the suitability of the propagation rate law in Equation (7) to describe at least qualitatively the kinetic behavior of the system.

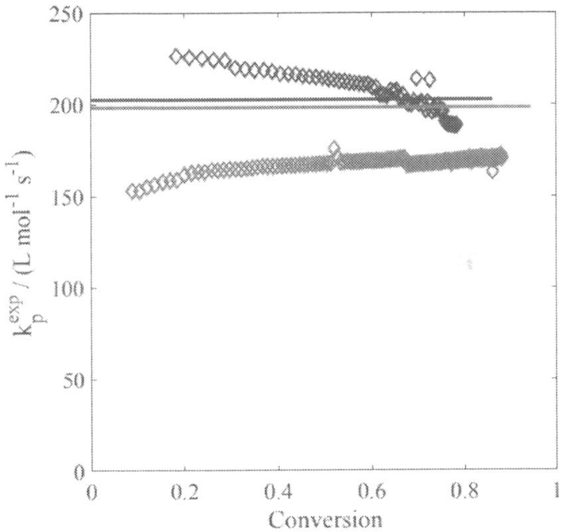

Figure 7: Propagation rate coefficient as a function of the monomer conversion calculated from the termination rate coefficient in Table 5 and the experimental data of conversion versus time of fully ionized MAA polymerization at constant ionic strength equivalent to 5 wt % MAA by addition of NaCl. Variation of monomer concentration 1 wt % (red) and 5 wt % (green, no NaCl addition): comparison with the simulated k_p using the model equations (lines).

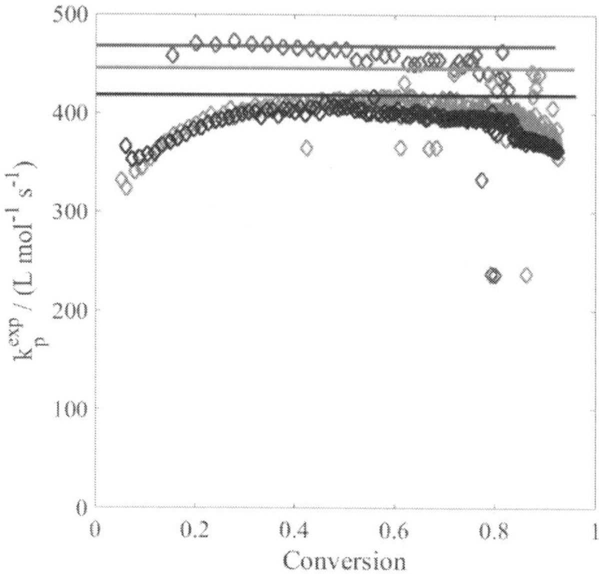

Figure 8: Propagation rate coefficient as a function of the monomer conversion calculated from the termination rate coefficient in Table 5 and the experimental data of conversion versus time of fully ionized MAA polymerization at constant ionic strength equivalent to 10 wt % MAA by addition of NaCl. Variation of monomer concentration 1 wt % (red), 5 wt % (green), and 10 wt % (blue, no NaCl addition): comparison with the simulated k_p using the model equations (lines).

The four parameters $k^0_{p,i}$, $k_{D,0}$, β, and B are estimated by fitting the model equations to the experimental data of conversion versus time for the fully ionized case. The ratio of the parameters $k^0_{p,i}\exp(-Bw_{M,0})$ and $k_{D,0}C_E^{\beta}$ is expected to be close to unity: this way, the diffusion limitation to k_p due to the electrostatic interactions should be effective inside the entire range of ionic strength values explored, as deduced from the experimental results. It should be noted that $k_{D,0}$ must be expressed in L mol^{-1} s^{-1} (kg mol^{-1}) in agreement with the definition of C_E in the model. The exponent β is dimensionless and it is expected to range between 0.1 and 10, while B is a measure of the influence of initial monomer concentration on the kinetics. As the value for non-ionized MAA has been determined as 5.3 [21], a similar value (between 1 and 10) is expected.

The parameters are estimated by minimizing the square of the error between experimental and simulated conversion versus time data. This minimization is carried out by a genetic algorithm which is implemented using Matlab. An optimization run was conducted to evaluate the four mentioned adjustable parameters and leads to the results and parameter values reported in Figure 9, Figure 10 and Figure 11 and Table 6, respectively. Figure 9 compares the influence of monomer concentration without salt on conversion: while the curve at 10 wt % is perfectly reproduced, at 5 wt % monomer, the model predicts a reaction rate which is slightly faster than that at 10 wt %, but while looking at the experimental data the rates should be exactly the same (i.e., the electrostatic and intrinsic kinetics effects on the propagation and termination rate coefficients should be perfectly balanced). Towards higher conversion, the model correctly predicts a decrease of the reaction rate at 5 wt %.

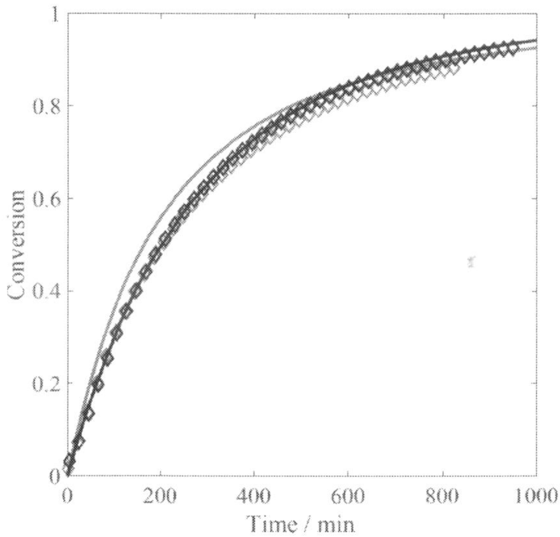

Figure 9: Monomer conversion versus time profiles of the radical batch polymerization of fully ionized MAA in aqueous solution at 50 °C, 0.1 wt % of V-50 initiator, and initial monomer concentration equal to 5 wt % (green) and 10 wt % (blue): comparison between experimental data by in-situ NMR (diamonds) and simulated curves (lines, using the parameter values of Table 6).

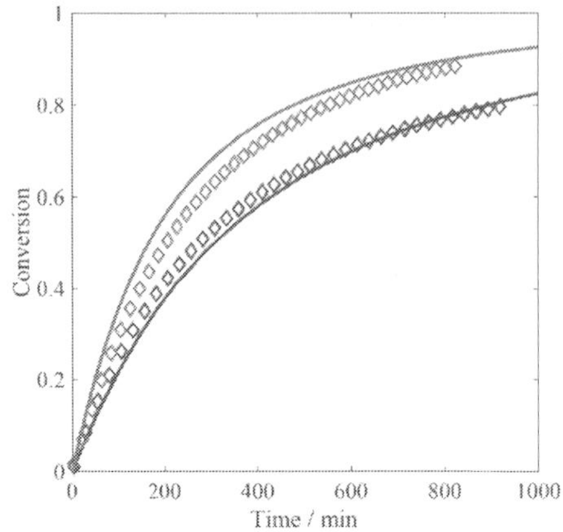

Figure 10: Monomer conversion versus time profiles of the radical batch polymerization of fully ionized MAA in aqueous solution at 50 °C, 0.1 wt % of V-50 initiator and constant ionic strength equivalent to 5 wt % MAA by the addition of NaCl. Variation of monomer concentration 1 wt % (red) and 5 wt % (green, no NaCl addition): comparison between experimental data by in-situ NMR (diamonds) and simulated curves (lines, using parameter values of Table 6).

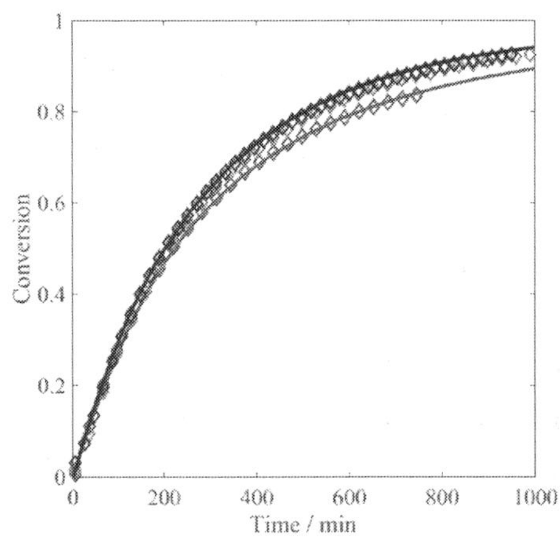

Figure 11: Monomer conversion versus time profiles of the radical batch polymerization of fully ionized MAA in aqueous solution at 50 °C, 0.1 wt % of V-50 initiator and constant ionic strength equivalent to 10 wt % MAA by addition of NaCl. Variation of monomer concentration 1 wt % (red), 5 wt % (green), and 10 wt % (blue, no NaCl addition): comparison between experimental data

by in-situ NMR (diamonds) and simulated curves (lines, using the parameter values of Table 6).

Table 6: Estimated parameter values of the parameter optimization carried out by fitting the model equations to the experimental data of conversion versus time (cf. Figure 9, Figure 10 and Figure 11) for the fully ionized MAA case

Parameter	Value
$k_{p,i}^0/(\mathrm{L\ mol^{-1}\ s^{-1}})$	758
$k_{D,0}/(\mathrm{L\ mol^{-1}\ s^{-1}\ (kg\ mol^{-1})^{\beta}})$	201
β	2.18
B	1.92

The influence of monomer concentration at an ionic strength equivalent to 5 wt % is shown in Figure 10. Qualitatively, the initial slopes at varying amounts of MAA are well reproduced, although the impact of monomer concentration on the initial rate is slightly too strong. The decrease in the reaction rate at conversions above 50% is very well described for both monomer concentrations.

A similar comparison is shown in Figure 11, where the ionic strength is set equivalent to 10 wt % MAA. As already mentioned in the discussion of Figure 5, the initial slope is the same for all MAA concentrations, which is well reproduced by the model, as well as the behavior at higher conversion. Since the model is able to correctly simulate the reduction at 1 wt %, this decrease of the reaction rate cannot be imputed to smaller local concentrations around the polymer chain. It is rather due to a conversion-dependent increase of the termination rate coefficient (caused by a decrease of M_n^R). Note that an effect of local concentration can still be present when considering the reaction at 1 wt % without salt addition, but since we lack an appropriate description of the termination rate at this ionic strength, we are unable to predict this situation.

The parameter values are provided in Table 6: the kinetic rates $k_{p,i}^0$ and $k_{D,0}$ are in the same order of magnitude and similar to the kinetic parameter obtained by Lacik et al. [20], whereas B is slightly smaller than two. It is therefore clearly smaller than the 5.3 obtained by Beuermann et al. [21], but since the reactivity of the ionized MAA is much smaller, it might be reasonable to assume that the effect of monomer concentration is also reduced. The parameter β is slightly larger than two, which is large enough to alter $k_{D,0}c_E^{\beta}$ in a way to qualitatively reproduce the experimental behavior at increasing ionic strength. The overall good balance between $k_{p,i}^{0}\exp(-Bw_{M,0})$ and $k_{D,0}c_E^{\beta}$ leads to a good description of the system: at low ionic strength, the propagation reaction is diffusion limited, whereas at ionic strengths larger than 5 wt % (either by addition of NaCl or MAA), the propagation seems to become more reaction controlled, although never completely (as shown by the experimental results in Figure 6). Moreover, the sensitivity of the model performances upon limited variations of the model parameter values is reported, indicating that small variations of each parameter (i.e., within 10 wt %) induce a very large change in the total error between experiments and simulations.

Eventually, in order to better describe what happens at ionic strengths below 5 wt % and above 10 wt %, a more generalized description of the termination rate is required. This knowledge is especially needed to elucidate the reason for the drastically reduced reaction rate at 1 wt % MAA without salt addition.

As a final remark, the knowledge of molecular weights would allow us to improve the reliability of the developed kinetic models, especially for the fully ionized case. Nevertheless, our choice of focusing on time-conversion data only in the analysis and modeling of the system kinetics was motivated by the following reasons: (i) the estimation of accurate MWD for MAA and generally for aqueous polymers is not trivial and requires ad hoc experimental setup and procedures; (ii) the accuracy and reliability of the kinetic data obtained by in-situ NMR is much higher compared to our current capability of measuring MWD; (iii) the availability of independent values of the termination rate coefficients from PLP of non-ionized MAA and, for specific reaction conditions, also for fully ionized MAA makes our approach reasonable for determining the propagation kinetics by focusing only on the measured polymerization rates.

CONCLUSIONS

Free radical aqueous polymerization reactions of non-ionized and fully ionized MAA were carried out in D_2O by in-situ NMR, with a focus on the effect of the monomer concentration (varied between 1 wt % and

10 wt %) on the polymerization rate. The experimental results revealed substantial differences in the behavior of the two systems. The degree of ionization of the monomer plays a major role in determining the overall reaction rate as well as its sensitivity to changes in monomer and electrolyte concentration.

In the non-ionized case, we were able to develop a kinetic model based on literature rate coefficients which is capable of nicely predicting the experimental data at the initial monomer concentration of 10 wt %. At lower monomer concentration and high conversion (ca. 0.5 wt % residual monomer concentration), a decrease in the reaction rate is observed, which cannot be explained by the current version of the model. Similar to what has also been previously highlighted for AA polymerization, interactions between MAA monomer molecules and their environment at high dilution are assumed to be responsible for this kinetic feature not contemplated by the currently developed polymerization models.

In the case of the fully ionized monomer, we improved the model for the non-ionized case by introducing a new rate law for propagation: namely, both electrostatic and non-electrostatic effects of monomer concentration on the reaction kinetics are explicitly considered. Due to the limited amount of termination data, we focused this development on the cases with an ionic strength relative to MAA concentrations 5 wt % and 10 wt %. The modeling of higher and lower monomer concentrations (or equivalent ionic strengths) would require the knowledge of either molecular weights or termination rate coefficients—both of which are very hard to determine. A reasonable agreement between model simulations and experimental results was obtained by considering together the effects of monomer concentration and salt addition on the polymerization kinetics. In particular, the model is capable of reproducing the effect of increasing reaction rate upon increasing the initial ionic strength.

A better characterization of the rate coefficients of propagation and termination of fully ionized MAA is supposed to substantially improve the elucidation of their combined impact on the polymerization rate, especially in view of the investigation of the effects of monomer and electrolyte concentration. With this respect, a promising method to access on-line the propagation and termination rate is represented by in-situ diffusion-ordered spectroscopy (DOSY) NMR [28], which might allow for measuring at the same time the residual monomer concentration and molecular weight. Furthermore, relevant information about the solution behavior of polar and ionizable molecules such as the MAA monomer and polymer as well as their interactions with the reaction environment (i.e., water and electrolytes) could be obtained by the application of computational chemistry, namely molecular dynamics simulations. Both strategies are currently under development in our research group.

ACKNOWLEDGMENTS

The financial support from the Swiss National Science Foundation with Grant Number 200021.153403/1 is gratefully acknowledged.

AUTHOR CONTRIBUTIONS

All three coauthors E.J.F., G.S. and D.C. designed and conceived the experiments, which were performed by E.J.F., whereas the resulting data was analyzed by E.J.F. and D.C. The paper was written by E.J.F. and D.C.

REFERENCES

1. Nemec, J.W.; Bauer, J.W. *Encyclopaedia of Polymer Science and Engineering*; Wiley: New York, NY, USA; Volume I, 1988.
2. Goin, J. Water Soluble Polymers. In *CEH Marketing Research Report 582.0000 D-E*; SRI International: Menlo Park, CA, USA, 1991.
3. Kumar, J.; Nalwa, H.S.; MacDiarmid, A.G. *Handbook of Polyelectrolytes and Their Applications*; Tripathy, S.K., Ed.; ASP: Stevenson Ranch, CA, USA, 2002.
4. Oosawa, F. *Polyelectrolytes*; Marcel Dekker: New York, NY, USA, 1971.
5. Dobrynin, A.V.; Rubinstein, M. Theory of Polyelectrolytes in Solutions and at Surfaces. *Prog. Polym. Sci.* 2005, *30*, 1049–1118.
6. Losada, R.; Wandrey, C. Non-Ideal Polymerization Kinetics of a Cationic Double Charged Acryl Monomer and Solution Behavior of the Resulting Polyelectrolytes. *Macromol. Rapid Commun.* 2008, *29*, 252–257.

7. Hahn, M.; Jaeger, W.; Wandrey, C.; Reinisch, G. Zur Kinetik der Radikalischen Polymerisation von Dimethyl-diallyl-ammoniumchlorid. IV. Mechanismus von Start-und Abbruchreaktion mit Persulfat als Initiator. *Acta Polym.* 1984, *35*, 350–358.
8. Buback, M. Solvent Effects on Acrylate k_p in Organic Media: A Response. *Macromol. Rapid Commun.* 2015, *36*, 1979–1983.
9. Kurenkov, V.F.; Nadezhdin, I.N.; Antonovich, O.A.; Lobanov, F.I. Phase Separation in Aqueous Solutions of Binary Copolymers of Acrylamide with Sodium 2-Acrylamido-2-Methylpropanesulfonate and Sodium Acrylate. *Russ. J. Appl. Chem.* 2004, *29*, 804–808.
10. Riahinezhad, M.; Kazemi, N.; McManus, N.; Penlidis, A. Effect of Ionic Strength on the Reactivity Ratios of Acrylamide/Acrylic Acid (sodium acrylate) Copolymerization. *J. Appl. Polym. Sci.* 2014, *131*, 40949.
11. Cuccato, D.; Storti, G.; Morbidelli, M. Experimental and Modeling Study of Acrylamide Copolymerization with Quaternary Ammonium Salt in Aqueous Solution. *Macromolecules* 2015, *48*, 5076–5087.
12. Beuermann, S.; Buback, M.; Hesse, P.; Lacik, I. Free-Radical Propagation Rate Coefficient of Nonionized Methacrylic Acid in Aqueous Solution from low Monomer Concentrations to Bulk Polymerization. *Macromolecules* 2006, *39*, 184–193.
13. Lacík, I.; Chovancova, A.; Uhelska, L.; Preusser, C.; Hutchinson, R.A.; Buback, M. PLP-SEC Studies into the Propagation Rate Coefficient of Acrylamide Radical Polymerization in Aqueous Solution. *Macromolecules* 2016, *49*, 3244–3253.
14. Drawe, P.; Buback, M.; Lacik, I. Radical Polymerization of Alkali Acrylates in Aqueous Solution. *Macromol. Chem. Phys.* 2015, *216*, 1333–1340.
15. Rintoul, I.; Wandrey, C. Polymerization of Ionic Monomers in Polar Solvents: Kinetics and Mechanism of the Free Radical Copolymerization of Acrylamide/Acrylic Acid. *Polymer* 2005, *46*, 4525–4532.
16. Preusser, C.; Ezenwajiaku, I.H.; Hutchinson, R.A. The Combined Influence of Monomer Concentration and Ionization on Acrylamide/Acrylic Acid Composition in Aqueous Solution Radical Batch Copolymerization. *Macromolecules*2016, *49*, 4746–4756.

17. Junkers, T.; Barner-Kowollik, C. The Role of Mid-Chain Radicals in Acrylate Free Radical Polymerization: Branching and Scission. *J. Polym. Sci. A* 2008, *46*, 7585–7605.
18. Preusser, C.; Chovancova, A.; Lacik, I.; Hutchinson, R.A. Modeling the Radical Batch Homopolymerization of Acrylamide in Aqueous Solution. *Macromol. React. Eng.* 2016, *10*, 490–501.
19. Wittenberg, N.F.G.; Preusser, C.; Kattner, H.; Stach, M.; Lacik, I.; Hutchinson, R.A.; Buback, M. Modeling Acrylic Acid Radical Polymerization in Aqueous Solution. *Macromol. React. Eng.* 2016, *10*, 95–107.
20. Lacik, I.; Ucnova, L.; Kukuckova, S.; Buback, M.; Hesse, P.; Beuermann, S. Propagation Rate Coefficient of Free-Radical Polymerization of Partially and Fully Ionized Methacrylic Acid in Aqueous Solution. *Macromolecules* 2009, *42*, 7753–7761.
21. Beuermann, S.; Buback, M.; Hesse, P.; Hutchinson, R.A.; Kukucova, S.; Lacik, I. Termination Kinetics of the Free-Radical Polymerization of Nonionized Methacrylic Acid in Aqueous Solution. *Macromolecules* 2008, *41*, 3513–3520.
22. Barth, J.; Buback, M. SP-PLP-EPR Study into the Termination Kinetics of Methacrylic Acid Radical Polymerization in Aqueous Solution. *Macromolecules* 2011, *44*, 1292–1297.
23. Buback, M.; Hesse, P.; Hutchinson, R.A.; Kasak, P.; Lacik, I.; Stach, M.; Utz, I. Kinetics and Modeling of Free-Radical Batch Polymerization of Nonionized Methacrylic Acid in Aqueous Solution. *Ind. Eng. Chem. Res.* 2008, *47*, 8197–8204.
24. Wittenberg, N.F.G.; Buback, M.; Stach, M.; Lacik, I. Chain Transfer to 2-Mercaptoethanol in Methacrylic Acid Polymerization in Aqueous Solution. *Macromol. Chem. Phys.* 2012, *213*, 2653–2658.
25. Wittenberg, N.F.G.; Buback, M.; Hutchinson, R.A. Kinetics and Modeling of Methacrylic Acid Radical Polymerization in Aqueous Solution. *Macromol. React. Eng.* 2013, *7*, 267–276.
26. Lacik, I.; Beuermann, S.; Buback, M. PLP-SEC Study into Free-Radical Propagation Rate of Nonionized Acrylic Acid in Aqueous Solution. *Macromolecules* 2003, *36*, 9355–9363.
27. Kattner, H.; Drawe, P.; Buback, M. Chain-Length-Dependent Termination of Sodium Methacrylate Polymerization in Aqueous

Solution Studied by SP-PLP-EPR. *Macromolecules* 2017, *50*, 1386–1393.

28. Rosenboom, J.-G.; Roo, J.D.; Storti, G.; Morbidelli, M. Diffusion (DOSY) 1H NMR as an Alternative Method for Molecular Weight Determination of Poly(ethylene furanoate) (PEF) Polyesters. *Macromol. Chem. Phys.* 2017, *218*, 1600436.

CHAPTER 8

MODEL-BASED REACTOR DESIGN IN FREE-RADICAL POLYMERIZATION WITH SIMULTANEOUS LONG-CHAIN BRANCHING AND SCISSION

Hidetaka Tobita

Department of Materials Science and Engineering, University of Fukui, Fukui 910-8507, Japan

ABSTRACT

Polymers are the products of processes and their microstructure can be changed significantly by the reactor systems employed, especially for nonlinear polymers. The Monte Carlo simulation technique, based on the random sampling technique, is used to explore the effect of reactor types on the branched polymer structure, formed through free-radical polymerization with simultaneous long-chain branching and scission, as in the case of low-density polyethylene synthesis. As a simplified model for a tower-type multi-zone reactor, a series of continuous stirred-tank reactors, consisting of one big tank and the same N-1 small tanks is considered theoretically. By simply changing the tank arrangement, various types of branched polymers, from star-like globular structure to a more randomly branched structure, can be obtained, while keeping the following properties of the final products, the monomer conversion to polymer, the average branching and scission densities, and the relationship between the mean-square radius of gyration and molecular weight.

Keywords: Branched; Monte Carlo simulation; polyethylene (PE); radical polymerization; reactor design

INTRODUCTION

Nonlinear polymer formation under a kinetically controlled condition is, in general, history-dependent, and the history of every polymer molecule determines the properties of final product polymers. The molecular architecture can be very complex; however, the history-dependence opens up the opportunities to control the nonlinear structure through various types of reactor operation.

In this article, free-radical polymerization that involves chain transfer to polymer, leading to long-chain branching and scission, as in the case of high-pressure polymerization of ethylene to produce low-density polyethylene [1,2], is considered. As shown in Figure 1, the birth time of the chains, C_1 and C_2 must be some time after that of A or A'. There is a definite time order for the chain connection statistics.

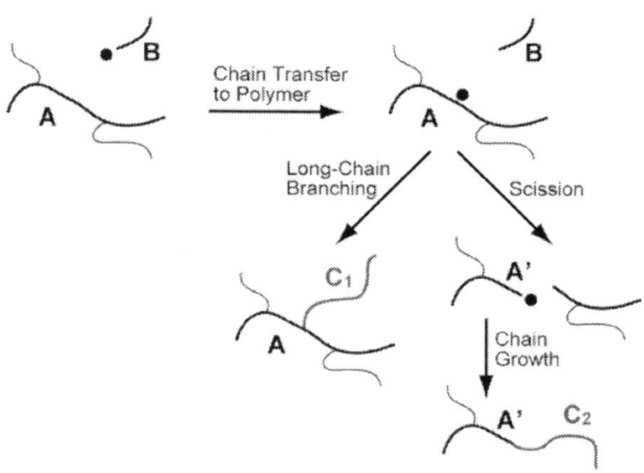

Figure 1: Schematic representation of the process of chain transfer to polymer, leading to long-chain branching and scission.

When the scission reaction is involved together with long-chain branching, the time sequence of branching and scission must be properly accounted for. This kind of reaction system cannot be fully represented by a simple set of population balance differential equations [3,4,5,6,7]. On the other hand, by application of Monte Carlo (MC) method, based on the random sampling technique [8,9], history-dependence of branching and scission can be fully accounted for [7,10,11,12]. In this MC simulation method, the structure of each polymer molecule can be observed directly on the computer screen, and very detailed structural information can be obtained. On the basis of such detailed structural information, it is possible to determine the viscoelastic properties of branched polymers [13].

This simulation method can be used to investigate the effect of reactor type on the formed branched structure. Following the fundamentals of chemical reaction engineering [14], consider how the difference of an ideal plug flow reactor (PFR) and an ideal continuous stirred-tank reactor (CSTR) affects the present polymerization system. Note that a PFR is equivalent to a batch reactor, in which the length in a PFR can be converted to time in a batch reactor. Figure 2 shows the MC simulation results [12] of the weight fraction distribution for the polymers synthesized in a PFR and in a CSTR. In the simulation, the final average branching density ($\bar{\rho}$=2.22×10−3), as well as the final scission density ($\bar{\eta}$=1.11×10−4), is set to be the same for both types of reactors. The independent variable shown in the figure is the logarithm of degree of polymerization (DP), $\log_{10}P$, as usually employed in the gel permeation chromatography (GPC) measurement. The high DP tail extends more significantly in a CSTR, and the weight-average DP is larger for the product in a CSTR, even though the average branching density level is set to be the same.

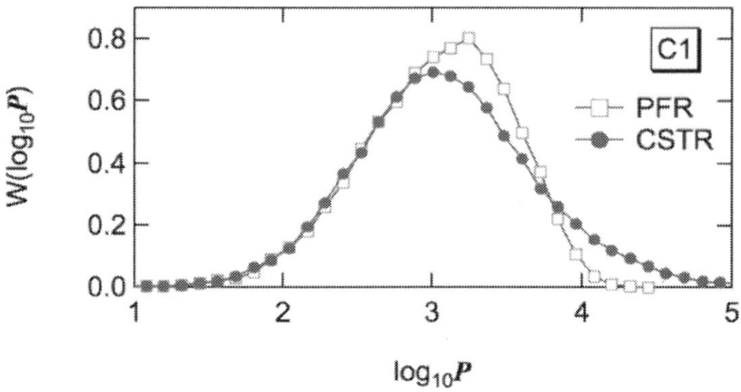

Figure 2: Weight fraction distribution of polymers formed in a PFR and in a CSTR, when the average branching density, as well as the average scission density, is set to be the same for both types of reactors [12]. The kinetic parameters used are the same as C1, shown later in Table 2 of this article.

Figure 3 shows the MC simulation results [12] for the relationship between the branching density ρ and the degree of polymerization P. The branching density is defined as the fraction of units having a branch point. In Figure 3a,b, each dot represents ρ and P of each polymer molecule simulated, and the solid blue curve with circular symbols shows the average ρ within small ΔP intervals, which is the estimate of the average branching density of polymers having degree of polymerization P, $\bar{\rho}$ (P). The dashed black line shows the average branching density of the whole system, $\bar{\rho}$. The value of $\bar{\rho}$ (P) increases with P, but reaches a constant limiting value $\rho_{P\to\infty}$, and $\rho_{P\to\infty}$ is larger than the average branching density of the whole system, $\bar{\rho}$. This is a rather general characteristic of the branched polymer systems [9], including the hyperbranched polymers [15]. Figure 3c shows the comparison of $\bar{\rho}$ (P) for a PFR and for a CSTR. The limiting branching density is larger for the CSTR.

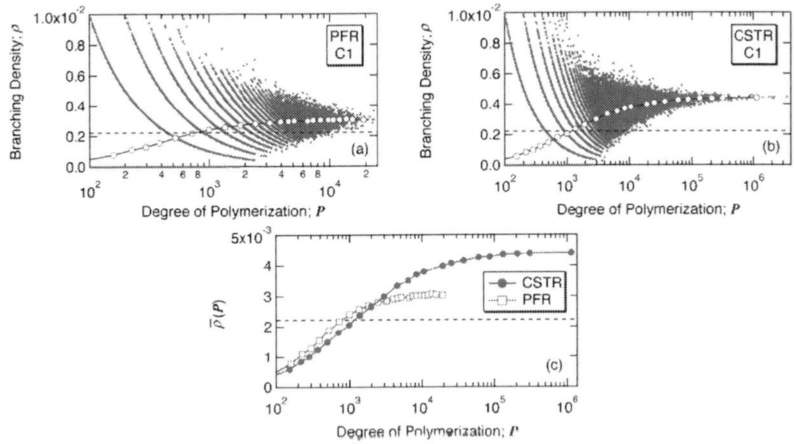

Figure 3: Relationship between branching density ρ and degree of polymerization P for the branched polymrs formed in (**a**) a PFR and in (**b**) a CSTR, and (**c**) $\bar{\rho}$ (P) for both types of reactors. The calculation condition is the same as in Figure 2.

Figure 4 shows the relationship between the mean square radius of gyration for the unperturbed chains $\langle s^2 \rangle_0$ and P. In the figure, b is a constant defined by $b = u/l^2$, where u is the number of monomeric units in a random walk segment, and l is the length of a random walk segment. Each dot in Figure 4 represents $\langle s^2 \rangle_0$ and P-value of each polymer molecule simulated, and the circular symbols show the average values of $\langle s^2 \rangle_0$ within small ΔP intervals, which are the estimates of the average $\langle s^2 \rangle_0$ of polymers having DP, P. For randomly branched polymers, the radii of gyration is known to be given by the Zimm-Stockmayer equation [16], represented by:

$$b\langle s^2 \rangle_0 = \frac{P}{6}\left[\left(1+\frac{m_P}{7}\right)^{0.5} + \frac{4m_P}{9\pi}\right]^{-0.5} \tag{1}$$

where m_p is the number of branch points with degree of polymerization P, and for the present case it can be estimated from:

$$m_P = P\bar{\rho}(P) \tag{2}$$

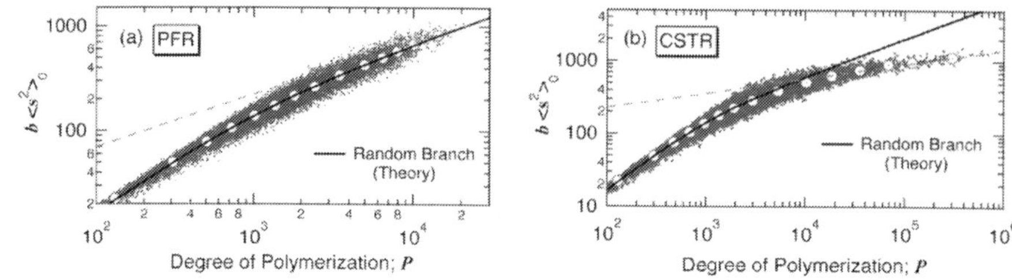

Figure 4: Relationship between the mean-square radius of gyration $\langle s^2\rangle_0$ and degree of polymerization P for the branched polymers formed in (**a**) a PFR and in (**b**) a CSTR [12]. In the figure, b is a constant defined by $b = u/l^2$, where u is the number of monomeric units in a random walk segment, and l is the length of a random walk segment. The black solid curve shows the relationship for random branched polymers. The slope of broken straight line (green) for (**a**) is 0.5, while that for (**b**) is 0.2.

In Figure 4, the radii of gyration of randomly branched polymers having the branching density, $\bar{\rho}$ (P) is shown by a black curve. For a PFR, the radius of gyration is essentially the same as for the random branched polymers, as shown in Figure 4a. In a PFR or a batch reactor, $\langle s^2\rangle_0$ is close to that for the random branched polymers, at least for low conversion regions, while the $\langle s^2\rangle_0$-values tend to become slightly larger than those for the random branched polymers at higher conversions [17]. On the other hand, for a CSTR, the $\langle s^2\rangle_0$-values for large polymers are significantly smaller than the random branched polymers, as shown in Figure 4b.

Incidentally, because the $\bar{\rho}$ (P)-value reaches a constant value at large P's, the $\langle s^2\rangle_0$ curve for the random branched polymer, represented by Equation (1), follows the power law with $\langle s^2\rangle_0 \propto P0.5$ for $P\to\infty$. In Figure 4a, the broken line shows the slope with 0.5. On the other hand, for a CSTR, it is not clear if the power law holds for large polymers, but the broken line drawn as a trial shows the slope with 0.2.

Table 1 summarizes the characteristics of PFR and CSTR for free-radical polymerization with chain transfer to polymer. Note that Table 1 applies for low scission frequency cases. Because the polymer transfer reaction is the reaction between a polymer and a polymer radical, larger polymer concentration promotes the chain transfer reaction. With a CSTR, the polymer concentration is high throughout the polymerization.

If the final conversion is set to be the same for both types of reactors, the average branching density is larger for the polymers synthesized in a CSTR. In the present example, because the final average branching density is set to be the same, the final conversion must set to be larger for a PFR. It is interesting to note that the weight-average DP, $\overline{P}_w$ is larger for a CSTR, even though the average branching and scission densities are the same for both types of reactors. As shown in Figure 2, the molecular weight distribution of polymers formed in a CSTR is broader.

Table 1: Comparison of PFR and CSTR, under condition where the final average branching density is set to be the same for both types of reactors

Property of product polymer	PFR (or Batch)	CSTR
Final conversion	>	
Weight-average DP, $\overline{P}_w$	<	
Branching density of large polymers, $\rho_{P\to\infty}$	<	
$\langle s^2\rangle_0$, compared with random branched polymers	Almost the same. (May become larger at higher conversions.)	Significantly smaller for large polymers.

The branching density of large polymers, represented by $\rho_{P\to\infty}$ is larger for a CSTR, as shown in Figure 3. For the polymers synthesized in a PFR, the radius of gyration of polymers having the degree of polymerization P is essentially the same as that for the random branched polymers. It would be reasonable to consider that the branched structure formed in a PFR is close to randomly branched. On the other hand, as Figure 4b shows, the polymers formed in a CSTR possess much smaller radius of gyration compared with that for the random branched polymers with the same P-value.

As the textbook describes [14], when the multiple CSTRs are connected in series, the residence time distribution approaches to that of a PFR, as the number of tanks increases. The tanks-in-series model was applied to the present reaction system [18] to confirm that as the number of tanks increases the produced polymers approach to those formed in a PFR. Figure 5 shows how the relationship between $\langle s^2\rangle_0$ and P changes by increasing the total number of tanks.

The MC simulation method employed in refs. [7,12,18] can be used to investigate the effect of reaction environment change on the formed branched polymer structure, which may lead to a discovery of novel type of reactor systems. In this article, a series of continuous stirred-tank reactors, consisting of one big tank and the same N-1 small tanks is discussed by using the data reported in reference [7] and newly created MC simulation data. The possibility of model-based reactor design for the nonlinear polymerization system is explored.

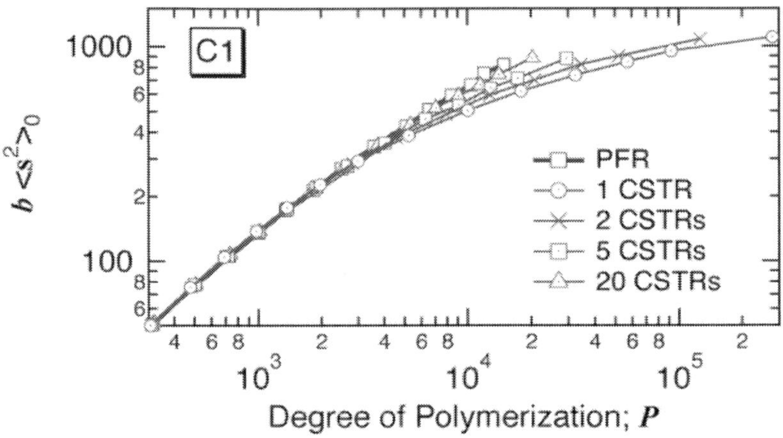

Figure 5: Effect of the number of CSTRs in the tanks-in-series model on the relationship between the mean-square radius of gyration $<s^2>_0$ and degree of polymerization P. The MC simulation data were taken from [18].

SIMULATION METHOD

The random sampling technique [8,9] is used to estimate the properties of final product polymers. In this method, a polymer molecule is selected randomly from the product polymers, and the molecular structure is reconstructed by following the history of this particular polymer molecule, as schematically represented by Figure 6.

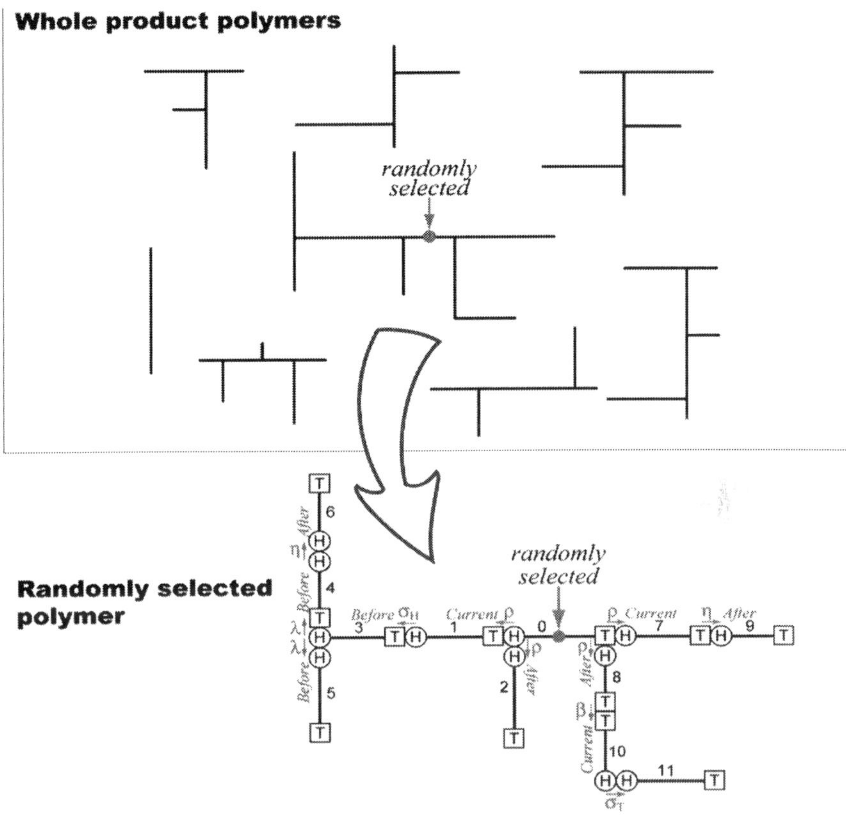

Figure 6. Schematic representation of the concept of random sampling technique.

When the random sampling technique is applied to the Monte Carlo method, a large number of polymers are sampled, and the statistical properties of the whole system are determined effectively. In this method, the system size considered is infinitely large, and therefore, the system boundary problem does not occur. In addition, the amount of calculation required does not increase significantly by increasing the number of CSTRs in series. This MC simulation does not proceed in the order of reaction time, as in the most of other MC methods [19]. Looking from the given chain, the chains with different birth times are connected back and forth by following the reaction history that particular polymer molecule has experienced.

The selection of polymer molecules can be done both on the number and the weight basis [9]. In the case shown in Figure 6, a polymer molecule is chosen by selecting one monomeric unit randomly, which is the selection on the weight basis. For example, the weight-average degree of polymerization, $\bar{P}_w$ can be obtained by simply taking the arithmetic average of the sample. Incidentally, the analytic representation of $\bar{P}_w$ can always be obtained by using the random sampling technique [20,21,22,23,24].

The elementary reactions considered here are as follows. Initiation (rate represented by R_I), propagation (R_p), chain transfer to small molecules including monomer, solvent and chain transfer agents (R_f), bimolecular termination by disproportionation (R_{td}) and by combination (R_{tc}), and chain transfer to polymer (R_{fp}) leading to long-chain branching (R_b) and chain scission (R_s), with $R_{fp} = R_b + R_s$.

The backbiting reaction to form short-chain branching, typically consisting of several carbon atoms, is not considered explicitly, because it has negligible effects on the formed molecular weight distribution and the radii of gyration, which are the major topics of the present investigation. However, the hydrogen abstraction reaction is much more significant for the tertiary carbon atom, rather than the secondary carbon, and therefore, the polymer transfer reactions would be promoted by the backbiting reactions. Because the location of the tertiary carbon atoms along the chain could be considered random, the overall rate coefficients for the branching and the scission reactions k_b and k_s given below are employed [7,12,18].

$$R_b = k_b[R^{\cdot}]Q_1 \qquad (3)$$

$$R_s = k_s[R^{\cdot}]Q_1 \qquad (4)$$

where $[R^{\cdot}]$ is the total radical concentration, and Q_1 is the first moment of the polymer distribution, representing the total number of monomeric units in polymer.

$$Q_1 = \sum_{P=1}^{\infty} P[P_P] \qquad (5)$$

where $[P_P]$ is the concentration of polymer having degree of polymerization, P.

In a strict sense, the long-chain branching reaction shown in Figure 1 is a second order reaction between an internal radical and a monomer, while the scission reaction is the first order reaction of the internal radical, and relative contribution would change during polymerization. However, it was found earlier that such kinetic differences could be neglected [25] for the cases with the final conversion smaller than ca. 0.25, which is normally satisfied for the commercial low-density polyethylene production processes. Actual MC simulation method for the tanks-in-series model is discussed in detail in references [7,18]. In this article, a series of N CSTRs, consisting of one big tank placed as the L-th tank and the same N-1 small tanks as shown in Figure 7 is considered. In the present reaction system, what is important is the magnitude of ξ, defined by the following equation, rather than the volume of reactor [7].

$$\xi_i = k_{p,i}[R^\bullet]_i \overline{t_i} \tag{6}$$

where kp,i and [R·]i are the propagation rate constant and the radical concentration both in the ith tank, and $\overline{t_i}$ is the mean residence time of the ith tank, i.e., $\overline{t_i} = V_i/v$. Here, V_i is the volume of the ith tank, and v is the volumetric flow rate. By neglecting the density change, v can be considered as a constant.

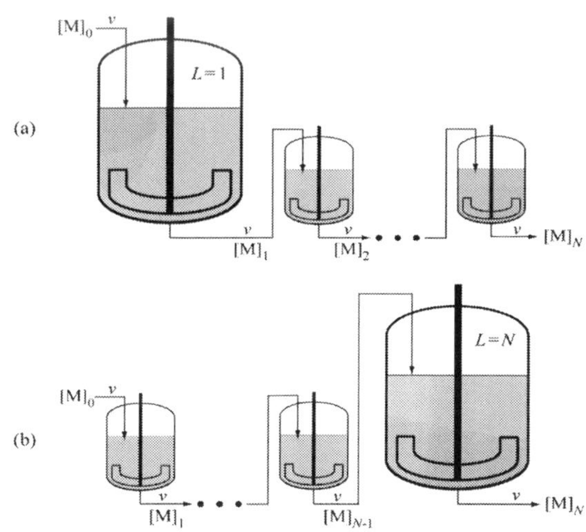

Figure 7: Illustrative representation of the N CSTRs in series, with the arrangement of (**a**) the first large tank L = 1; and (**b**) the last large tank L = N.

From the mathematical point of view, the cases where one large ξ-value, ξ_L with $\xi_i = \xi$ for i≠L are discussed in this article. The value of ξ can be controlled by changing kp,i and/or [R·]i, however by assuming constant values for kp,i and [R·]i, the control factor is the volume of reactor V, as schematically represented by Figure 7.

Under the condition where only the L-th tank is large with the same volume for the other tanks, because every fluid element must flow through all the tanks, the residence time distribution is kept the same irrespective of the tank arrangement. In the present article, the case where the volume of one large tank is equal to the sum of other tanks, $V_L = (N-1)V$, is considered. In this case, the residence time distribution is shown in Figure 8 [7], where $\theta = t/\bar{t}$, and $\bar{t}$ is the mean residence time of the whole reactor system. The long tail is formed due to a single large CSTR that takes up a half of total reactor volume.

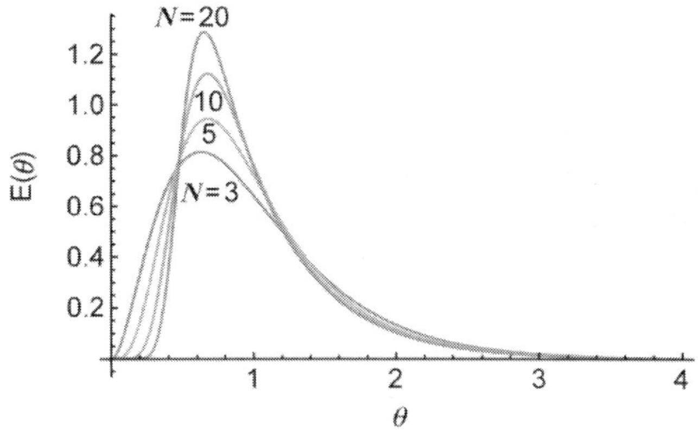

Figure 8: Calculated residence time distribution for the cases with the total number of tanks, N = 3, 5, 10 and 20 [7].

In the present reactor system, the final conversion x_N, the average branching density of the product polymer $\bar{\rho}_N$, and the average scission density of the product polymer $\bar{\eta}_N$ are given, respectively, by [7]:

$$x_N = 1 - \frac{1}{(1-\xi_L)(1-\xi)^{N-1}} \quad (7)$$

$$\bar{\rho}_N = \frac{C_b}{x_N(1+\xi_L)}\left[(\xi_L)^2 + \xi\sum_{i=1}^{N-1}\left\{(1+\xi_L) - \left(\frac{1}{1+\xi}\right)^i\right\}\right] \quad (8)$$

$$\bar{\eta}_N = \frac{C_s}{x_N(1+\xi_L)}\left[(\xi_L)^2 + \xi\sum_{i=1}^{N-1}\left\{(1+\xi_L) - \left(\frac{1}{1+\xi}\right)^i\right\}\right] \quad (9)$$

where C_b and C_s are defined respectively by $C_b = k_b/k_p$ and $C_s = k_s/k_p$.

Equations (7)–(9) do not involve the large tank number L, which means that x_N, $\bar{\rho}_N$ and $\bar{\eta}_N$ do not change irrespective of the tank arrangement. The present system of CSTRs in series makes it possible to investigate the effect of tank arrangement, while keeping the following properties the same; the final conversion, the final average branching and scission densities, and the residence distribution of the whole reactor system.

RESULTS AND DISCUSSION

The parameters used in the present article are shown in Table 2. The values of rate ratios, τ and β may be different depending on the tank number, however, they are assumed to be constant in the present investigation. The polymer transfer constant, $C_{fp} = k_{fp}/k_p$ is equal to the sum of C_b and C_s, i.e., $C_{fp} = C_b + C_s$. For C3, combination termination is included. With combination termination, the crosslinks are formed, and possibility of gelation needs to be considered [24]. With C3, the value of C_s is increased to 0.005, in order to prevent gelation.

Table 2: Parameters used in the present investigation

Parameter	C1	C3
$\tau = (R_{td} + R_f)/R_p$	0.002	0.002
$\beta = R_{tc}/R_p$	0	0.001
$C_b = k_b/k_p$	0.02	0.02
$C_s = k_s/k_p$	0.001	0.005

The MC simulations are conducted to generate 5×10^5 polymer molecules to determine the statistical properties of the product polymers. The total number of tanks, N investigated in this study is N = 5 and 10.

Figure 9 shows how the conversion increases with the progress of tank number for N = 5. The final conversion is set to be $x_N = 0.2$. For L = 1, the conversion goes up in the first tank to $x_1 = 0.1035$ and increases gradually to $x_5 = 0.2$. For L = 2, the conversion increases significantly in the second tank. Figure 10 shows the comparison with N = 10.

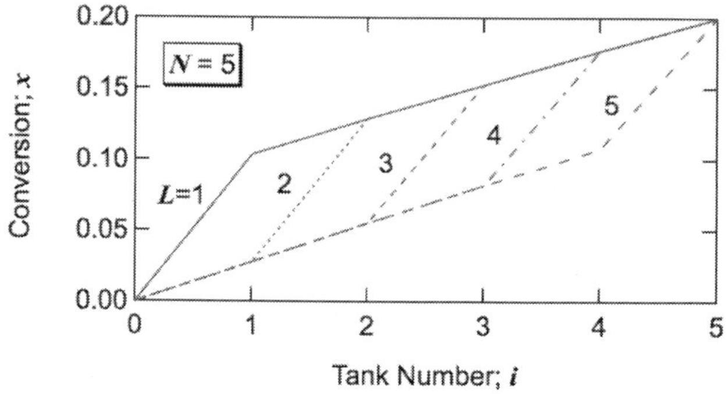

Figure 9: Development of monomer conversion to polymer x with the progress of tank number i for N = 5.

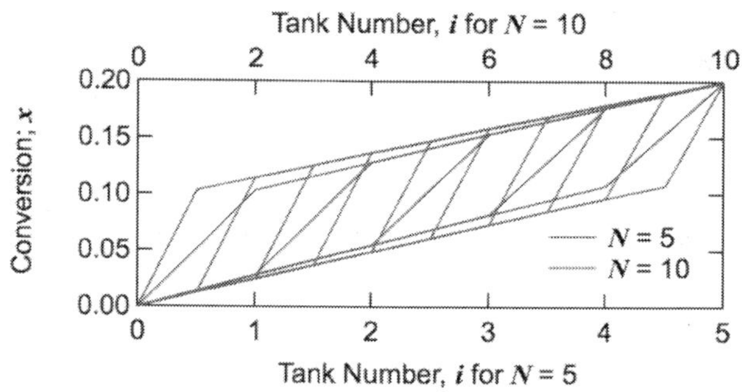

Figure 10: Development of monomer conversion to polymer x with the progress of tank number i for N = 5 (blue) and N = 10 (red).

Figure 11 shows the development of average branching density $\bar{\rho}$, with the progress of tank number for N = 5 and 10. The final average branching density is slightly higher for N = 5, although the final conversion is the same. The small difference in $\bar{\rho}_N$ is caused by the series of small tanks part. With N = 5, there are only four small tanks, while there are nine small tanks for N = 10. Larger number of the same-sized tanks makes the behavior closer to a PFR. When the conversion level is set to be the same, a PFR leads to smaller average branching density than a series of CSTRs, as discussed in Introduction.

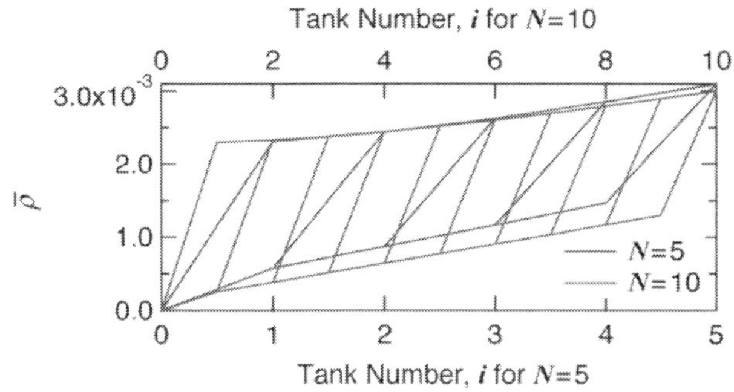

Figure 11: Development of average branching density $\bar{\rho}$ with the progress of tank number i for N = 5 (blue) and N = 10 (red).

Bearing the developments of conversion and average branching density shown in Figure 10 and Figure 11 in mind, let us investigate the effect of tank arrangement on the formed branched structure.

Molecular Weight Distribution

Figure 12 shows the final average degree of polymerization formed in each reactor system. The symbols are MC simulation results. The red lines and symbols are for N = 5, and the blue ones are for N = 10. The weight-average degree of polymerization, $\bar{P}_w$ is the largest for the first large tank case (L = 1), and decreases with larger L-values.

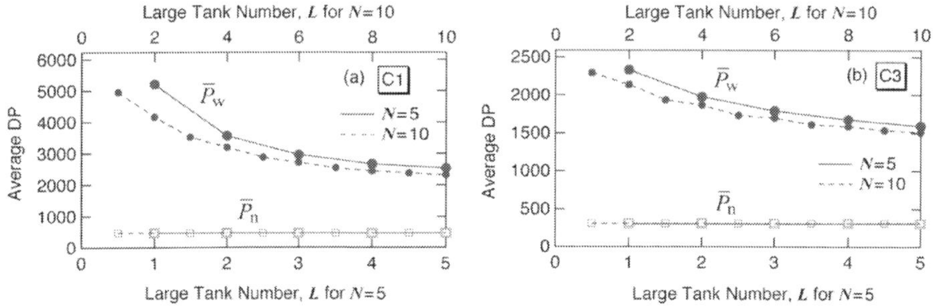

Figure 12: MC simulation results for the number ($\bar{P}_n$) and weight-average ($\bar{P}_w$) degree of polymerization of the polymers formed for conditions (**a**) C1 and (**b**) C3.

The number-average degree of polymerization, $\bar{P}_n$ is given theoretically by [18]:

$$\bar{P}_n = \frac{1}{\tau + \beta/2 + \bar{\eta}_N} \qquad (10)$$

Because the final average scission density $\bar{\eta}$, as well as the values of τ and β, is the same irrespective of tank arrangement, $\bar{P}_n$ does not change with the large tank number L, as Equation (10) and the MC simulation results show.

Figure 13 shows the full weight fraction distribution for N = 10. The cases with N = 5 can be found in ref. [7]. The high molecular weight tail prevails in L = 1.

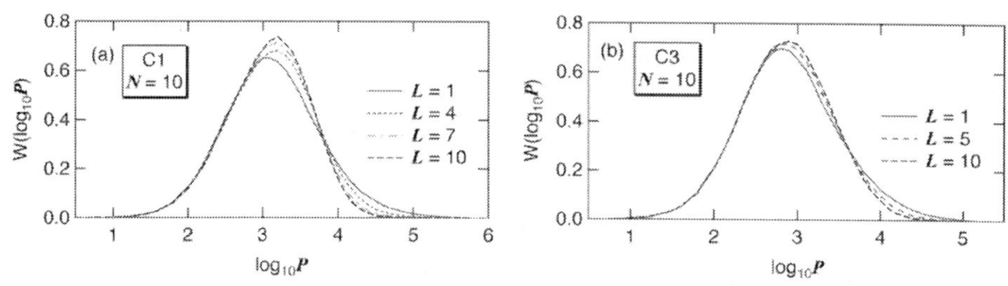

Figure 13: MC simulation results for the weight fraction distribution, plotted as a function of $\log_{10}P$, $W(\log_{10}P)$ of the polymers formed for conditions (**a**) C1 and (**b**) C3, with N = 10.

Now, consider the qualitative explanation why the first big tank case gives the largest $\bar{P}_w$. Two important characteristics need to be reminded. The long-chain branches are formed through the polymer transfer reaction, and larger polymer molecules have a better chance of being attacked by a polymer radical. Therefore, (1) larger molecules grow faster than the smaller molecules in the present reaction system. Note that the frequency of long-chain branching is higher than that of chain scission, $C_b > C_s$ in the present reaction system. (2) A CSTR produces broader molecular weight distribution (MWD) than that for a PFR, as shown in Figure 2. In a CSTR, the polymer molecules whose residence time is large tend to grow significantly.

Now, consider the case with L = 1. In this case, large polymer molecules are formed in the first big CSTR. After that all the polymers flow through a series of small CSTRs, which resembles a PFR. According to Item (1), larger polymer molecules grow faster in a PFR, and very large polymer molecules can be formed for the case with L = 1.

On the other hand, for the case with the last big tank, L = N, the first part is a PFR. With a PFR, the MWD surely becomes broader during polymerization, but not very much compared with a CSTR. In the final big CSTR, the MWD becomes broader, however, large polymer molecules formed before entering the last big CSTR do not necessarily stay long time in this CSTR and may not grow significantly. Therefore, the last CSTR is not very effective to make larger molecules even larger. As a result, $\bar{P}_w$ is small for the case with the last big tank, L = N.

Branching Density

Figure 14 shows the MC simulation results for the relationship between the branching density ρ and the degree of polymerization P for N = 10. Figure 14a–c are for condition C1, and Figure 14d–f are for condition C3. The MC simulation results for N = 5 can be found in ref. [7], and the fundamental characteristics are the same as the present simulation results for N = 10. In Figure 14, each dot represents ρ and P of each polymer molecule simulated, and the solid blue curve with circular symbols shows the average branching density of polymers having degree of polymerization P, $\bar{\rho}$ (P). The dashed black line shows the average branching density of the whole system $\bar{\rho}$, which is $\bar{\rho}$=0.003 for all cases shown in Figure 14. As in the cases of other branched polymer systems, the value of $\bar{\rho}$ (P) increases with P but reaches a constant limiting value $\rho_{P\to\infty}$, and $\rho_{P\to\infty}$ is larger than the average branching density of the whole system, $\bar{\rho}$. Because the $\bar{\rho}$ -value, shown by the broken line is the same, one notices that $\rho_{P\to\infty}$ becomes larger as the value of L increases, i.e., as the big tank moves backward.

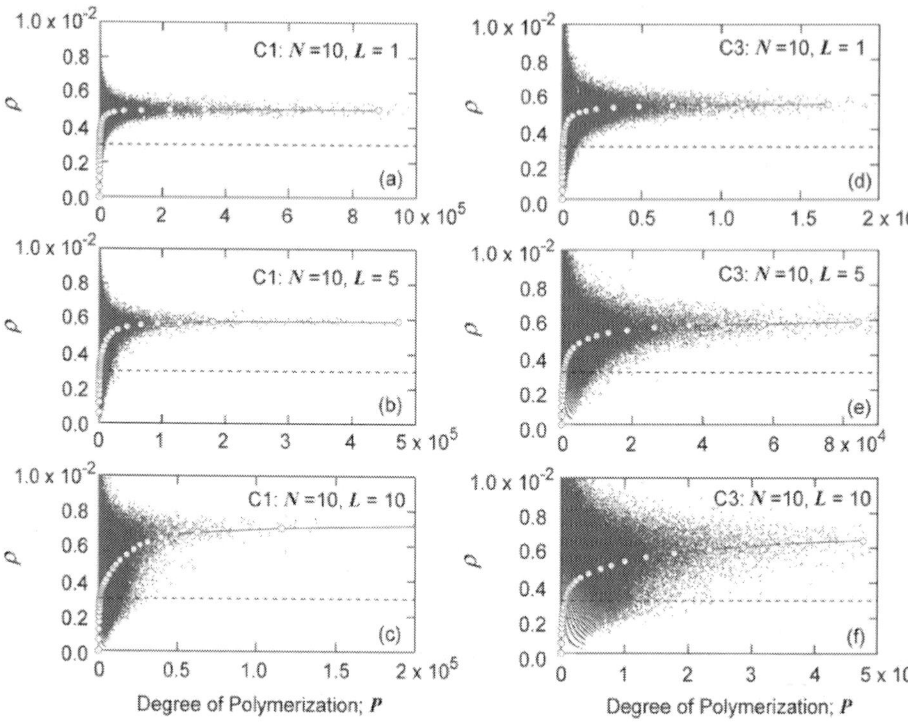

Figure 14: Relationship between branching density ρ and degree of polymerization P, obtained from the MC simulation for N = 10: (**a**) C1, L = 1; (**b**) C1, L = 5; (**c**) C1, L = 10; (**d**) C3, L = 1; (**e**) C3, L = 5; and (**f**) C3, L = 10. The black broken line shows the average branching density of the whole reaction system, $\bar{\rho}$=0.003 for all cases.

As shown in Figure 12, the weight-average DP, $\bar{P}_w$ becomes smaller as the value of L increases. According to the present series of simulation results, $\rho_{P\to\infty}$ is larger for the cases that gives smaller $\bar{P}_w$. This is a completely opposite result, compared with the relationship between a PFR and a CSTR, as shown in Figure 2 and Figure 3. A CSTR gives larger $\rho_{P\to\infty}$ and larger $\bar{P}_w$, compared with a PFR. Now, think of the qualitative explanation for this seemingly strange conflict.

First, consider the $\bar{\rho}$ (P) curves of a CSTR and a PFR shown in Figure 3c. A larger branching density for a given degree of polymerization (P) simply means that the average chain length of the primary chains that make up a branch polymer molecule having P is smaller for a CSTR. In free-radical polymerization, each primary chain is formed within a very small time interval, and therefore, one can conveniently define the instantaneous chain length distribution [8,9]. The number-average chain length of a linear polymer radical formed through propagation starting from the time of radical generation until the dead chain formation is given by [9]:

$$\bar{P}'_{n,\text{linear}} = \frac{R_p}{R_{td} + R_f + R_{tc} + R_{fp}} = \frac{1}{\tau + \beta + R_{fp}/R_p} \tag{11}$$

In a PFR, the final term, R_{fp}/R_p is 0 at x = 0 because the polymer concentration is 0, and increases to the value at the final conversion, x_f. For a CSTR, R_{fp}/R_p is a constant, and is given by:

$$\frac{R_{fp}}{R_p} = C_{fp}\frac{x_f}{1-x_f} \tag{12}$$

where C_{fp} is the polymer transfer constant, and is equal to $C_{fp} = C_b + C_s$. Equation (12) represents the maximum value for a PFR reached at x_f.

The primary chain length is smaller for a CSTR. A larger number of branch points are required to form a branched polymer molecule having a given degree of polymerization, P. This is the reason for the larger $\bar{\rho}$ (P) curve of a CSTR, compared with a PFR, shown in Figure 3c.

Next, consider the CSTRs in series, with L = 1 and L = N. As shown in Figure 10, about one half of polymer by weight is formed in the large tank, and therefore, the L-th CSTR has the most important effect on the formed branched structure. The value of R_{fp}/R_p is $C_{fp}x_1/(1-x_1)$ for L = 1, and is $C_{fp}x_N/(1-x_N)$ for L = N. Because $x_N > x_1$, R_{fp}/R_p is larger for L = N, leading to give smaller primary chains for L = N. Therefore, a larger number of branch points are required to form branched polymers having a given P. The $\rho_{P\to\infty}$ is larger for L = N that gives smaller $\bar{P}_w$.

Radius of Gyration

Figure 15 shows the MC simulation results for the relationship between the mean-square radius of gyration $<s^2>_0$ and degree of polymerization P of the polymer molecules formed in various reactor placements with N = 10. Similar figures for N = 5 can be found in ref. [7]. Each dot in Figure 15 represents $<s^2>_0$ and P-value of each polymer molecule simulated, and the blue curve with circular symbols shows the average $<s^2>_0$ of polymers having DP, P. The black curve shows the radius of gyration of the random branched polymers given by the Zimm-Stockmayer equation [16], i.e., Equation (1). The radius of gyration is smaller than that for random branched polymers, but the degree of deviation becomes smaller as the value of Lincreases.

As was shown in Figure 4b, much more compact polymers are formed in a CSTR. In a CSTR, the primary polymer molecules whose residence time is large are expected to possess a larger number of branch points. These primary chains tend to form a core region of a star-like structure. This is the reason for forming compact branched polymers, especially for large polymers. On the other hand, Figure 4a shows that more randomly branched structure is tend to be formed in a PFR.

In the CSTRs in series with L = 1, compact star-like polymers are formed in the first big CSTR. After the first tank, branches are attached to the star-like polymers rather randomly, and the star-like structure is preserved, leading to form polymers with smaller $<s^2>_0$, compared with that for random branched polymers.

Now, consider the case with the last big tank, L = N. In the first stage of N-1 small tanks, the reactor characteristics are more like a PFR, and polymers with relatively random branched structure are expected to be formed. In the final big tank, the branched polymer molecules whose residence time is large will connect many branch chains. However, looking from an entering branched polymer molecule, the branch points formed on the polymer molecule would be distributed rather randomly, and a core region will not be formed. This is the reason for obtaining polymers having relatively randomly branched polymers.

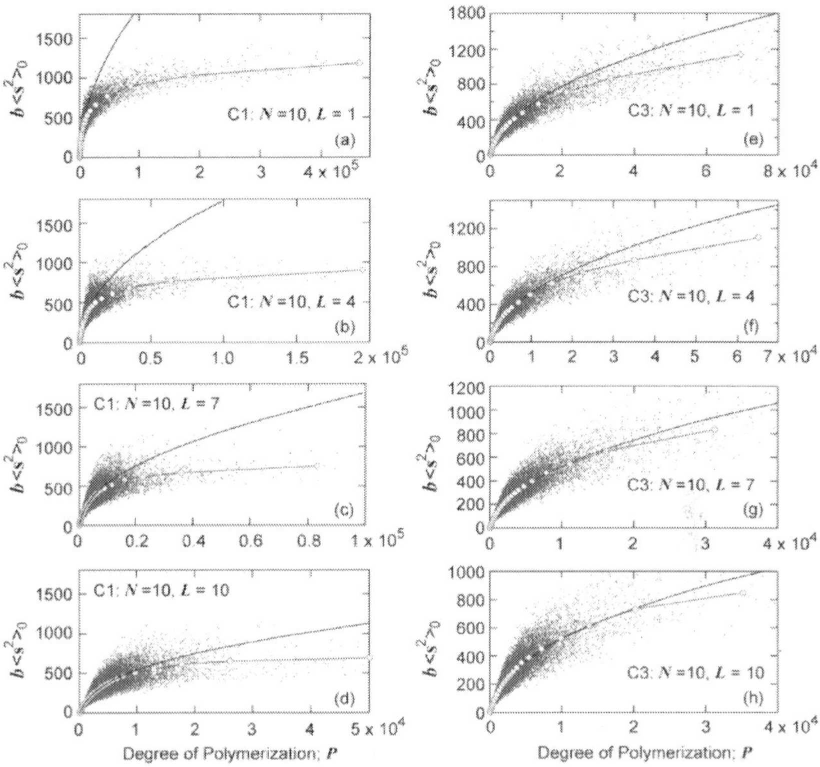

Figure 15: Relationship between the mean-square radius of gyration $<s^2>_0$ and degree of polymerization P of the polymer molecules formed in various reactor placements with N = 10: **(a)** C1, L = 1; **(b)** C1, L = 4; **(c)** C1, L = 7; **(d)** C1, L = 10; **(e)** C3, L = 1; **(f)** C3, L = 4; **(g)** C3, L = 7; and **(h)** C3, L = 10. The black curve shows the relationship for random branched polymers calculated from Equation (1).

With L = 1, the branching density, $\rho_{P\to\infty}$ is small, as shown in Figure 14a, but much more compact polymers are formed, compared with the random branched polymers, as shown in Figure 15a. On the other hand, with L = N, the branching density, $\rho_{P\to\infty}$ is large, as shown in Figure 14c, but the branched structure is more random and $<s^2>_0$ is large for the given branching density level, as shown in Figure 15d. As a result, the $<s^2>_0$-value for the given P becomes essentially the same, irrespective of the tank arrangement, as shown in Figure 16. Similarly as reported in reference [7], there seem to exist small differences for C1, but the differences are very small for C3.

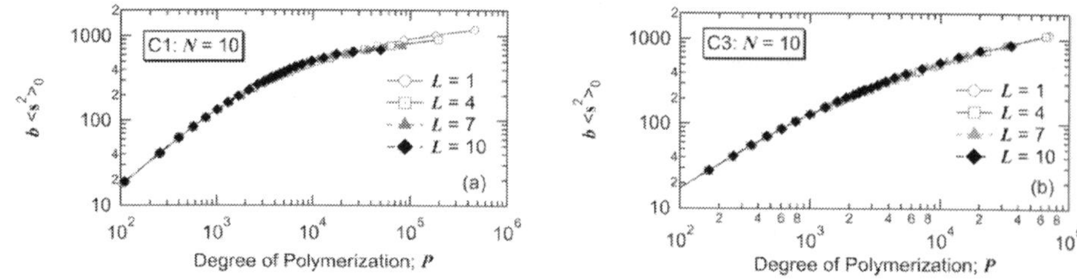

Figure 16: Relationship between the mean-square radius of gyration $<s^2>_0$ and degree of polymerization P of the polymer molecules formed with N = 10: **(a)** C1 and **(b)** C3.

CONCLUSIONS

In this article, a series of continuous stirred-tank reactors, consisting of one big tank and the same N-1 small tanks is investigated to design a new reactor system for free-radical polymerization with simultaneous long-chain branching and minor contribution of scission. Table 3 summarizes the results obtained in this study. When the final conversion is set to be the same, the present reactor configuration does not change the final average branching and scission densities, irrespective of the tank arrangement. The residence time distribution does not change with the tank arrangement, and therefore, the differences are not caused by the residence time distribution, but by the different history of branched polymer formation.

Table 3: N CSTRs in series with one large tank, when the final conversion, x_N, is set to be the same

Large Tank Number, $L = 1...N$	
Final average branching density	Unchanged
Final average scission density	Unchanged
Residence time distribution	Unchanged
Weight-average degree of polymerization, $\bar{P}_w$	>
Branching density for large polymers, $\rho_{P\to\infty}$	<
Branched structure	Star-like ↔ Random
Relationship between $<s^2>_0$ and P	Essentially the same

The weight-average DP, $\bar{P}_w$ is the largest for the first big tank case L = 1, and decreases as the big tank moves backward, i.e., as L increases.

The branching density of large polymers, $\rho_{P\to\infty}$ is the smallest for L = 1, and becomes larger as L increases. With L = 1, polymers with star-like globular structure are formed. By increasing L, the polymer structure changes to a more randomly branched structure.

With L = 1, the branching density of large polymers ($\rho_{P\to\infty}$) is small, while the structure is star-like and $<s^2>_0$-value is small for the given branching density. On the other hand, with L = N, the branching density of large polymers ($\rho_{P\to\infty}$) is large, while the structure is more like randomly branched and $<s^2>_0$-value is large for the given branching density level. What is interesting is that the relationship between $<s^2>_0$ and P is essentially unchanged by the tank arrangement.

The present investigation shows that by changing the tank arrangement, various types of branched polymers, from star-like globular structure to a more randomly branched structure, can be obtained, while keeping the followings are the same: (1) final conversion; (2) final average branching and scission densities; and (3) the relationship between $<s^2>_0$ and P. In the case of a tower-type multi-zone reactor, the branched structure could be controlled by changing the location of division plates. On the basis of the present investigation, it is expected that the first big zone case will tend to produce polymers with star-like globular structure, while the case with the last big zone arrangement will tend to form more randomly branched polymers.

Conventionally, most mathematical models have been used to describe the behavior in the reactor that has been applied already in industry. The simulation has been used rather passively. The present type of positive use of simulation may lead to propose novel reactor systems. Table 4 highlights the characteristics of the positive model use adopted in this article.

Table 4: Positive model use proposed in this article

Passive Model Use	Positive Model Use
Aiming at improving the process already used in industry	Aiming at creating a new process
Emphasizing the quantitative agreement with the experimental data	Emphasizing the novelty of the proposed processes
Straightforward cause-and-effect logic	Divergence through try-and-error and convergence through reasoning
Initiative by chemists	Initiative by process engineers

ACKNOWLEDGMENTS

I would like to express my sincere thanks to Professor Masoud Soroush, Drexel University, for kind invitation to this special issue.

REFERENCES

1. Ehrlich, P.; Mortimer, G.A. Fundamentals of the free-radical polymerization of ethylene. Adv. Polym. Sci. 1970, 7, 386–448.
2. Goto, S.; Yamamoto, K.; Furui, S.; Sugimoto, M. Computer model for commercial high-pressure polyethylene reactor based on elementary reaction rates obtained experimentally. J. Appl. Polym. Sci. Appl. Polym. Symp. 1981, 36, 21–40.
3. Giudici, R.; Hamielec, A.E. A simulation study on random scission of branched chains. Polym. React. Eng. 1996, 4, 73–101.
4. Tobita, H. Random degradation of branched polymers. 2. multiple branches. Macromolecules 1996, 29, 3010–3021.
5. Iedema, P.D.; Wulkow, M.; Hoefsloot, C.J. Modeling molecular weight and degree of branching distribution of low-density polyethylene. Macromolecules 2000, 33, 7173–7184.
6. Iedema, P.D. Predicting MWD and branching distribution of terminally branched polymers undergoing random scission. Macromol. Theory Simul. 2012, 21, 166–186.
7. Tobita, H. Continuous tanks-in-series process for free-radical polymerization with long-chain branching and scission: Effect of the order of a large tank. Macromol. React. Eng. 2015.
8. Tobita, H. Random sampling technique to predict the molecular weight distribution in nonlinear polymerization. Macromol. Theory Simul. 1996, 5, 1167–1194.
9. Tobita, H. Polymerization processes, 1. Fundamentals. In Ullmann's Encyclopedia of Industrial Chemistry; Wiley-VCH: Weinheim, Germany, 2015.
10. Tobita, H. Simultaneous long-chain branching and random scission: 1. Monte Carlo simulation. J. Polym. Sci. Part B: Polym. Phys. 2001, 39, 391–403.

11. Tobita, H.; Kawai, H. Simulation of size exclusion chromatography for branched polymers formed by simultaneous long-chain branching and random scission. E-Polymers 2002, 2, 688–700.
12. Tobita, H. Free-radical polymerization with long-chain branching and scission in a continuous stirred-tank reactor. Macromol. React. Eng. 2013, 7, 181–192.
13. Read, D.J.; Auhl, D.; Das, C.; den Doelder, J.; Kapnistos, M.; Vittorias, I.; McLeish, T.C.B. Linking models of polymerization and dynamics to predict branched polymer structure and flow. Science 2011, 333, 1871–1874. [PubMed]
14. Levenspiel, O. Chemical Reaction Engineering; John Wiley & Sons: New York, NY, USA, 1972.
15. Tobita, H. Markovian approach to self-condensing vinyl polymerization: Distributions of molecular weights, degrees of branching, and molecular dimensions. Macromol. Theory Simul. 2015, 24, 117–132.
16. Zimm, B.H.; Stockmayer, W.H. The dimensions of chain molecules containing branches and rings. J. Chem. Phys. 1949, 17, 1301–1314.
17. Tobita, H. Dimensions of branched polymers formed in simultaneous long-chain branching and random scission. J. Polym. Sci.: Part B: Polym. Phys. 2001, 39, 2960–2968.
18. Tobita, H. Continuous free-radical polymerization with long-chain branching and scission in a tanks-in-series model. Macromol. Theory Simul. 2014, 23, 182–197.
19. Brandao, A.L.; Soares, J.B.P.; Pinto, J.C.; Alberton, A.L. When Polymer reaction engineers play dice: Application of monte carlo models in PRE. Macrtomol. React. Eng. 2015, 9, 141–185.
20. Tobita, H. Molecular weight distribution in random branching of polymer chains. Macromol. Theory Simul. 1996, 5, 129–144.
21. Tobita, H.; Zhu, S. Statistical crosslinking of heterochains. Polymer 1997, 38, 5431–5439.
22. Tobita, H. General matrix formula for the weight-average molecular weights of crosslinked polymer systems. J. Polym. Sci.: Part B: Polym. Phys. 1998, 36, 2423–2433.

23. Tobita, H. Markovian approach to nonlinear polymer formation: Free-radical crosslinking copolymerization. Macromol. Theory Simul. 1998, 7, 675–684.
24. Tobita, H. Free-radical polymerization with long-chain branching and scission: Markovian solution of the weigh-average molecular weight. Macromol. Theory Simul. 2014, 23, 477–489.
25. Tobita, H. Markovian approach to free-radical polymerization with simultaneous long-chain branching and scission: Effect of branching and scission kinetics. Macromol. React. Eng. 2015, 9, 245–258.

CHAPTER 9

DETERMINING THE EFFECT OF SIDE REACTIONS ON PRODUCT DISTRIBUTIONS IN RAFT POLYMERIZATION BY MALDI-TOF MS†

Antonio Veloso , Wendy García , Amaia Agirre , Nicholas Ballard , Fernando Ruipérez , José C. de la Cal and José M. Asua

POLYMAT, University of the Basque Country UPV/EHU, Joxe Mari Korta R&D Ctr, Avda. Tolosa-72, Donostia-San Sebastián 20018, Spain

ABSTRACT

Reversible Addition–Fragmentation chain Transfer (RAFT) polymerization has emerged as one of the most versatile reversible deactivation radical polymerization techniques and is capable of polymerizing a wide range of monomers under various conditions. One of the most important factors governing the success of a RAFT polymerization is the fraction of living chains at the end of the reaction, which can be maximized by using a low amount of initiator. From the point of view of the process, it is tempting to perform the polymerization in solution, which allows a better mixing. However, in this work it is shown that this choice may be negative for the quality of the polymer. Detailed analysis using Matrix Assisted Laser Desorption Ionization Time of Flight Mass Spectrometry (MALDI-TOF MS) of poly(n-butyl acrylate) (pBA) obtained at high conversion in the RAFT solution polymerization revealed that in addition to the polymer chains, formed by the RAFT mechanism, there were two distinct populations resulting from chain transfer to solvent and transfer to polymer followed by β-scission. Complementary results from Size Exclusion Chromatography coupled with Multi Angle Light Scattering detector (SEC/MALS), quantum chemical calculations, and a mathematical model that predicts product distributions, were also used to further confirm the assigned structures. The results highlight the scope and limitation on the living fraction of chains due to chain transfer events using RAFT polymerization and reversible deactivation radical polymerizations in general, and furthermore, yielded information about the fate of midchain radicals formed by intramolecular transfer to polymer.

INTRODUCTION

The large commercial demand for acrylic polymers has prompted research toward improving processes for the polymerization of acrylates. Most poly(acrylates) are produced by free radical polymerization that, in comparison to anionic and cationic polymerizations, provides poor control of the polymerization.[1] Reversible deactivation radical polymerization techniques[2–8]allow a much better control of this microstructure. Most of these methods utilize a dynamic equilibrium between the growing radicals and dormant species. Among these techniques, reversible-addition–fragmentation chain transfer (RAFT) polymerization is attractive because it has the advantage of being compatible with a wide range of monomers,[9] notably including acidic monomers, and it can be performed in an extensive range of solvents and under a broad range of conditions. In addition, recent work has highlighted the potential of RAFT polymerization for the synthesis of multiblock copolymers due to the high living fraction of polymer chains that can be obtained by exploiting the degenerative transfer mechanism of the polymerization.[10,11] In RAFT polymerization, thiocarbonylthio compounds with generic formula R–S–(C=S)–Z (referred to as RAFT agents) are employed as highly active chain transfer agents which allow for growth of all chains throughout the polymerization.

Standard free radical initiators are used to generate radicals which react with the thiocarbonylthio compound to generate an intermediate radical, as can be seen in Scheme 1. This intermediate radical then fragments via a β-scission process which results in reformation of the thiocarbonylthio group at the dormant chain end and an active, propagating radical. Through this reversible addition–fragmentation process an equilibrium between dormant and active chains is established and control over molecular weight and end group functionality can be achieved.

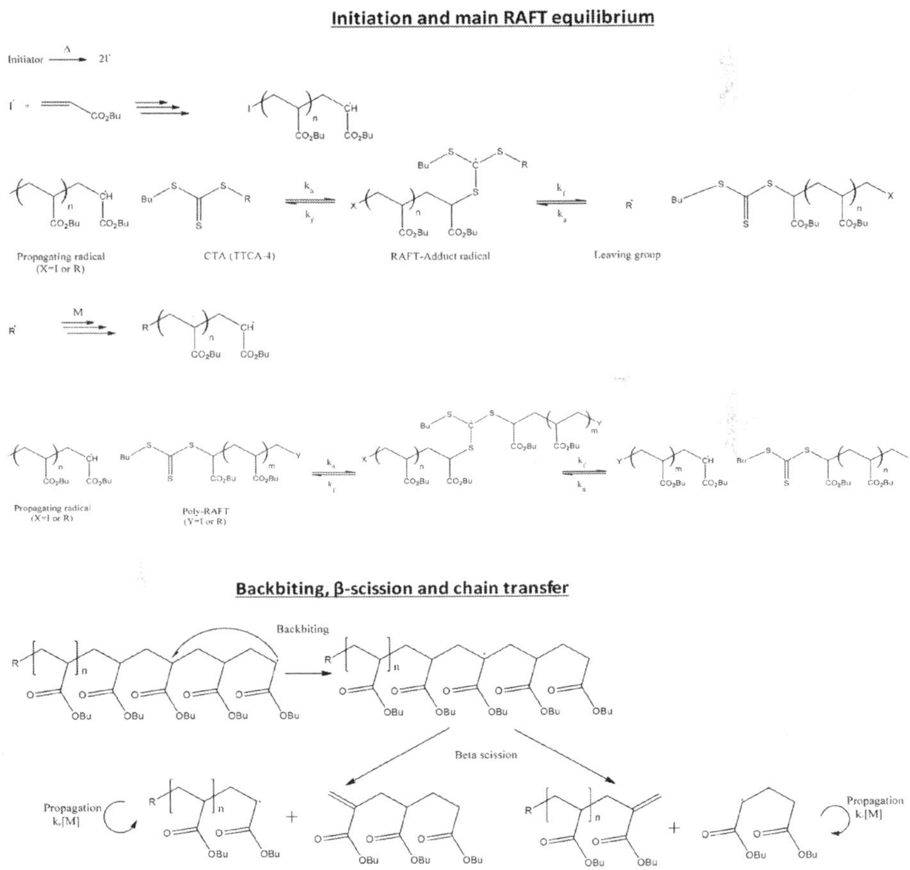

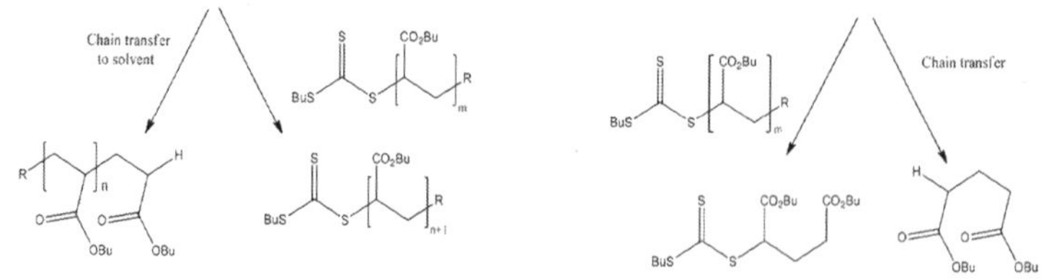

Scheme 1: Mechanism of the RAFT polymerization of pBA using trithiocarbonate as CTA (e.g. TTCA-4) and products formed from of backbiting and β-scission.

The factors governing the success of a RAFT polymerization are the fraction of living chains at the end of the reaction and the achievement of a narrow molecular weight distribution (MWD). Under conditions where termination reactions lead to a significant number of dead chains, the MWD can be narrowed and the fraction of living chains increased by simply lowering the [initiator]/[RAFT agent] ratio.

The choice of RAFT agent and initiator can also play a key role in the control of the polymers structure and can avoid undesired products which can result from reactions of the RAFT intermediate radicals.[8,9,12–15]

From the point of view of the process, it is beneficial to perform the reaction in solution, particularly for higher T_g polymers, which allows for a better mixing during the reaction. However, solution polymerization may present two major drawbacks in comparison to bulk polymerization. Firstly, it increases the possibility of chain transfer to solvent, which increases the fraction of dead chains. Secondly, in solution polymerization of acrylic monomers, the lower monomer concentration promotes intramolecular transfer to polymer (backbiting)[16,17] and this results in a large fraction of tertiary radicals which can subsequently react to form branch points, or undergo β-scission to yield macromonomers, thus impacting on the kinetics, molecular weight distribution and end group functionality of the polymerization (see Scheme 1).[18–20]

In order to ascertain the extent to which the product distribution is affected by the side reactions influenced by the process conditions, a comprehensive analysis of the product mixture is needed. Matrix assisted laser desorption ionization time of flight mass spectrometry (MALDI-TOF MS)[21–24] is the most promising technique for such analysis.

Despite the widespread use of acrylic polymers in RAFT polymerization and the importance associated with maintaining the living nature of the polymeric RAFT agents, relatively little has been published regarding the detailed characterization of pBA-RAFT polymers by MALDI-TOF MS[25,26] especially at higher conversion, which is the region of interest for most applications. In addition, most work to date has focused on the reactivity of the RAFT intermediate radical and potential side products arising from various modes of termination. In the RAFT polymerization of BA using cumyl dithiobenzoate Venkatesh et al.[26] used MALDI-TOF as a tool in order to explore the effects of termination of the RAFT intermediate during monomer-free, model experiments and observed the formation of 3- and 4-arm star structures. Similarly, Ah Toy et al.[14] studied the RAFT polymerization of methyl acrylate and observed significant number of species arising from oxidation of the RAFT agent as well as two armed stars which were thought to arise by disproportionation of the RAFT intermediate radical. Chirowodza et al.[27] synthesized polymer/clay hybrid materials by grafting polymer chains from the surface of LAPONITE® clay via surface-initiated RAFT polymerization of BA, and the grafted polymer chains were analyzed using MALDI-TOF MS. The end-group structures of the free and grafted polymers were studied. Their work expands the application range for MALDI-TOF MS, as it shows that the analysis of polymers attached to a solid surface via electrostatic interactions is possible without detaching them.

In order to gain a greater insight into the effects of chain transfer reactions and intermediate radicals on the product distribution in RAFT polymerization of acrylates, in this work we study in detail poly(n-butyl acrylate) (pBA) produced using a trithiocarbonate RAFT agent in solution and in bulk by means of MALDI-TOF MS combined with other techniques including Nuclear Magnetic Resonance (NMR) and Size Exclusion Chromatography coupled with a Multi Angle Light Scattering detector (SEC/MALS). The results are analyzed by means of a detailed mathematical model, highlighting the importance of the process conditions on the fraction of living chains as well as on the individual molecular weight distributions of the distinct species formed in the polymerization.

To begin, we present results for the RAFT polymerization of butyl acrylate in ethanol, a solvent that is particularly useful in synthesis of multiblock polymers made out of monomers of widely different polarity such as BA and acrylic acid (AA)[28–30] whose effectiveness is hindered by chain transfer to solvent. After showing the huge impact that transfer reactions can have on the living fraction of chains we subsequently present results under conditions where transfer reactions are considered to have less impact using the same RAFT agent, 2-[(butylsulfanyl) carbonothioyl sulfanyl] propanoic acid (TCCA-4). Solution polymerization are conducted at 20% monomer content in benzene, where solvent transfer is low but transfer to polymer and β-scission reactions remain significant, at 70% in dioxane, as a typical reaction medium, and in bulk where both transfer to polymer and transfer to solvent are low.

MATERIALS AND METHODS

Chemicals

Butyl acrylate (BA, Quimidroga), 4,4′-azobis(4-cyanopentanoic acid) (V-501, Fluka), and 2,2′-azobis(2-methylbutyronitrile) (AMBN, Sigma-Aldrich) were used as received. Ethanol, dioxane, benzene (Scharlab) and HPLC grade tetrahydrofuran (THF, Scharlab) were used as solvents. 2-[(butylsulfanyl) carbonothioyl sulfanyl] propanoic acid (TTCA-4) was prepared according to Ferguson et al.[31] MALDI matrices: α-cyano-4-hydroxycinnamic acid (CHCA, Aldrich) and trans-2-[3-(4-tert-butylphenyl)-2-methyl-2-propenylidene]malononitrile (DCTB, Aldrich). Cationization agents: sodium iodide (NaI, Aldrich), sodium trifluoroacetate (NaTFA, Aldrich) and potassium trifluoroacetate (KTFA, Aldrich).

Synthesis of pBA by RAFT polymerization

Solution and bulk RAFT polymerizations were carried out using TTCA-4 as RAFT agent. Scheme 1 shows the reaction of the synthesis. The solution polymerization was carried out by mixing BA, TTCA-4 and V-501 in different solvents at molar ratios 15 : 1 : 0.1 and total solids content of 20%. Ethanol, dioxane and benzene were used as solvents. The solution was placed in a 0.10 L glass bottle, sealed and purged for 10 min under N_2 atmosphere. The reaction was carried out by placing the bottle in a rotating device and immersed in a thermostatic bath at 60 °C by 24 h. Bulk polymerization was performed at 60 °C in a round bottomed flask on a scale of 2 g to alleviate the possibility of any exotherm. The molar ratio of reactants and the reaction temperature was the same as the solution polymerizations but the initiator was replaced with AMBN due to low solubility of V-501 in butyl acrylate. The resulting homopolymer (pBA) was dried in a vacuum oven for 40 °C in order to remove all the residual monomer and the solvent. Then, it was analyzed by MALDI-TOF MS, NMR and SEC/MALS. The theoretical molar mass was calculated according to:

$$\bar{M}_{n,calculated} = \left(X \frac{[BA]}{[RAFT]} M_{wBA} \right) + M_{wRAFT}$$

where [BA] and [RAFT] are the initial monomer and RAFT agent concentrations, M_{wBA} and M_{wRAFT} are the molar masses of butyl acrylate and RAFT agent, and X is the monomer conversion determined gravimetrically.

Size exclusion chromatography (SEC) coupled with a multi angle light scattering (MALS) and refractive index (RI) detectors

The average molar mass of the pBA was analyzed by SEC/MALS. The equipment was composed by a LC20 pump (Shimadzu) coupled to a miniDAWN Treos multiangle (3 angles) light scattering laser photometer equipped with an He–Ne laser (λ = 658 nm), an Optilab Rex differential refractometer (λ = 658 nm) (all from Wyatt Technology Corp., USA). Separation was carried out using three columns in series (Styragel HR6, Styragel HR4 and Styragel HR2, with pore sizes of 10^6, 10^4 and 10^2 Å, respectively). Filtered toluene (HPLC-grade from Sigma-Aldrich) was used for the calibration of the 90° angle scattering intensity. The detectors at angles other than 90° in the MALS instrument were normalized to the 90° detector using a standard (PS 28 770 g mol^{-1}, Polymer Labs), which is small enough to produce isotropic scattering, at a flow rate of THF through the detectors of 1 mL min^{-1}. In addition, the same standard and conditions were used to perform the alignment (interdetector delay volume) between concentration and light scattering detectors and the band broadening correction for the sample dilution between detectors.

The analysis was performed at 35 °C and THF was used as mobile phase at a flow rate of 1 mL min^{-1}. The dried polymer was diluted in HPLC grade THF at concentrations of about 4 mg mL^{-1} and then was injected into the equipment. The dn dc^{-1} used for the molar mass calculation was the one corresponding to pBA (dn dc^{-1} = 0.064 mL g^{-1}).[32] Due to the low molecular weight of the polymers synthesized herein, this dn dc^{-1} value will be an approximation of the true value and may result in some error in the calculation of molecular weights. The SEC/MALS data was analyzed by using the ASTRA software version 6.0.6. (Wyatt Technologies, USA). The absolute molar mass was calculated from the MALS/RI data using the Debye plot (with 1st order Zimm formalism).

Mass spectrometry

MALDI-TOF MS measurements were performed on a Bruker Autoflex Speed system (Bruker, Germany) instrument equipped with a 355 nm Nd:YAG laser. All spectra were acquired in the positive-ion reflectron mode (accelerating voltage 20 kV, pressure 5×10^{-6} mbar). Samples were prepared with CHCA and DCTB. CHCA matrix was dissolved at concentration of 10 g L^{-1} and DCTB at 40 g L^{-1} in THF.[26] NaI, NaTFA and/or KTFA were added as cationic ionization agents (approximately 10 g L^{-1} dissolved in THF). The matrix, salt and polymer solutions were premixed in the ratio 10 : 1 : 10 (matrix : salt : sample).

Density functional theory calculations

All the density functional theory calculations (DFT) have been carried out with the Gaussian 09[33] suite of programs using the M06-2X[34] functional. Structure optimizations were performed in gas phase by using the 6-31+G(d) basis set[35] and harmonic vibrational frequencies were obtained, at the same level of theory, by analytical differentiation of the gradients in order to determine whether the structures were minima or transition states. These frequencies were then used to evaluate the zero-point vibrational energy (ZPVE) and the thermal corrections, at T = 298 K, to the enthalpy and Gibbs free energy in the harmonic oscillator approximation. The electronic energy was refined by single-point calculations using the 6-311++G(d,p) basis set.[36]

Nuclear magnetic resonance

The degree of polymerization (DP_{BA}) was determined by ^{1}H-NMR spectra recorded on a Bruker nuclear magnetic resonance instrument (400 MHz Spectrometer, Germany) using $CDCl_3$ as the solvent and tetramethylsilane (TMS) as internal standard. Molar mass was calculated according to:

$$\bar{M}_{n,NMR} = DP_{BA} M_{wBA} + M_{wRAFT}$$

RESULTS AND DISCUSSION

In order to demonstrate the effect of transfer to solvent on the livingness and molecular weight distribution of polymers produced by RAFT we first synthesized a low molecular weight pBA polymer in ethanol, a known transferring solvent which has been used in the synthesis of multiblock BA/AA copolymers.[28–30] The synthesis of pBA by RAFT polymerization in ethanol solution proceeded to 98% conversion. Table 1 presents the number and weight-average molar masses $(\bar{M}_n, \bar{M}_w)$, the dispersity (Đ) and the polymerization degree (DP_n) obtained by the different techniques (MALDI-TOF MS, SEC/MALS and ^{1}H-NMR) together with the theoretical values. It should be noted that, as expected from RAFT polymerization, a low dispersity was achieved and the degree of polymerization is also in good agreement with the theoretical value. For MALDI-TOF MS, the species with the highest ionization efficiency observed in the spectra were considered for molar mass calculations. The slightly higher molecular weight values obtained by SEC/MALS are related to the low-light scattering signal at high elution times (small molecular weights). It is worth noting that in SEC/MALS chromatography, in order to accurately calculate the molar mass, the signal of the detectors should be sufficient in the whole elution range, and at small elution volumes SEC/MALS signal is more sensitive than refractive index (RI), and the opposite happens at large elution times. A consequence is that dispersity is likely underestimated.

Table 1: Number and weight-average molar masses ($\bar{M}_n$, $\bar{M}_w$), dispersity (Đ) and polymerization degree (DP_n) obtained by different techniques (MALDI-TOF MS, SEC/MALS and ^{1}H-NMR) together with the theoretical values for the pBA synthesized in ethanol solution (98% conversion)

Label	$\bar{M}_n$ (g mol^{-1})	$\bar{M}_w$ (g mol^{-1})	*Đ*	DP_n
Theoretical	2102	—	—	15
SEC/MALS	2900	3000	1.025	21
MALDI	2393	2509	1.048	17
^{1}H-NMR	2289	—	—	16

Taking into account the different possible reactions of the radicals showed in Table 1S (ESI†), 18 different structures were considered for the assignment of the peaks in the spectra (Fig. 1). Moreover, for each structure, three possible adducts (proton, $[M + H]^+$; sodium, $[M + Na]^+$, $[M + 2Na–H]^+$;[37] and potassium, $[M + K]^+$) were considered, therefore, more than 100 possibilities were evaluated in this work.

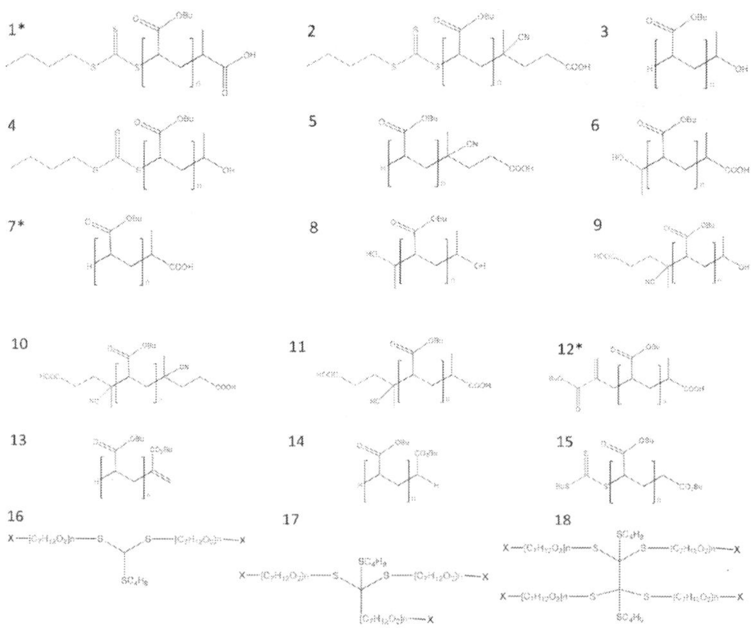

Figure 1: Possible structures derived from the reaction of n-pBA using TTCA-4 as RAFT agent, and initiated with V-501 in ethanol. Structures from **1** to **15** are formed due to the RAFT equilibrium, bimolecular termination by combination

or chain transfer to solvent, and 2-, 3- and 4-arm star structures (from **16** to **18**) (being X the initiator radical and/or R group) that result from termination of the intermediate RAFT radicals with various radical species. These 2-, 3- and 4-arm star structures have been reported in the literature[26] using dithiobenzoate as RAFT agent. * Detected structures **1**, **7** and **12**.

The MALDI-TOF mass spectra of pBA produced by solution polymerization in ethanol using DCTB as MALDI matrix and NaTFA salt as cationization agent[26] are shown in Fig. 2. A CHCA matrix and NaI salt was also used, but better resolution of the spectral peaks was obtained with the aforementioned system. The mass spectrum of this sample consists of numerous groups of peaks (or series of peaks). Fig. 2 presents these groups of peaks in enlargements made at different mass ranges (A) 1500–1580, (B) 1880–1960, and (C) 2520–2600 Da. It can be seen that although the relative intensity of the peaks varies with the mass range, the main peaks appeared in all ranges. It is observed that the molar mass difference between two consecutive peaks of the same kind, e.g., between two consecutive S1 peaks, is 128.08 Da, which corresponds to the mass of one BA monomer unit. Peaks S1 and S3 were assigned to structure **1** in Fig. 1, which is formed in the main equilibrium of the RAFT agent and contains an R group and a RAFT group at either end of the chain. The difference is that S1 corresponds to $[M + Na]^+$ adduct (M being the polymer) and S3 to the $[M + 2Na–H]^+$ adduct. The formation of these adducts has been also reported by Arakawa et al.[37] for acrylic acid polymers. A good agreement between the theoretical and experimental isotopic distribution was achieved (Fig. 3).

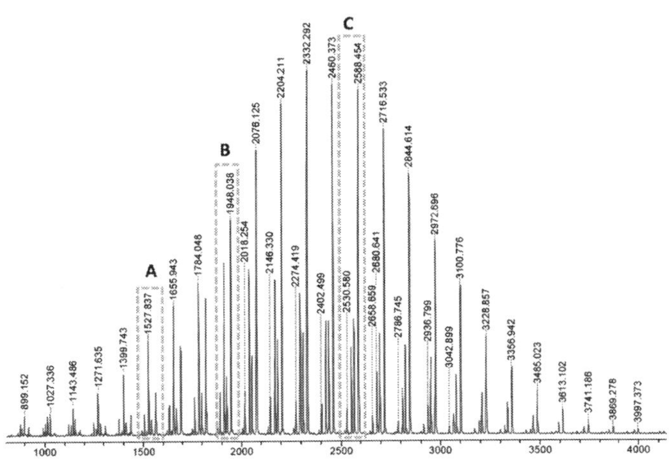

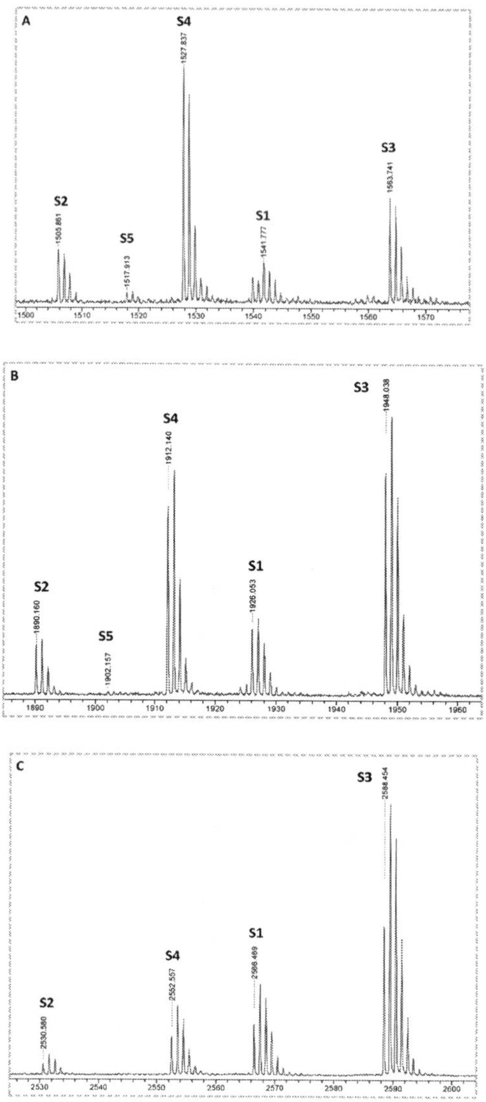

Figure 2: MALDI-TOF mass spectra of RAFT polymerization of BA in ethanol solution (98% conversion) using DCTB matrix and NaTFA cationization agent. Complete spectra in 700–4000 Da mass range (left). Enlargements (right) in different mass ranges (A) 1500–1580, (B) 1880–1960, and (C) 2520–2600 Da show the different detected series from S1 to S5. See Table 2S (ESI†) for details.

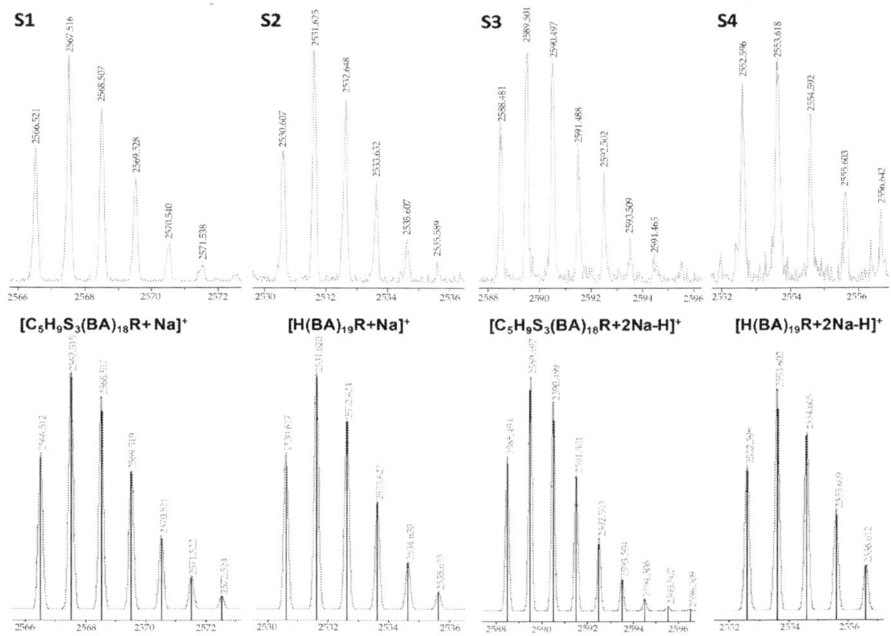

Figure 3: MALDI-TOF mass spectra in reflectron mode of pBA on the basis of the isotopic mass distribution: observed (above) and theoretical (below). The series S1 and S3 were assigned to structure **1** from Fig. 1, and for S2 and S4 were assigned to species **7**from Fig. 1 due to chain transfer to solvent. S1 and S2 correspond to sodiated species $[C_5H_9S_3–(C_7H_{12}O_2)_{18}–C_3H_5O_2 + Na]^+$ (S1) and $[H–(C_7H_{12}O_2)_{19}–C_3H_5O_2 + Na]^+$ (S2), while S3 and S4 form different adduct $[C_5H_9S_3–(C_7H_{12}O_2)_{18}–C_3H_5O_2 + 2Na–H]^+$ (S3) and $[H–(C_7H_{12}O_2)_{19}–C_3H_5O_2 + 2Na–H]^+$ (S4).[37]

The S2 and S4 series were attributed to a hydrogen terminated product containing an R terminal group, labeled species **7** in Fig. 1, with $[M + Na]^+$ and $[M + 2Na–H]^+$ adducts, respectively. The structure can be further confirmed by the good agreement between the theoretical isotopic distribution and the experimental one (Fig. 3). In order to explain species **7**, two mechanisms may be proposed: (a) fragmentation of the polymer during the ionization process inside the spectrometer, which could allow the generated radical $\dot{P}_m$ to take a proton from the acidic medium of the matrix preparation; and (b) proton transfer from the solvent to the radical species $\dot{P}_m$.

Characterizing by MALDI and electrospray ionization (ESI) a poly(methyl methacrylate) synthesized by RAFT polymerization, Jiang et al.[38] demonstrated that labile end-groups, such as the dithioester groups, can be lost in MALDI-TOF MS experiments during the ionization process. In order to investigate if this may occur in the present case, a study of the energetics of the process using density functional theory (DFT) calculations was performed. This consists of breaking different bonds and reacting with the protons of the medium. The calculations showed that the C–S bond cleavage during MALDI ionization with the laser energy is feasible since the BuS(C=S) $\dot{S}$ group may be detached from the polymer during the desorption process, forming a new carbanion ($\Delta H = 51.94$ kcal mol^{-1}) which is able to remove a proton from the matrix to form the hydrogenated species ($\Delta H = -188.19$ kcal mol^{-1}). The whole process is thermodynamically favorable ($\Delta H = -129.32$ kcal mol^{-1}). This point should not be underestimated since in many types of reversible deactivation radical polymerization, the control agent is relatively labile and during MALDI ionization may cause the presence of hydrogen terminated products.[39] However, if fragmentation of the polymer during the ionization process were the cause of the presence of species **7**, the ratio (S1 + S3)/(S2 + S4) should be constant along the MWD because fragmentation during MALDI ionization is proportional to the number of C–S bonds, normally one per chain. Therefore, the fragmentation should be the same for all lengths. However, Fig. 2 shows that this is not the case and the ratio (S1 + S3)/(S2 + S4) increased with the molecular weight. This suggests that chain transfer to solvent is responsible for the hydrogen terminated species (S2 + S4).

The explained results about the end-groups S1, S2, S3 and S4 were confirmed by the use of other MALDI matrices and cationization agents: CHCA/NaI, DCTB/NaTFA and DCTB/KTFA (Fig. 1S, ESI†). Venkatesh et al[26] selected the DCTB/NaTFA system to characterize pBA polymers synthesized using cumyl dithiobenzoate as RAFT agent and, by generating polymeric radicals using ATRP type chemistry in the presence of RAFT agent, were able to detect 2-, 3- and 4-arm star structures due to the relatively high stability of the intermediate radical. In our case, after comparing the mass spectral peaks obtained using the same matrix system (Fig. 2) and using CHCA and NaI (not shown), it could be confirmed that none of these arm star structures was present, so the trithiocarbonate type RAFT agent used here seems to be more selective and appropriate to avoid undesired products resulting from termination of the RAFT intermediate radical and additionally to avoid oxidation of the RAFT agent.

Fig. 2 shows the presence of small amounts of species S5 that can only be observed at moderate molecular weights. This species was attributed to the product of intramolecular transfer to polymer, followed by β-scission (Scheme 1) leading to a terminally unsaturated species, labeled structure **12** in Fig. 1. In order to verify the backbiting reaction, the low masses were analyzed and the small fragments arising from these reactions were detected as proton and potassium adducts as KTFA salt was used (Fig. 4). Both the low molecular weight radical formed by β-scission in one direction (observed as the corresponding radical having terminated by disproportionation) and the low molecular weight macromonomer formed by scission in the other direction were observed. Interestingly, no higher molecular weight analogues of this compound were observed.

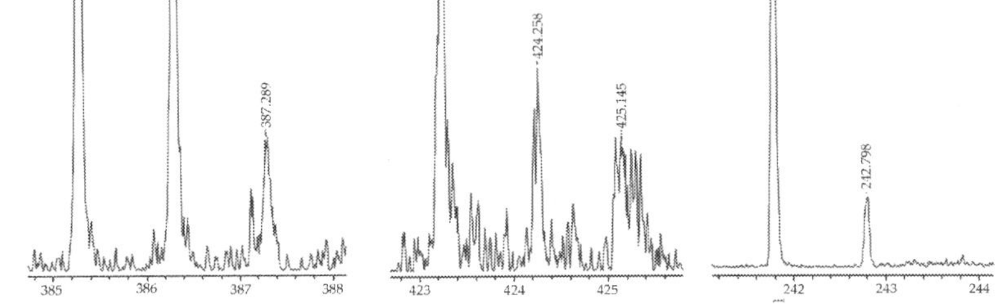

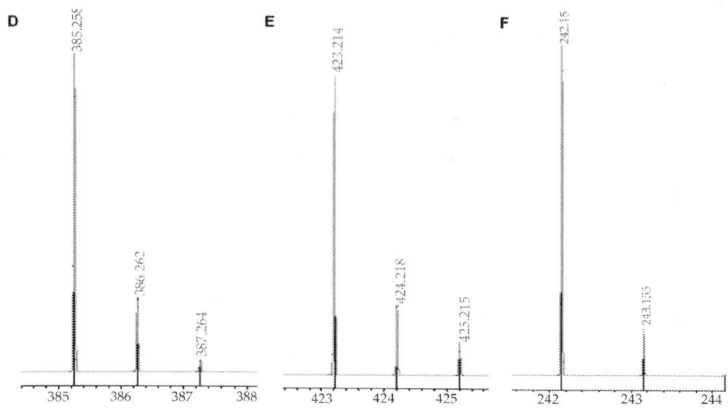

Figure 4: Enlargements of MALDI-TOF mass spectra in reflectron mode of two generated molecules after backbiting and β-scission reactions (Scheme 1), and the corresponding adducts (blue, upper trace). (A) and (B) represent proton and potassium adducts of $C_{21}H_{36}O_6$ (M_1, m/z = 384.25 Da) and (C) represents the $C_{13}H_{22}O_4$ structure without forming any adduct (M_2, m/z = 242.15 Da). Comparison of each adduct with theoretical isotopic mass distribution is also shown (green, below) (D, E and F).

It has previously been shown that migration of the midchain radical can result in a size selective molecular weight distribution[39,40] of this macromonomer in monomer free experiments designed for macromonomer synthesis. The fact that this cannot be seen in this case would suggest that the migration process is negligible in the presence of competing propagation reactions. This result will impact on expected distributions of short and long chain branches in synthesis of acrylic polymers and would suggest that only short arm branches (of two monomer units) would be expected under standard reaction conditions.

In order to determine whether the proposed structures were compatible with the accepted mechanisms for their formation we constructed a mathematical model that takes into account propagation, activation and deactivation reactions, chain transfer to solvent, backbiting and β-scission, and is capable of yielding the complete molecular weight distribution of each individual species (see ESI† for details). Due to the known experimental observations of a reduction in products arising from intramolecular transfer to polymer both in the presence of hydrogen bonding solvents[41,42] and under conditions of reversible deactivation,[43,44]the rate constant for backbiting obtained from radical polymerization in hydrocarbon solvents was reduced accordingly (see ESI†). The cause of the reduction in the rate coefficient under these conditions is still uncertain and therefore the amount by which it is reduced was estimated from previous literature data. Fig. 5 shows a comparison between the spectra generated from the mathematical model (above) and the most intense series of the MALDI spectra (below) representing the peaks related to series S1 (+S3), S2 (+S4) and S5. Both spectra show similar MWD for each series in the same mass range. The profile of the series S2 is also similar to the experimental one, and displays the experimental trend of a shift to lower masses compared to S1. The shift to lower masses is to be expected if the peak is the result of chain transfer to solvent because this event results in a dead chain whose mass can no longer increase while the main chain distribution will continue to extend throughout the polymerization due to the living nature of the reaction. This is in contrast to what one would expect if the peak was simply due to loss of the labile RAFT group upon MALDI ionization. The comparison of these spectra reinforce the idea that a significant amount of chain transfer to solvent takes during RAFT polymerization of pBA under these conditions. This also highlights that hydrogen terminated species which result from ionization, can be distinguished from products resulting from chain transfer simply by the looking at the relative mass distributions.[45,46]

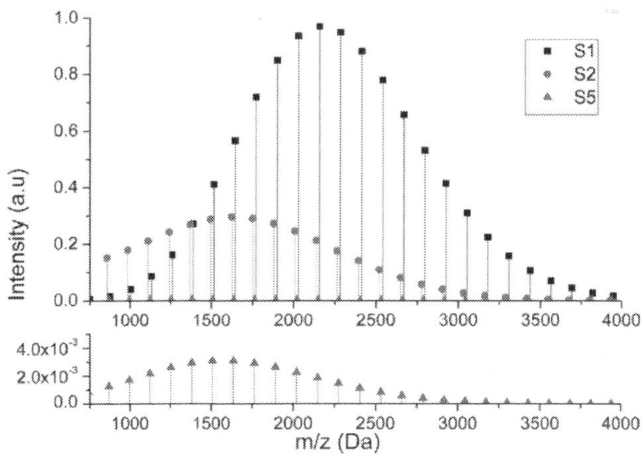

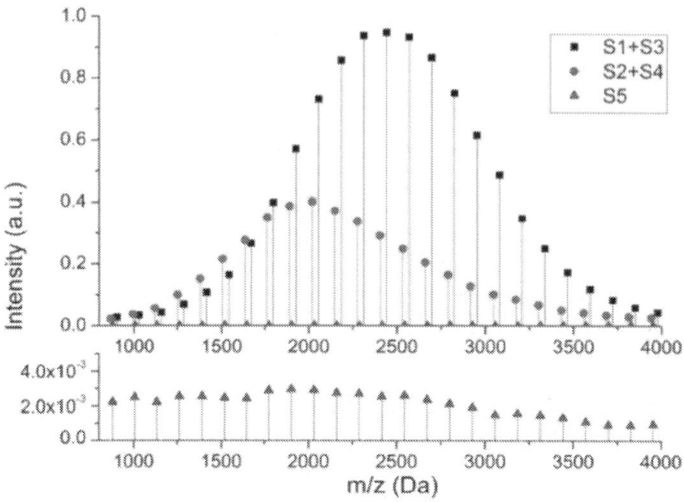

Figure 5: Comparison between the spectra generated from mathematical model (above) and the most intense series of the MALDI spectra (below) for RAFT polymerization of BA in solution of ethanol with TCCA-4 and V-501, representing the three different detected species. In MALDI spectrum the experimental peaks represented as black squares and red circles are the sum of the two adducts for each one. Below each spectrum an enlargement to see S5 behaviour.

Similarly, good agreement between the model and experimental data for the molecular weight distribution for the product of β-scission, S5, can be observed with a greater intensity at lower molecular weights. This higher intensity at lower molecular weights is due to the decrease in molecular weight with respect to the main chain caused by the β-scission reaction, which fragments the polymer, creating a lower molecular weight macromonomer. Although this macromonomer is technically a reactive species,[47] due to the lower rate constant for propagation and (mainly) the low concentration the probability of this molecule reacting further is low. Thus, while the main chain distribution is constantly growing, any macromonomer formed tends to remain at lower molecular weight. This tendency becomes more obvious as target chain length increases since the chains grow to a greater extent.

One important conclusion from this analysis is that even at moderate temperatures a significant number of side products can arise that can affect not only the molecular weight distribution but also the livingness of the polymerization. This is particularly important for obtaining high molecular weight polymers and for recently developed processes for synthesis of multiblock copolymers in one-pot systems where conversion is taken to high values between each sequential monomer addition stage.[48–50] Formation of multiblocks is only possible if after each extension step (which typically proceeds to high conversion) a high living fraction of chains remains. In RAFT it has been assumed that a high living fraction can be achieved simply by using a low [Initiator]/[RAFT] ratio[51,52] but the present results highlight that a substantial fraction of dead chains can arise from alternative reactions.

In order to quantify this effect, the fraction of living chains with RAFT groups attached can be calculated by summing the concentration chains formed from the RAFT agent and of new chains created by transfer to solvent $[C_s]$, β-scission $[C_\beta]$ and the concentration of terminated chains according to

$$L = \frac{[\mathrm{RAFT}]}{[\mathrm{RAFT}] + [\mathrm{C_S}] + [\mathrm{C}_\beta] + 2f[\mathrm{I}]_0(1 - e^{-k_d t})\left(1 - \frac{\delta}{2}\right)}$$

where δ is the fraction of total chains terminated that terminate by combination as opposed to disproportionation. The concentration of dead chains formed by chain transfer to solvent is given by

$$[\mathrm{C_S}] = -\mathrm{C_{ct}}[\mathrm{S}]\ln(1 - X)$$

and from β scission

$$[\mathrm{C_\beta}] = \frac{K_{\mathrm{bb}}}{1 + K_{\mathrm{p3}}K_{\mathrm{bb}}}\ln\left(1 + \frac{1 + K_{\mathrm{p3}}K_{\mathrm{bb}}}{1 + K_{\mathrm{p3}}[\mathrm{M}]_0 + K_{\mathrm{p3}}K_{\mathrm{bb}}}\frac{X}{1 - X}\right)$$

where $C_{ct} = k_{ct}/k_p$, [S] is the solvent concentration, $K_{bb} = k_{bb}/k_p$, $K_{p3} = k_{p3}/k_\beta$, and k_{ct}, k_p, k_{bb}, k_{p3}, k_β are the rate coefficients of chain transfer to solvent, propagation, backbiting, propagation of the tertiary radical and β-scission respectively. The derivation of these equations along with that to calculate chain transfer to monomer is given in the ESI.† From this equation it can be seen that if any chain transfer process generates a significant number of chains with respect to the initial RAFT agent concentration then the living fraction of chains will be affected and it follows that the number average degree of polymerization will decrease compared to the ideal case. This is therefore particularly important at high target degrees of polymerization (low $[RAFT]_0$) where small numbers of chains generated will affect the living fraction to a greater extent. This effect applies equally to RAFT as to other reversible deactivation radical polymerization techniques such as NMP and ATRP.

Fig. 6 shows the fraction of dead chains accumulated as a result of chain transfer to solvent for the solution polymerization of butyl acrylate with varying solvents and from the fraction of dead chains arising from β scission reactions of midchain radicals at various temperatures calculated using the parameters given in Table 3S of the ESI.† It can be seen that the process of chain transfer to solvent is less sensitive to changes in conversion than β-scission as may be predicted from the strong effect of monomer concentration on formation of tertiary radicals, and the subsequent scission reaction. With regards to chain transfer to solvent, it can be seen that the solvent choice impacts hugely on the livingness of the polymerization and that even for the short chain lengths obtained with the $[M]_0/[RAFT]_0$ ratio used in the simulations, a solvent with very low chain transfer rate coefficient must be used in order to obtain a polymer with high living fraction. For chains arising from β-scission of midchain radicals a large effect of temperature on the number of new chains created can be observed. This is expected due to the high activation energies of backbiting and particularly β-scission and highlights the need to polymerize at low temperatures to avoiding the degradative effects of β-scission.

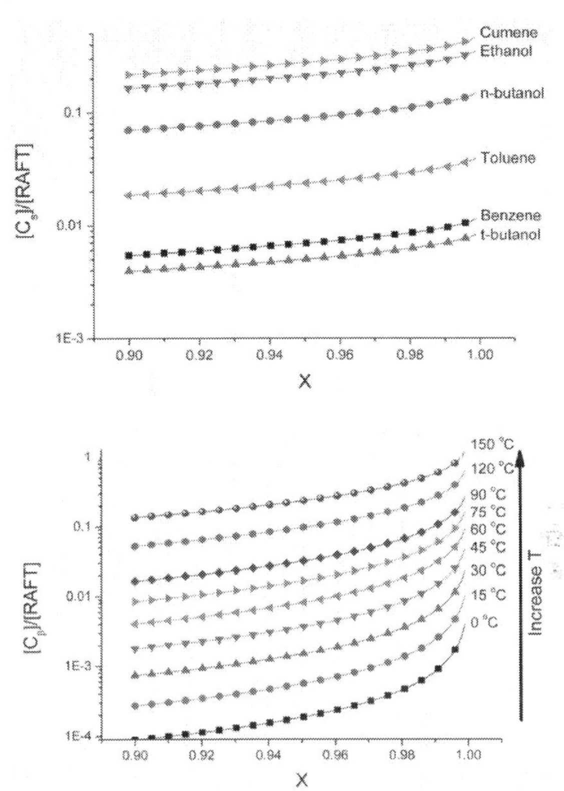

Figure 6: (Above) Modelled effect of solvent and conversion on fraction of dead chains formed by chain transfer to solvent compared to living chains $[C_s]/[RAFT]_0$ for polymerization of butyl acrylate at 60 °C ($[M]_0 = 1.2$ M, ($[RAFT]_0 = 0.08$ M). Values for chain transfer constants were obtained from literature ($C_{ct} = k_{ct}/k_p = 0.17 \times 10^{-4}$ (t-butanol), 0.22×10^{-4} (benzene), 0.9×10^{-4}(toluene), 2.91×10^{-4} (n-butanol), 4.38×10^{-4} (ethanol), 13.8×10^{-4} (cumene), 3.9×10^{-4} (dioxane)).[53,54] (Below) Modelled effect of temperature and conversion on fraction of dead chains formed by β-scission compared to living chains $[C_s]/[RAFT]_0$ for polymerization of butyl acrylate ($[M]_0 = 1.2$ M, ($[RAFT]_0 = 0.08$ M). No adjustment was made for any potential reduction of the rate of backbiting under conditions of reversible deactivation or with hydrogen bonding solvents.

These results emphasize the importance of solvent choice and temperature on the potential for multiblock polymers. To date, such syntheses have only been shown to be successful in solvents with low transfer to solvent rate coefficients such as dimethyl sulfoxide (DMSO) and water. Unfortunately, we could not find any rate coefficients to demonstrate the effect of DMSO in Fig. 6. Furthermore, synthesis of such multiblock copolymers consisting of acrylic monomers, which are highly influenced by chain transfer to polymer events, have been shown to be successful only at low temperatures where backbiting and subsequent scission of the midchain radical can be sufficiently neglected. In this respect the success of the synthesis of a given multiblock copolymer is highly subject to the polymerization conditions.

The results above show that significant deviation from the synthetic conditions that have been reported in the literature (low temperature, low C_{ct} solvents) and the use of monomers which have a greater tendency to undergo chain transfer will result in high fraction of dead chains and limit the potential for multiple chain extension steps. It is also important to note that these chain transfer events will lead to a lower living fraction of polymer chains, and that, for multiblock copolymers, the cumulative effect of chain transfer for each additional block formed will limit the synthetic scope of the procedure if conditions are not selected with great care.

In order to further highlight the importance of avoiding chain transfer and its effects on product distributions, a series of RAFT polymerizations were carried out under conditions where transfer reactions are less favoured. A reaction was conducted at 20 wt% in benzene using identical conditions to those employed for the solution polymerization in ethanol above and led to slightly lower conversion (93%). Fig. 2S† shows the MALDI spectrum for this reaction in which it can be seen that due to the lower conversion, macromonomer formation (species S5) is less prevalent. The use of benzene prevented the formation of any dead chains due to transfer to solvent due to the extremely low transfer constant, so no species S2 was detected. Subsequently, a polymerization was carried out under more "standard" RAFT conditions in dioxane at 70 wt% (MALDI also included in Fig. 2S, ESI†). In this case, chains arising from both transfer to solvent (species S2) and β-scission (species S5) could be observed although at lower concentration than those of the reaction conducted in ethanol due to the higher monomer concentration and lower tendency for dioxane to undergo transfer. Although, in both cases the dead chains were more prevalent at low molecular weight in agreement with the other experiments conducted herein, in dioxane 70 wt% there is higher conversion value and therefore, more β-scission species (S5) are detected. It should also be noted that in both cases the major chain stopping event is not termination. This highlights the importance of both transfer to solvent and transfer to polymer followed by β-scission, which must be considered when accounting for the living fraction of polymers produced by reversible deactivation polymerizations. The MWD values obtained using MALDI and SEC/MALS and the chromatograms are shown in Table 4S and Fig. 3S of the ESI.†

Finally, a reaction was carried out under bulk conditions where both transfer to monomer and intramolecular transfer to polymer are expected to be low. The molar ratio of the reactants was the same as for solution polymerization, but AMBN was used due to the low solubility of V-501 in butyl acrylate. Fig. 7 shows a comparison between the spectra from pBA created in ethanol solution and in bulk. In the complete spectra (left) we can observe fewer peaks in the bulk reaction. In the enlargements of the spectra (right), it is clear that the most intense peaks are due to the RAFT main mechanism, corresponding to species S1 and S3. The species S2 and S4 disappear totally and the intensity of the structure S5 diminishes drastically as expected since transfer to polymer is largely averted and essentially no hydrogen terminated product, which could be formed due to either termination by disproportionation or unavoidable effects of transfer to monomer could be observed.

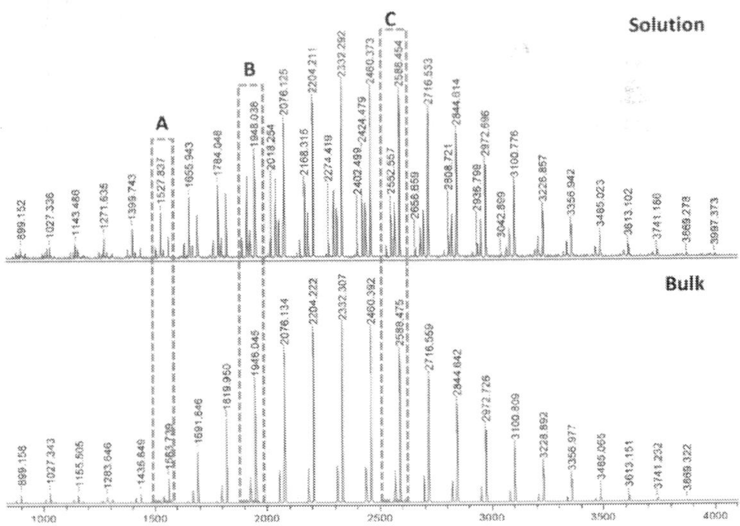

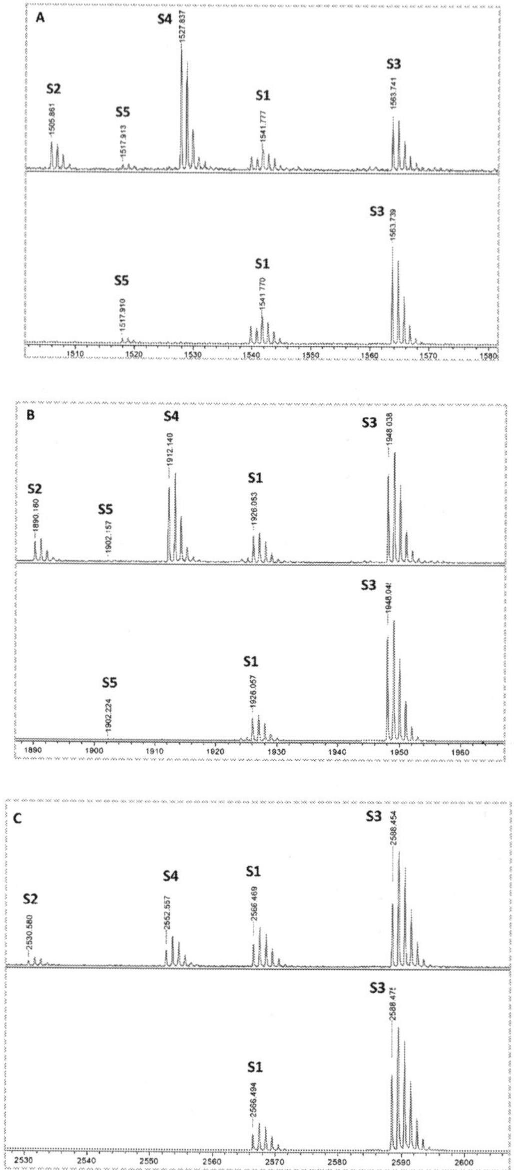

Figure 7: Comparison between MALDI-TOF spectra of the same RAFT reaction in ethanol solvent (98% conversion) and in bulk (99% conversion). Complete spectra in 800–4000 Da mass range (left). Enlargements (right) in different mass ranges (A) 1500–1580, (B) 1880–1960, and (C) 2520–2600 Da show the different detected series from S1 to S5.

In order to confirm that the decrease in hydrogen terminated products arising from chain transfer to solvent (monomer in this case) and terminal double bonds resulting from β-scission of midchain radicals is directly related to the polymerization conditions the theoretical mass distribution was calculated for this reaction using the mathematical model (see equations from ESI†). Fig. 8 shows a comparison between the spectra generated from the mathematical model (above) and the most intense series of the MALDI spectra (below) representing the peaks related to series S1 and S5. In this case, comparison between Fig. 5and 8 clearly shows that in bulk polymerization, the amount of species due to chain transfer disappear totally. This confirms that transfer to monomer does not have any substantial effect on the molcecular weight distribution and confirms that the choice of solvent is critical if high living fraction of chains is necessary. This also further demonstrates that S2 and S4 species were created due to transfer to solvent and not due to the rupture of C–S bond during MALDI ionization process. One point of interest is that the experimental concentration of the S5 species is higher in the case of bulk polymerization than in the solution polymerization. It would be expected that due to the solution polymerization being conducted at lower monomer concentration the reverse would be the case. However, there are several additional factors that this supposition does not take into account. Firstly, from an experimental point of view, although the reaction was performed under conditions designed to minimize any effect arising from the exothermic nature of the reaction in bulk, local temperature rises are inevitable and can lead to increased rate of both backbiting and β scission relative to propagation. Secondly, as the solution polymerization is conducted in ethanol, hydrogen bonding may result in a decrease in the amount of backbiting relative to the bulk polymerization.[41,42] Finally, the potential of the midchain radical to undergo chain transfer to solvent for the reaction conducted in ethanol, which has not been considered herein, may also lead to a reduction in the products arising from the midchain radical as has been demonstrated to be important for reactions carried out in the presence high concentrations of chain transfer agent.[55,56]

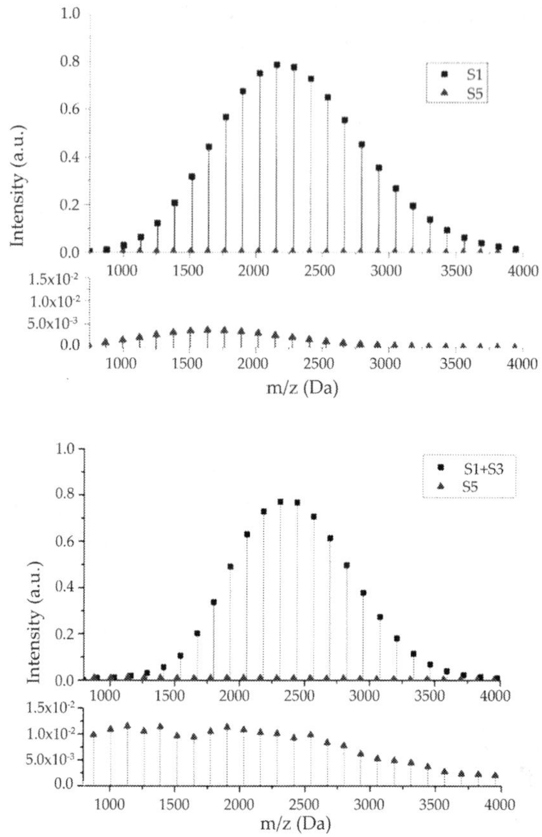

Figure 8: Comparison between the spectra generated from mathematical model (above) and the most intense series of the MALDI spectra (below) for bulk RAFT polymerization butyl acrylate using TCCA-4, representing the three different detected species. In MALDI spectrum the experimental peaks represented as black squares are the sum of the two adducts. In order to see behaviour of S5 species, the enlargements are shown below.

CONCLUSIONS

In conclusion, through the synthesis of low molar mass poly(butyl acrylate) by RAFT polymerization under varying conditions and detailed analysis of the resulting polymers by MALDI-TOF we have shown the critical effect of chain transfer events on the living fraction of chains at high conversion in reversible deactivation radical polymerizations. Under conditions where chain transfer is heightened (dilute solution polymerization in ethanol) three products with distinct molecular weight distributions were present in the final reaction mixture. The main population corresponded to polymer chains formed according to the RAFT polymerization mechanism described in Scheme 1. The second population corresponds to dead chains arising from chain transfer to solvent and the third species results from intramolecular transfer to polymer followed by fragmentation of the polymer chain by β-scission. No evidence of detrimental effects of the intermediate RAFT radical leading to 2-, 3-, or 4-arm star products which have been detected by other groups was found. In contrast, polymerization in conditions at which transfer events do not significantly affect the reaction (bulk polymerization at 60 °C) resulted in almost exclusively a single product formed according to the main RAFT equilibrium. Chain transfer to solvent was shown to result in a significant number of dead chains across a wide range of conversion thus limiting the solvent choice for synthesis of multi-block copolymer and high molecular weight polymers where high chain end fidelity is required. Furthermore, from analysis of the products of the β-scission reaction it can be concluded that intramolecular transfer to polymer occurs almost exclusively via a six membered ring transition state and that migration of the midchain radical along the polymer backbone did not occur under the present reaction conditions.

ABBREVIATIONS

MALDI	Matrix assisted laser desorption ionization
TOF	Time of flight
MS	Mass spectrometry
MWD	Molar mass distribution
SEC/MALS	Size exclusion chromatography/multi angle light scattering
BA	Butyl acrylate
CRP	Controlled/living polymerization
RAFT	Reversible addition–fragmentation chain transfer
ATRP	Atom transfer radical polymerization
NMP	Nitroxide-mediated controlled free-radical polymerization
ITP	Iodine transfer polymerization
RITP	Reverse iodine transfer polymerization
V-501	4,4′-Azobis(4-cyanopentanoic acid)

THF	Tetrahydrofuran
CHCA	α-Cyano-4-hydroxycinnamic acid
TTCA-4	2-[(Butylsulfanyl) carbonothioyl sulfanyl] propanoic acid
DMSO	Dimethyl sulfoxide
AMBN	2,2′-Azobis(2-methylbutyronitrile
NaI	Sodium iodide
KTFA	Potassium trifluoroacetate
SA	Sinapinic acid
MBT	2-Mercaptobenzothiazole
DHB	2,5-Dihydroxy benzoic acid
DCTB	trans-2-[3-(4-tert-Butylphenyl)-2-methyl-2-propenylidene]malononitrile.

ACKNOWLEDGEMENTS

Diputación Foral de Gipuzkoa, University of Basque Country (UFI 11/56), Basque Government (GVIT373-10), and Ministerio de Economía y Competitividad (CTQ2011-25572) and POLYMAT Industrial Liaison Program (Akzo Nobel, Allnex, Arkema, BASF, Foresa, Nuplex Resins, Solvay, Stahl, Synthomer, Vinavil, and Wacker) are gratefully acknowledged for their financial support. The SGI/IZO-SGIker UPV/EHU is gratefully acknowledged for generous allocation of computational resources.

NOTES AND REFERENCES

1. K. Matyjaszewski, Y. Nakagawa and C. Jasieczek, Macromolecules, 1998, 31, 1535–1541 .
2. C. J. Hawker, Acc. Chem. Res., 1997, 30, 373–382 .
3. J. Chiefari, Y. Chong, F. Ercole, J. Krstina, J. Jeffery, T. Le, R. Mayadunne, G. Meijs, C. Moad, G. Moad, E. Rizzardo and S. Thang, Macromolecules, 1998, 31, 5559–5562 .
4. G. Moad, J. Chiefari, Y. Chong, J. Krstina, R. Mayadunne, A. Postma, E. Rizzardo and S. Thang, Polym. Int., 2000, 49, 993–1001 .
5. M. Dietrich, M. Glassner, T. Gruendling, C. Schmid, J. Falkenhagen and C. Barner-Kowollik, Polym. Chem., 2010, 1, 634–644 .
6. K. Matyjaszewski, Macromolecules, 2012, 45, 4015–4039 .

7. E. Rizzardo and D. H. Solomon, Aust. J. Chem., 2012, 65, 945–969 .
8. D. Keddie, G. Moad, E. Rizzardo and S. Thang, Macromolecules, 2012, 45, 5321–5342 .
9. C. Barner-Kollowik, Handbook of RAFT Polymerization, WILEY-VCH Verlag GmbH & Co. KGaA, Weinheim, 2008, pp. 5–49 .
10. G. Gody, T. Maschmeyer, P. B. Zetterlund and S. Perrier, Nat. Commun., 2013, 4, 2505–2514 .
11. P. B. Zetterlund, G. Gody and S. Perrier, Macromol. Theory Simul., 2014, 23, 331–339 PubMed .
12. K. Matyjazewski, Controlled and Living Radical Polymerizations: Progress in ATRP, NMP and RAFT, American Chemical Society, 1st edn, 2000 .
13. S. Perrier, C. Barner-Kowollik, J. F. Quinn, P. Vana and T. P. Davis, Macromolecules, 2002, 35, 8300–8306 .
14. A. Ah Toy, P. Vana, T. P. Davis and C. Barner-Kowollik, Macromolecules, 2004, 37, 744–751 .
15. G. Moad, E. Rizzardo and S. H. Thang, Polymer, 2008, 49, 1079–1131 PubMed .
16. C. Plessis, G. Arzamendi, J. R. Leiza, H. A. S. Schoonbrood, D. Charmot and J. M. Asua, Macromolecules, 2000, 33, 5041–5047 .
17. A. N. Nikitin, R. A. Hutchinson, G. A. Kalfas, J. R. Richards and C. Bruni, Macromol. Theory Simul., 2009, 18, 247–258 .
18. C. Plessis, G. Arzamendi, J. R. Leiza, H. A. S. Schoonbrood, D. Charmot and J. M. Asua, Macromolecules, 2000, 33, 5041–5047 .
19. R. X. E. Willemse, A. M. van Herk, E. Panchenko, T. Junkers and M. Buback, Macromolecules, 2005, 38, 5098–5103 .
20. T. Junkers and C. Barner-Kowollik, J. Polym. Sci., Part A: Polym. Chem., 2008, 46, 7585–7605 .
21. C. Schilli, M. G. Lanzendörfer and A. H. E. Müller, Macromolecules, 2002, 35, 6819–6827 .
22. F. D'Agosto, R. Hughes, M. Charreyre, C. Pichot and R. Gilbert, Macromolecules, 2003, 36, 621–629 .

23. J. Loiseau, N. Doerr, J. Suau, J. Egraz, M. Llauro and C. Ladaviere, Macromolecules, 2003, 36, 3066–3077 .
24. A. Favier, C. Ladaviere, M. Charreyre and C. Pichot, Macromolecules, 2004, 37, 2026–2034 .
25. J. T. Lai, D. Filla and R. Shea, Macromolecules, 2002, 35, 6754–6756 .
26. R. Venkatesh, B. Staal, B. Klumperman and M. Monteiro, Macromolecules, 2004, 37, 7906–7917 .
27. H. Chirowodza, P. Hartmann and H. Pasch, Macromol. Chem. Phys., 2012, 213, 847–857 .
28. M. Jacquin, P. Muller, G. Lizarraga, C. Bauer, H. Cottet and O. Théodoly, Macromolecules, 2007, 40, 2672–2682 .
29. A. Krieg, C. Pietsch, A. Baungaertel, M. D. Hager, C. R. Becer and U. S. Schubert, Polym. Chem., 2010, 1, 1669–1676 .
30. X. Zhang, F. Boisson, O. Colombani, C. Chassenieux and B. Charleux, Macromolecules, 2014, 47, 51–60 .
31. C. Ferguson, R. Hughes, D. Nguyen, B. Pham, R. Gilbert, A. Serelis, C. Such and B. Hawkett, Macromolecules, 2005, 38, 2191–2204 .
32. S. Podzimeck, Light Scattering, Size Exclusion Chromatography and Asymmetric Flow Field Fractionation: Powerful Tools for the Characterization of Polymers, Proteins and Nanoparticles, Wiley, Canada, 2011 .
33. M. J. Frisch, G. W. Trucks, H. B. Schlegel, G. E. Scuseria, M. A. Robb, J. R. Cheeseman, G. Scalmani, V. Barone, B. Mennucci, G. A. Petersson, H. Nakatsuji, M. Caricato, X. Li, H. P. Hratchian, A. F. Izmaylov, J. Bloino, G. Zheng, J. L. Sonnenberg, M. Hada, M. Ehara, K. Toyota, R. Fukuda, J. Hasegawa, M. Ishida, T. Nakajima, Y. Honda, O. Kitao, H. Nakai, T. Vreven, J. A. Montgomery, Jr., J. E. Peralta, F. Ogliaro, M. Bearpark, J. J. Heyd, E. Brothers, K. N. Kudin, V. N. Staroverov, R. Kobayashi, J. Normand, K. Raghavachari, A. Rendell, J. C. Burant, S. S. Iyengar, J. Tomasi, M. Cossi, N. Rega, J. M. Millam, M. Klene, J. E. Knox, J. B. Cross, V. Bakken, C. Adamo, J. Jaramillo, R. Gomperts, R. E. Stratmann, O. Yazyev, A. J. Austin, R. Cammi, C. Pomelli, J. W. Ochterski, R. L. Martin, K. Morokuma, V. G. Zakrzewski, G. A. Voth, P. Salvador, J. J. Dannenberg, S. Dapprich, A. D. Daniels, Ö. Farkas, J. B.

Foresman, J. V. Ortiz, J. Cioslowski and D. J. Fox, Gaussian 09, Revision A.1, Gaussian, Inc., Wallingford CT, 2009 .

34. Y. Zhao and D. G. Truhlar, Theor. Chem. Acc., 2008, 120, 215 .
35. W. Hehre, R. Ditchfield and J. A. Pople, J. Chem. Phys., 1972, 56, 2257 .
36. R. Krishnan, J. S. Binkley, R. Seeger and J. A. Pople, J. Chem. Phys., 1980, 72, 650 .
37. R. Arakawa, S. Egami and S. Okuno, J. Mass Spectrom., 2006, 41, 549–550 .
38. X. Jiang, P. Schoenmakers, J. van Dongen, X. Lou, V. Lima and J. Brokken-Zijp, Anal. Chem., 2003, 75, 5517–5524 .
39. J. Vandenbergh and T. Junkers, Macromolecules, 2012, 45, 6850–6856 .
40. J. Vandenbergh and T. Junkers, Macromolecules, 2013, 46, 3324–3331 .
41. K. Liang, R. A. Hutchinson, J. Barth, S. Samrock and M. Buback, Macromolecules, 2011, 44, 5843–5845 .
42. K. Liang and R. A. Hutchinson, Macromol. Rapid Commun., 2011, 32, 1090–1095 .
43. W. Wang, A. N. Nikitin and R. A. Hutchinson, Macromol. Rapid Commun., 2009, 30, 2022–2027 .
44. N. Ballard, S. Rusconi, E. Akhmatskaya, D. Sokolovski, J. C. de la Cal and J. M. Asua, Macromolecules, 2014, 47, 6580–6590 .
45. M. Bednarek, T. Biedron and P. Kubisa, Macromol. Chem. Phys., 2000, 201, 58–66 .
46. L. Charles, Mass Spectrom. Rev., 2014, 33, 523–543 .
47. A. Postma, T. P. Davis, G. Li, G. Moad and M. S. O'Shea, Macromolecules, 2006, 39, 5307–5318 .
48. A. H. Soeriyadi, C. Boyer, F. Nyström, P. B. Zetterlund and M. R. Whittaker, J. Am. Chem. Soc., 2011, 133, 11128–11131 .
49. A. Anastasaki, C. Waldron, P. Wilson, C. Boyer, P. B. Zetterlund, M. R. Whittaker and D. Haddleton, ACS Macro Lett., 2013, 2, 896–900 .

50. G. Gody, T. Maschmeyer, P. B. Zetterlund and S. Perrier, Macromolecules, 2014, 47, 3451–3460 .
51. G. Gody, T. Maschmeyer, P. B. Zetterlund and S. Perrier, Macromolecules, 2014, 47, 639–649 .
52. J. Vandenbergh and T. Junkers, Macromolecules, 2014, 47, 5051–5059 .
53. J. Brandrup and E. H. Immergut, Polymer Handbook, Wiley-Inteience, New York, 4th edn, 1999 .
54. J. Loiseau, N. Doërr, J. M. Suau, J. B. Egraz, M. F. Llauro and C. Ladavière, Macromolecules, 2003, 36, 3066–3077 .
55. M. Gaborieau, S. P. S. Koo, P. tignolles, T. Junkers and C. Barner-Kowollik, Macromolecules, 2010, 43, 5492–5495 .
56. N. Ballard, J. C. de la Cal and J. M. Asua, Macromolecules, 2015, 48, 987–993 .

CHAPTER 10

COPPER MEDIATED CONTROLLED RADICAL COPOLYMERIZATION OF STYRENE AND 2-ETHYLHEXYL ACRYLATE AND DETERMINATION OF THEIR REACTIVITY RATIOS

Bishnu P. Koiry and Nikhil K. Singha

Rubber Technology Centre, Indian Institute of Technology Kharagpur, Kharagpur, India

ABSTRACT

Copolymerization is an important synthetic tool to prepare polymers with desirable combination of properties which are difficult to achieve from the different homopolymers concerned. This investigation reports the copolymerization of 2-ethylhexyl acrylate (EHA) and styrene using copper bromide (CuBr) as catalyst in combination with N,N,N′,N″,N″-pentamethyldiethylenetriamine (PMDETA) as ligand and 1-phenylethyl bromide (PEBr) as initiator. Linear kinetic plot and linear increase in molecular weights vs. conversion indicate that copolymerization reactions were controlled. The copolymer composition was calculated using ^{1}H NMR studies. The reactivity ratio of styrene and EHA (r_1 and

r_2) were determined using the Finemann–Ross (FR), inverted Finemann–Ross (IFR), and Kelen–Tudos (KT) methods. Thermal properties of the copolymers were also studied by using TGA and DSC analysis.

Keywords: controlled radical polymerization, ATRP, polystyrene, poly(2-ethylhexyl acrylate), copolymers, reactivity ratio, thermal properties

INTRODUCTION

The homopolymer and copolymers of 2-ethylhexyl acrylate (EHA) have very good film formation characteristics and have very good low temperature flexibility, because of the presence of branched and longer alkyl pendant group in the EHA. They have also low volume shrinkage. So the copolymers of EHA are widely used in paints, coating and adhesive applications (Skeist, 1977; Plessis et al., 2001; Webster and Crain, 2002). Copolymerization is an important synthetic tool which can control the thermal and mechanical properties of the polymers (Kavousian et al., 2004). In the copolymers of styrene and EHA, the properties of the copolymers can be monitored by controlled incorporation of the respective comonomers styrene and EHA. Polystyrene possess high glass transition temperature, T_g (~100°C) where as poly(2-ethylhexyl acrylate) (PEHA) has low T_g of −60°C. Copolymers of styrene and EHA are important components for hard coating (Plessis et al., 2001) and also are used as blend compatibilizers (Haldankar, 2001). Conventional free radical polymerization (FRP) leads to uncontrolled molecular weight and broad dispersity (*Đ*). There is also gel formation tendency, because of the several side reactions during FRP. This makes them difficult to apply for paints and coating material application owing to high viscosity (Solomon and Moad, 1995; Odian, 2004). Since 1990s there have been spectacular advances in the field of controlled radical polymerization (CRP). There were several CRP techniques namely, atom transfer radical polymerization (ATRP) (Kamigaito et al., 2001; Matyjaszewski and Jia, 2001; Kavitha and Singha, 2008), nitroxide-mediated polymerization (NMP) (Harth et al., 2001; Hawker et al., 2001), and reversible addition–fragmentation chain transfer (RAFT) (Chiefari et al., 1998; Moad et al., 2000; Moad, 2006; Barner-Kowollik, 2008). Among the different CRP techniques, ATRP is applicable to polymerize wide range of monomers and can be carried out at wide range of temperature (−20 to 200°C) (Haloi et al., 2009). ATRP has been successfully applied to synthesize a wide range of polymers with varied molecular weights, different architectures, functionalities etc. Transition metal catalyzed CRP known as ATRP is an important CRP method (Matyjaszewski, 2000) which is carried out in presence of an active alkyl halide using a transition metal halide as catalyst in combination with a suitable ligand (Matyjaszewski and Jia, 2001). A wide variety of copolymers can be prepared via ATRP with controlled molecular weight, functionality, and low dispersity. There are several reports on the copolymerization of different acrylate monomers using ATRP and determination of their reactivity ratios. The reactivity ratios are important parameters for a set of monomer. The reactivity ratios of monomers predict the copolymer composition as well as the sequence distribution of the comonomers. It also predicts the properties of the copolymer. ATRP provides random copolymer with similar chain compositions which is very much different from FRP (Matyjaszewski, 2002). During the polymerization reaction, the polymer chains grow simultaneously and thus all the polymer chains have same composition. However, in FRP macromolecular chains start growing at different times during the polymerization and the monomer composition continuously changes. As a result, in FRP the different chains will have different compositions in the end (Solomon and Moad, 1995). The reactivity ratios of co-monomers in a CRP are somewhat different from the same in FRP. This is because in CRP processes there is intermittent activation-deactivation of the active species which results in different rates of consumption of comonomers (Matyjaszewski, 2002; Braunecker and Matyjaszewski, 2007). For example, Mignard et al. reported the reactivity ratios of the copolymerization of styrene and butyl acrylate (BA) via NMP at 120°C in solution. They reported the reactivity ratios for styrene and BA within the range of 0.60–1.2 ($r_{styrene}$) and 0.16–0.29 (r_{BA}) respectively (Mignard et al., 2004). Arehart and Matyjaszewski reported the reactivity ratios of styrene and BA prepared via ATRP at 110°C in solution. They reported the reactivity ratios for styrene and BA as $0.68 < r_{styrene} < 0.82$ and $0.22 < r_{BA} < 0.26$ respectively (Arehart and Matyjaszewski, 1999). Chambard et al. reported the copolymerization of styrene and BA prepared in bulk via FRP. The reactivity ratios of styrene and BA prepared via FRP at 90°C were reported to be 0.95 and 0.20 respectively (Chambard et al., 1999). Ziegler and Matyjaszewski reported the variation in reactivity ratios of MMA and BA with the change in ligand from 4,4′-di(5-nonyl)-2,2′-bipyridine (dNbpy) (r_{MMA} = 2.52, r_{BA}= 0.26) to N,N,N′,N″,N″-pentamethyldiethylenetriamine (PMDETA) (r_{MMA} = 3.15, and r_{BA} = 0.37) (Ziegler and Matyjaszewski, 2001). Lessard et al. reported the reactivity ratios of styrene and *tert*-butyl acrylate(*t*-BA) prepared via NMP at 115°C in bulk as $r_{t\text{-}BA}$ = 0.09–0.12 and $r_{Styrene}$ = 0.40–0.49 (Lessard et al., 2007). We reported the ATRP of furfuryl methylacrylate and methyl methacrylate (Kavitha and Singha, 2007), 2-ethylhexyl acrylate and glycidyl methacrylate (Haloi et al., 2009). However, there is no report on the ATRP of styrene and EHA and to determine their reactivity ratios. The objective of this investigation is to study the copolymerization of styrene and EHA via ATRP and to determine their reactivity ratios. The reactivity ratios of styrene and EHA were calculated using Finemann–Ross (FR), inverted Finemann–Ross (IFR), and Kelen–Tudos (KT) methods (Fineman and Ross, 1950; Kelen et al., 1980; Makrikosta et al., 2005).

MATERIALS AND METHODS

The monomers, 2-ethylhexyl acrylate (EHA) (Aldrich, USA) and styrene (Jyoti Chemicals, Mumbai) were purified by vacuum distillation. CuBr (Aldrich, USA) was purified by washing with glacial acetic acid, and then it was washed thoroughly with diethyl ether and was finally dried under vacuum. Phenyl ethylbromide (PEBr) (97%) and N,N,N′,N″,N″-pentamethyldiethylenetriamine (PMDETA) (97%) were purchased from Aldrich, USA and were used as received.

Characterization

Number average molecular weight (Mn, GPC) and dispersity (*Đ*) were determined by Gel Permeation Chromatography (GPC). GPC analysis was carried out at room temperature using a Viscotek GPC equipped with a refractive index detector (Model VE3580), two ViscoGEL GPC columns (model GMHHR-M # 17392) connected in series. GPC analysis was carried out using tetrahydrofuran as eluent at a flow rate of 1 ml/min. Linear and narrow disperse polystyrene was used as calibration standard and Viscotek OMNI-01 software was used for data processing.

^{1}H NMR spectra of the polymers were recorded on a 200 MHz Brucker NMR spectrometer using $CDCl_3$ as solvent which had a small amount of tetramethylsilane (TMS) as an internal standard.

Differential scanning calorimetry (DSC) analysis was carried out by using TA Instrument (DSC Q100 V8.1 Build 251) under nitrogen atmosphere at a heating rate of 10°C/min within a temperature range of −100°C to +150°C. The baseline calibration was done by scanning the temperature domain with the help of an empty pan. The enthalpy was calibrated by using indium standard and the heat capacity was calibrated by using the sapphire disc that was supplied by TA instrument. The glass transition temperature (T_g) was determined from the plot of heat flow vs. temperature in the second heating scan in the DSC analysis.

Thermogravimetric analysis (TGA) was carried out by using a TA Instrument (Q50) at a heating rate of 20°C/min in the temperature range of 30–600°C in nitrogen atmosphere. TGA analyzer consists of high precision balance with a pan which was placed in a small electrically heated oven. The temperature was measured accurately with the help of a thermocouple. From the plot of weight percent vs. temperature the polymer degradation temperature was determined.

Synthesis of Copolymers of Styrene and EHA via Atom Transfer Radical Copolymerization (ATRcP)

The polymerization reaction was carried out in a Schlenk tube. In a typical ATRP reaction EHA (4.82 g, 26.1 mmol), styrene (0.909 g, 8.7 mmol) and CuBr (0.050 g, 0.35 mmol) were accurately weighed and transferred to the Schlenk tube. The PMDETA ligand (0.0604 g, 0.35 mmol) was then added to the reaction tube. Oxygen was removed from the reaction mixture by passing nitrogen through the reaction tube. The polymerization was started by adding PEBr (0.0646 g, 0.35 mmol) and was carried out at 90°C. Aliquot samples were taken out at different time intervals and were used to calculate the conversion by gravimetric method. The samples were also used to find out the molecular weight by GPC. The final product was diluted with THF and was purified by passing through alumina column to remove the copper catalyst. The same procedure was adopted for other feed ratios of ATRcP (atom transfer radical copolymerization) of EHA and styrene.

Homopolymerization of Styrene via Atom Transfer Radical Polymerization (ATRP)

The homopolymerization of styrene was carried out in bulk in a Schlenk tube equipped with silicone septum and magnetic stirring bar. In the Schlenk tube styrene (4.5 g, 43.2 mmol), CuBr (0.031 g, 0.21 mmol) and PMDETA (0.037 g, 0.21 mmol) were weighed and degassed by passing nitrogen gas for 15 min. The reaction was started by adding PEBr (0.040 g, 0.21 mmol) in the mixture. The reaction was carried out at 110°C for 6 h. ^1H NMR ($CDCl_3$, 200 MHz): δ (in ppm) = 6.4–7.2 (phenyl protons of polystyrene) and 1.4–2.2 ppm ($-CH_2-$ and >CH− protons).

Synthesis of Diblock Copolymer of Polystyrene with EHA via ATRP

The homopolymer of polystyrene (i.e., PS-Br) was used as macroinitiator for the synthesis of diblock copolymer of styrene and EHA. The macroinitiator, PS-Br (1.0 g, 0.066 mmol) was taken in a Schlenk tube and was dissolved in THF solvent into which CuBr (0.009 g, 0.069 mmol) and PMDETA (0.017 g, 0.10 mmol) were added followed by EHA (1 g, 5.4 mmol). The polymerization reaction was carried out at 90°C for 6 h. The polymer obtained was dissolved in THF and was purified by passing through basic alumina column and then precipitated in methanol.

RESULTS AND DISCUSSION

Copolymerization of styrene and 2-ethylhexyl acrylate was carried out via ATRcP (shown in Scheme 1) at different feed ratios by using phenylethylbromide (PEBr) as initiator, CuBr as catalyst in combination with PMDETA as a ligand. From the kinetic plot of ln(1/1-X) (where X is the percent conversion of monomer) vs. time, it was observed that the value of ln(1/1-X) was linearly increased with time (Figure 1). This linear dependency is the characteristics of the controlled polymerization reaction which follows the first order kinetics. Figure 1 showed that with increase in styrene content in the feed the rate of polymerization increased (Jianying et al., 2006). Figure 2 infers that there was linear increase in molecular weight with conversion and dispersity (*Đ*) was relatively narrow. GPC traces of poly(styrene-co-EHA) (50:50). It indicates the controlled nature of the copolymerization reaction. Table 1summarizes the feed ratio as well as the copolymer composition of the different copolymerization reactions.

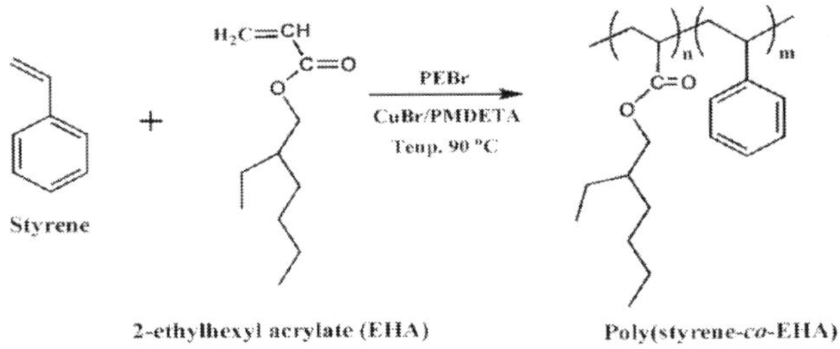

Scheme 1: Copolymerization of styrene and EHA via atom transfer radical polymerization (ATRP).

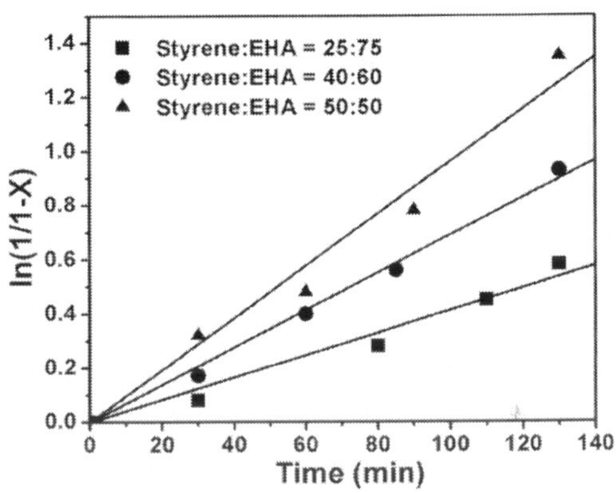

Figure 1: Kinetic plot of ln[1/(1-X)] vs time for copolymerization of styrene and EHA. [PEBr]: $[M]_o$:[CuBr]:[PMDETA] = 1:100:1:1, at 90°C.

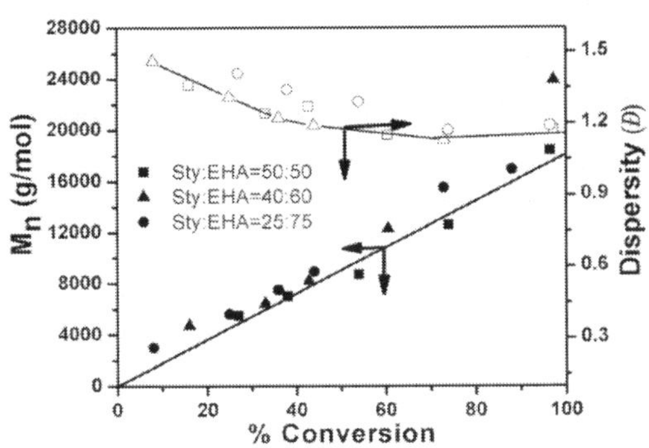

Figure 2: Plot of M_n **and** *Đ* **vs. conversion (%) for** ATRcP of Styrene and EHA in bulk. [PEBr]: $[M]_o$:[CuBr]:[PMDETA] = 1:100:1:1, at 90°C.

Table 1: Copolymerization of styrene and EHA in bulk at 90°C

Sl. No.	Feed composition (%) styrene: EHA	Copolymer composition≠ (%) styrene: EHA	Conversion (%)	$M_{n,GPC}$ (g/mol)	Đ
1	25:75	27:73	98.1	16,900	1.27
2	40:60	44:56	88.0	24,000	1.15
3	50:50	56.6:43.4	97.8	18,500	1.17

[PEBr]: [M]₀:[CuBr]:[PMDETA] = 1:100:1:1. Reaction time = 2 h 20 min.
≠Compositions were calculated by ^{1}H NMR spectroscopy.

STRUCTURAL CHARACTERIZATION AND COPOLYMER COMPOSITION

The structural characterization and copolymer composition were determined by ^{1}H NMR spectroscopy. Figure 3 shows the ^{1}H NMR spectra of poly(styrene-*co*-EHA) of 40:60 feed ratio. The resonance at $\delta = 0.8$ ppm is due to the $-CH_3$ protons of PEHA part. The broad resonances at $\delta = 1.0$ to 2.1 ppm are due to the different $-CH_2-$ and >CH– protons of pendant group of PEHA part as well as those of the main chain backbone of the copolymer. The resonances at $\delta = 6.5$–7.1 ppm are due to the different aromatic protons of polystyrene part. The resonances at 3.8 ppm are due to $-OCH_2-$ protons of PEHA part. The distinct resonances at 6.5–7.3 for five aromatic protons of styrene and 3.8 ppm for two protons of EHA were used to calculate the composition of the copolymer of styrene and EHA. The copolymer composition was determined by the equation (1) as shown below

$$F_{EHA} = \frac{5A}{5A + 2B} \times 100 \quad (1)$$

where, A and B represent the integral area at $\delta = 3.7$ and $\delta = 6.5$–7.3 ppm for $-OCH_2-$ protons in PEHA and aromatic protons of polystyrene unit respectively.

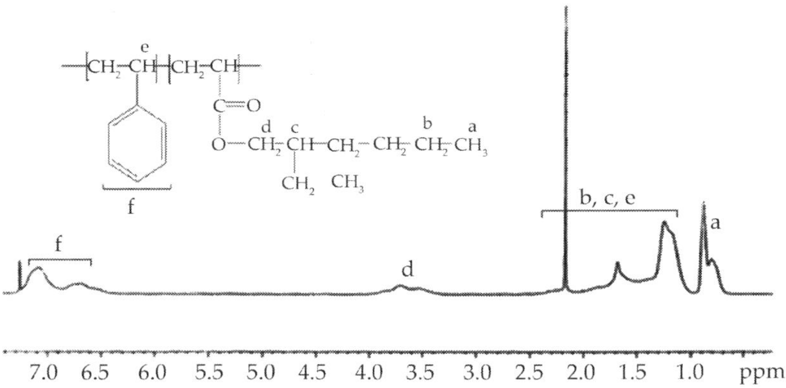

Figure 3: ^{1}H NMR spectrum of poly(styrene-*co*-EHA) with feed molar composition 50:50.

REACTIVITY RATIO DETERMINATION

In the copolymerization of styrene (1) and EHA (2) the reactivity ratio r_1 is defined as k_{11}/k_{12}, where k_{11} is the rate constant of the reaction between the growing polymer chain carrying free radical of styrene as the terminal unit and styrene (homo propagation) and k_{12} is the rate constant of the reaction between the same reactive chain end and the EHA monomer. Similarly the reactivity ratio r_2 is also defined as k_{22}/k_{21}, where k_{22} is the rate constant of homo propagation reaction between the growing macromolecular chain having EHA active radical as the terminal unit and EHA and k_{21} is the rate constant of the reaction between EHA active radical and styrene monomer (cross propagation). For determining the r_1 and r_2, copolymerization of styrene and EHA was carried out at different feed ratio of styrene (1) and EHA (2). In this case copolymerization was carried out at low conversion (~10%) and its molar composition was determined by ^{1}H NMR spectroscopy. Composition of the low conversion copolymer was used for the determination of monomer reactivity ratios. In this case FR, IFR, and KT methods were used to determine the reactivity ratio of the monomers.

In the FR method the following equation was used

$$M - \frac{M}{P} = -r_2 + r_1 \frac{M^2}{P}$$

where, M = molar feed ratio (M_1/M_2) and P = copolymer composition (m_1/m_2)

$$\text{or, } G = -r_2 + r_1 H$$

where, G = M−M/P and H = M^2/P

The plot of G vs. H gives the straight line (Figure 4). From this the slope and intercept were calculated to be r_1 = 1.24 and r_2 = 0.71 respectively.

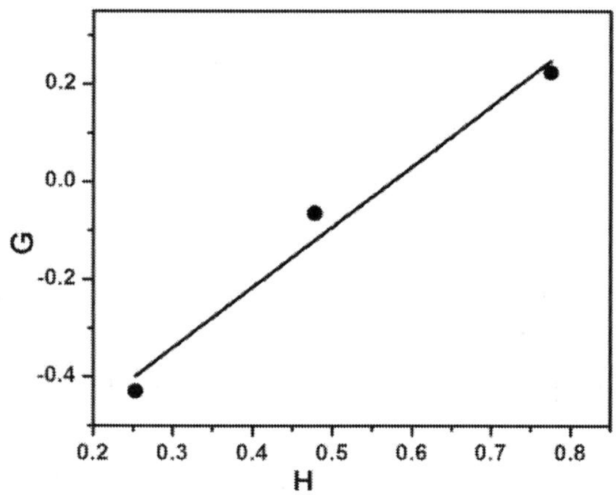

Figure 4: Finemann–Ross plot for copolymerization of styrene with EHA.

In the IFR method the equation used is

$$\frac{G}{H} = -r_2 \frac{1}{H} + r_1$$

So, from the plot of G/H and 1/H (Figure 5) the reactivity ratios, r_1 and r_2 were calculated as 1.34 and 0.76 from the intercept and the slope respectively.

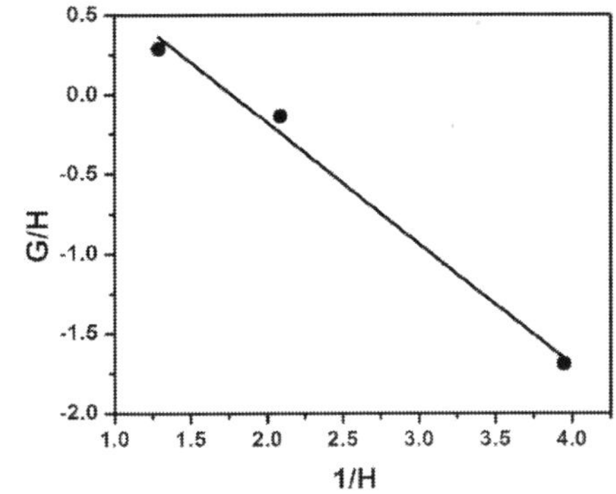

Figure 5: Inverted Finemann–Ross plot for copolymerization of styrene with EHA.

In KT method the reactivity ratio was determined by using the equation

$$\eta = \left(r_1 + \frac{r_2}{\propto}\right)\xi - \frac{r_2}{\propto}$$

where, $\eta = G/(\propto+H)$ and $\xi = H/(\propto+H)$ and $\propto = (H_{min} \cdot H_{max})^{1/2}$, $H_{min} = 0.253$ and $H_{max} = 0.775$. H_{min} and H_{max} values are taken from Table 2.

Table 2: Finemann–Ross, Inverted Finemann–Ross and Kelen–Tudos parameter for the copolymer of styrene and EHA in bulk*

Monomer feed ratio styrene: EHA	$M = M_1/M_2$	$^{\#}P = m_1/m_2$	*M/P*	P-1	G	H	G/H	1/H	∝+H	η	ξ
25:75	0.33	0.43	0.767	−0.57	−0.430	0.253	−1.69	3.95	0.695	−0.618	0.364
40:60	0.66	0.91	0.725	−0.09	−0.065	0.478	−0.135	2.09	0.920	−0.070	0.519
50:50	1.0	1.29	0.775	0.29	0.224	0.775	0.289	1.29	1.217	0.184	0.636

*[PEBr]: $[M]_0$:[CuBr]:[PMDETA] = 1:100:1:1, at 90°C (where, ∝ = 0.442).

\# Copolymer composition at 10% conversion.

In the plot of η vs. ξ, (Figure 6) the slope gives the value of $(r_1 + r_2/\propto)$ and intercept provides $r_2/\propto$. From these two values r_1 and r_2 were calculated as $r_1 = 1.30$ and $r_2 = 0.73$ respectively. All the parameters used for the three methods are given in Table 2.

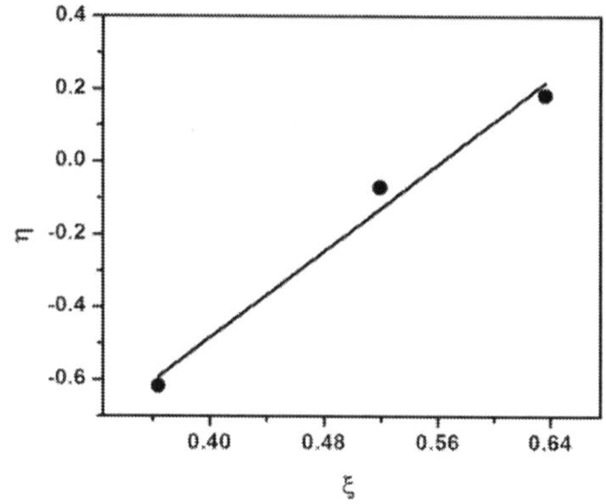

Figure 6: Kelen-Tudos plot for copolymerization of styrene with EHA.

The reactivity ratios, r_1 and r_2 calculated by the different methods are tabulated in Table 3. They are quite comparable. The values; $r_1 > 1$, $r_2 < 1$ and $r_2 < r_1$ indicate that the styrene has much influence on the copolymer formation during the reaction (Jianying et al., 2006). Srivastava et al. (Srivastava and Rai, 1995) reported the copolymerization of styrene and EHA initiated by azobisisobutyronitrile (AIBN) in bulk in the presence of anhydrous $ZnCl_2$. They reported the reactivity ratios for styrene and EHA as 0.10 and 0.175 respectively. Moreover, Kavousian et al. reported the copolymerization of styrene and EHA via conventional radical polymerization (Kavousian et al., 2004). They reported the reactivity ratio of styrene and EHA as 0.926 and 0.238 respectively. There is a difference in the polymerization mechanism of ATRP and non-ATRP processes. In ATRP, atom transfer from an organic halide to a transition metal complex occurs to generate the active radical species, which are then quickly "deactivated" by back transfer of the atom from the transition metal to the radical species (Matyjaszewski, 2002; Braunecker and Matyjaszewski, 2007). So, there is difference in reactivity ratio values in comparison to FRP. Barim et al. studied the FRP and ATRP of phenoxycarbonylmethyl methacrylate (PCMMA) and styrene at 110°C. They reported the reactivity ratios of PCMMA and styrene prepared via ATRP were 0.33 and 0.96 respectively and the same prepared via FRP were 0.47 and 1.16 respectively (Barim et al., 2007). The reactivity ratio varies with the polymerization temperature (Chambard et al., 1999; McManus et al., 2002). We did the polymerization reaction via ATRP in bulk at

90°C. However, the reactivity ratios calculated in this work follow the same trend ($r_{EHA} < r_{styrene}$) as reported by the other authors. In addition, in ATRcP technique the product of the reactivity ratios is less than one. It shows the tendency of random copolymer formation, where the chances of incorporation of styrene is more in comparison to EHA.

Table 3: Reactivity ratio of styrene (r_1) and EHA (r_2) determined by three different models

Methods used	r_1 (styrene)	r_2 (EHA)	$r_1.r_2$
FR	1.24	0.71	0.88
IFR	1.34	0.76	1.01
KT	1.30	0.73	0.91

The mean sequence length (l) of the copolymers were determined by using the equations (Pazhanisamy et al., 1997)

$$l_1 = r_1 \frac{M_1}{M_2} + 1$$

and

$$l_2 = r_2 \frac{M_2}{M_1} + 1$$

where, r_1 (styrene) = 1.29 and r_2 (EHA) = 0.73

The results of the mean sequence length in the copolymer are shown in Table 4. It indicates that the length of EHA increases as its content in the monomer feed increases.

Table 4: The monomer composition and sequence length ratio

Copolymer feed ratio styrene: EHA	M_1/M_2	M_2/M_1	l_1	l_2	$l_1: l_2$
25/75	0.33	3.00	1.42	3.19	4:9
40/60	0.66	1.50	1.85	2.09	4:5
50/50	1.00	1.00	2.29	1.73	4:3

THERMAL PROPERTIES

The glass transition temperature (T_g) of the copolymer was determined by DSC analysis (shown in Figure 7). All the copolymers showed a single T_g and they are shown in Table 5. It indicates that as the EHA (T_g = −60°C) content increases the T_g decreases. The T_g for the copolymers determined by DSC analysis was compared with the same ($T_{g,\ F\text{-}F}$) determined by Flory-Fox equation (Lijia et al., 1997). There is consistent difference between the T_g values at different content of two comonomers (Table 5). Table 5 shows that there is some discrepancy between the experimental T_g and T_g calculated by Flory-Fox equation. The Flory-Fox model is based on the free volume theory. The discrepancy in the T_g values is due to the fact that the effect of the chemical nature and organization of the monomers on the mobility of a polymer chain was not considered (Fernandez-Garcia et al., 1999).

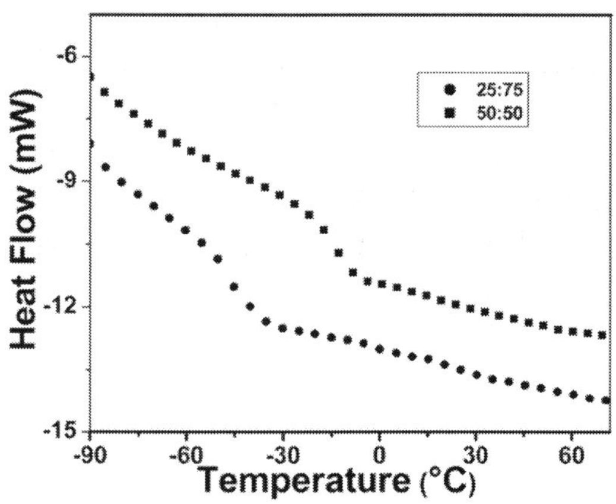

Figure 7: DSC thermogram of poly(styrene-*co*-EHA).

Table 5: Thermal properties of copolymers of styrene and EHA

Sl. no.	Feed composition styrene: EHA	Copolymer composition styrene: EHA	T_g (°C)	* $T_{g,F\text{-}F}$ (°C)	# T_{onset} (°C)	T_{max} (°C)
1	25:75	27:73	−47	−36	381	409
2	40:60	44:56	−27	−18	389	414
3	50:50	56.6:43.4	−14	−5	385	415
4	Polystyrene-b-PEHA	85: 15	+123 and −65	–	–	–

*$T_{g,F\text{-}F}$ was calculated by using Flory–Fox equation.

#Temperature at 10% weight loss was taken as T_{onset}.

T_{max} represents the maximum degradation temperature at which the polymer back-bone starts degrading. It was determined from the DTG thermogram.

The block copolymer of styrene and EHA (PS-*b*-PEHA) was prepared by using polystyrene-Br as macroinitiator. The shift of GPC traces of the diblock copolymer toward lower elution volume indicated the successful preparation of block copolymer. This block copolymer showed two T_gs, PEHA block at −64°C and polystyrene at +123°C (Figure 8). However, the copolymers, poly(styrene-*co*-EHA) showed only one T_g indicating the copolymers were not blocky in nature. Thermal stability of the copolymer was studied by TGA (Figure 9). It is clear that as the styrene content increases, there is a slight increase in T_{onset}. However, there was no significant change in T_{max} as shown in Table 5.

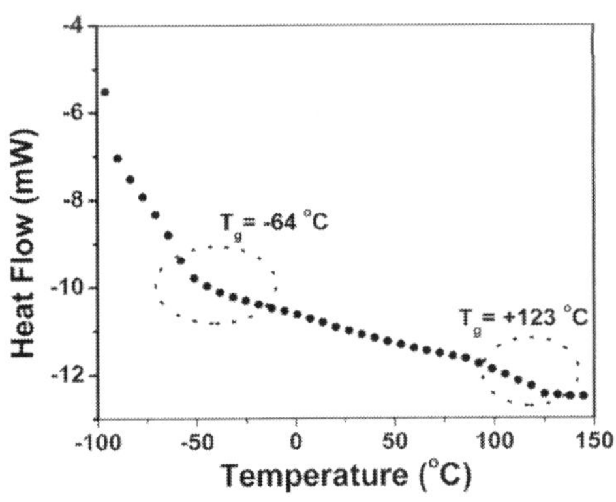

Figure 8: DSC thermogram of the polystyrene-*b*-PEHA.

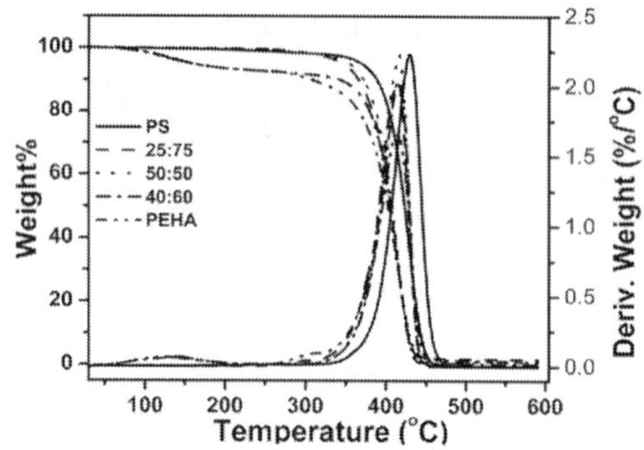

Figure 9: TGA thermogram of homopolymer and copolymers of styrene and EHA.

CONCLUSIONS

Copolymers of styrene and 2-ethylhexyl acrylate were synthesized successfully in bulk by using atom transfer radical copolymerization (ATRcP). The chemical composition was studied by ^{1}H NMR spectroscopy and the reactivity ratios of the two monomers were calculated by using FR, IFR, and KT methods. The reactivity ratios of styrene and EHA were somewhat different from the polymerization reaction of styrene and EHA using the FRP system. DSC analysis showed that the T_g of the copolymer increases on increasing styrene content.

CONFLICT OF INTEREST STATEMENT

The authors declare that the research was conducted in the absence of any commercial or financial relationships that could be construed as a potential conflict of interest.

ACKNOWLEDGMENT

Bishnu P. Koiry gratefully acknowledges the fellowship from IIT, Kharagpur, India.

REFERENCES

1. Arehart, S. V., and Matyjaszewski, K. (1999). Atom transfer radical copolymerization of styrene and n-butyl acrylate. *Macromolecules* 32, 2221–2231. doi: 10.1021/ma981693v
2. Barim, G., Demirelli, K., and Coskun, M. (2007). Conventional and atom transfer radical copolymerization of phenoxycarbonylmethyl methacrylate-styrene and thermal behavior of their copolymers. *Express Polym. Lett.* 1, 535–544. doi: 10.3144/expresspolymlett.2007.76
3. Barner-Kowollik, C. (2008). *Handbook of RAFT Polymerization.* Weinheim: Wiley-VCH Verlag GmbH and Co. KGaA. doi: 10.1002/9783527622757
4. Braunecker, W. A., and Matyjaszewski, K. (2007). Controlled/living radical polymerization: features, developments, and perspectives. *Prog. Polym. Sci.* 32, 93–146. doi: 10.1016/j.progpolymsci.2006.11.002
5. Chambard, G., Klumperman, B., and German, A. L. (1999). Dependence of chemical composition of styrene/butyl acrylate copolymers on temperature and molecular weight. *Polymer* 40, 4459–4463. doi: 10.1016/S0032-3861(98)00690-9
6. Chiefari, J., Chong, Y. K., Ercole, F., Krstina, J., Jeffery, J., Le, T. P. T., et al. (1998). Living free-radical polymerization by reversible addition-fragmentation chain transfer: the RAFT process. *Macromolecules* 31, 5559–5562. doi: 10.1021/ma9804951
7. Fernandez-Garcia, M., Cuervo–Rodriguez, R., and Madruga, E. L. (1999). Glass transition temperatures of butyl acrylate–methyl methacrylate copolymers. *J. Polym. Sci. B Polym. Phys.* 37, 2512–2520. doi: 10.1021/ma001182k
8. Fineman, M., and Ross, S. D. (1950). Liner method for determining monomer reactivity ratios in copolymerization. *J. Polym. Sci.* 5, 259–262. doi: 10.1002/pol.1950.120050210
9. Haldankar, G. S. (2001). *Compatibilizer for Polymer Blends.* Frankfort, IL: The Sherwin-Williams Co. US Patent 6,541,571.
10. Haloi, D. J., Roy, S., and Singha, N. K. (2009). Copper catalyzed atom transfer radical copolymerization of glycidyl methacrylate

and 2-ethylhexyl acrylate. *J. Polym. Sci. A Polym. Chem.* 47, 6526–6533. doi: 10.1002/pola.23695

11. Harth, E., Horn, B. V., and Hawker, C. J. (2001). Acceleration in nitroxide mediated 'living' free radical polymerizations. *Chem. Commun.* 9, 823–824. doi: 10.1039/b102145c
12. Hawker, C. J., Bosman, A. W., and Harth, E. (2001). New polymer synthesis by nitroxide mediated living radical polymerization. *Chem. Rev.* 101, 3661–3688. doi: 10.1021/cr990119u
13. Jianying, H., Jiayan, C., Jiaming, Z., Yihong, C., Lizong, D., and Yousi, Z. (2006). Some monomer reactivity ratios of styrene and (meth)acrylates in the presence of TEMPO. *J. Appl. Polym. Sci.* 100, 3531–3535. doi: 10.1002/app.22512
14. Kamigaito, M., Ando, T., and Sawamoto, M. (2001). Metal-catalyzed living radical polymerization. *Chem. Rev.* 101, 3689–3745. doi: 10.1021/cr9901182
15. Kavitha, A., and Singha, N. K. (2007). Atom-transfer radical copolymerization of furfuryl methacrylate (FMA) and methyl methacrylate (MMA): a thermally-amendable copolymer. *Macromol. Chem. Phys.* 208, 2569–2577. doi: 10.1002/macp.200700239
16. Kavitha, A., and Singha, N. K. (2008). High temperature resistant tailor-made poly(meth)acrylates bearing adamantyl group via atom transfer radical polymerization. *J. Polym. Sci. A Polym. Chem.* 46, 7101–7113. doi: 10.1002/pola.23015
17. Kavousian, A., Ziaee, F., Nekoomanesh, H., Leamen, M. J., and Penlidis, A. (2004). Determination of monomer reactivity ratios in styrene/2-ethylhexylacrylate copolymer. *J. Appl. Polym. Sci.* 92, 3368–3370. doi: 10.1002/app.20338
18. Kelen, T., Tudos, F., and Turcsanyi, B. (1980). Confidence intervals for copolymerization reactivity ratios determined by the Kelen-Tudos method. *Polym. Bull.* 2, 71–76. doi: 10.1007/BF00275556
19. Lessard, B., Graffe, A., and Maric, M. (2007). Styrene/t-Butyl acrylate random copolymers synthesized by nitroxide-mediated

polymerization: effect of free nitroxide on kinetics and copolymer composition. *Macromolecules* 40, 9284–9292. doi: 10.1021/ma071689w

20. Lijia, A. N., Dayong, H. E., Jlaokai, J., Zhigang, W., Donghong, Y. U., Jiang, B., et al. (1997). Effects of molecular weight and interaction parameter on the glass transition temperature of polystyrene mixture and its blends with polystyrene/poly(2,6-dimethyl-p-phenylene oxide). *Eur. Polym. J.* 33, 1523–1528. doi: 10.1016/S0014-3057(97)00089-X
21. Makrikosta, G., Georgas, D., Siakali-Kioulafa, E., and Pitsikalis, M. (2005). Statistical copolymers of styrene and 2-vinylpyridine with trimethylsilyl methacrylate and trimethylsilyloxyethyl methacrylate. *Eur. Polym. J.* 41, 47–54. doi: 10.1016/j.eurpolymj.2004.08.001
22. Matyjaszewski, K. (2000). "Controlled/Living radical polymerisation: progress in ATRP, NMP and RAFT," in *ACS Symposium Series* (New Orleans, LA). doi: 10.1021/bk-2000-0768
23. Matyjaszewski, K. (2002). Factors affecting rates of comonomer consumption in copolymerization processes with intermittent activation. *Macromolecules* 35, 6773–6781. doi: 10.1021/ma0208180
24. Matyjaszewski, K., and Jia, J. (2001). Atom transfer radical polymerization. *Chem. Rev.* 101, 2921–2990. doi: 10.1021/cr940534g
25. McManus, N. T., Penlidis, A., and Dube, M. A. (2002). Copolymerization of alpha-methyl styrene with butyl acrylate in bulk. *Polymer* 43, 1607–1614. doi: 10.1016/S0032-3861(01)00738-8
26. Mignard, E., Leblanc, T., Bertin, D., Guerret, O., and Reed, W. F. (2004). Online monitoring of controlled radical polymerization: nitroxide-mediated gradient copolymerization. *Macromolecules* 37, 966–975. doi: 10.1021/ma035589b
27. Moad, G. (2006). The emergence of RAFT polymerization. *Aust. J. Chem.* 59, 661–662. doi: 10.1071/CH06376

28. Moad, G., Chiefari, J., Chong, Y. K., Krstina, J., Mayadunne, R. T. A., Postma, A., et al. (2000). Living free radical polymerization with reversible addition–fragmentation chain transfer (the life of RAFT). *Polym. Int.* 49, 993–1001. doi: 10.1002/1097-0126(200009)49:9<993::AID-PI506>3.0.CO;2-6
29. Odian, G. (2004). *Principles of Polymerization*. New Jersey, NJ: Wiley-Interscience publication. doi: 10.1002/047147875X
30. Pazhanisamy, P., Ariff, M., and Anwaruddin, Q. (1997). Copolymers of α-methylstyrene with *N*-cyclohexylacrylamide: synthesis, monomer reactivity ratios, and mean sequence length. *J. Macromol. Sci. A Pure Appl. Chem.* A34, 1045–1054. doi: 10.1080/10601329708015009
31. Plessis, C., Arzamendi, G., Alberdi, J. M., Agnely, M., Leiza, J. R., and Asua, J. M. (2001). Intramolecular chain transfer to polymer in the emulsion polymerization of 2-ethylhexyl acrylate. *Macromolecules* 34, 6138–6143. doi: 10.1021/ma0018190
32. Skeist, I. (1977). *Handbook of Adhesives*. New York, NY: Chapman & Hall.
33. Solomon, D. H., and Moad, G. (1995). *The Chemistry of Free Radical Polymerization*. Oxford: Elsevier Science Ltd.
34. Srivastava, N., and Rai, J. S. P. (1995). Kinetics, mechanism, and rheological properties of the copolymers of 2-ethylhexylacrylate and styrene. *J. Macromol. Sci. A Pure Appl. Chem. A* 32, 2049–2062. doi: 10.1080/10601329508011044
35. Webster, D. C., and Crain, A. L. (2002). Synthesis of latexes containing diesters of 3-butene-1,2-diol. *Prog. Org. Coat.* 45, 43–48. doi: 10.1016/S0300-9440(02)00099-1
36. Ziegler, M. J., and Matyjaszewski, K. (2001). Atom transfer radical copolymerization of methyl, methacrylate and n-butyl acrylate. *Macromolecules* 34, 415–424. doi: 10.1021/ma001182k

CITATIONS

CHAPTER 1

Ates O (2015) Systems Biology of Microbial Exopolysaccharides Production. *Front. Bioeng. Biotechnol.* 3:200. doi: 10.3389/fbioe.2015.00200.

CHAPTER 2

Michele Greque de Morais, Bruna da Silva Vaz, Etiele Greque de Morais, and Jorge Alberto Vieira Costa, "Biological Effects of Spirulina (Arthrospira) Biopolymers and Biomass in the Development of Nanostructured Scaffolds," BioMed Research International, vol. 2014, Article ID 762705, 9 pages, 2014. doi:10.1155/2014/762705.

CHAPTER 3

Oladeji O. Ige, Lasisi E. Umoru, and Sunday Aribo, "Natural Products: A Minefield of Biomaterials," ISRN Materials Science, vol. 2012, Article ID 983062, 20 pages, 2012. doi:10.5402/2012/983062.

CHAPTER 4

Tizazu Mekonnen, Paolo Mussone, Hamdy Khalil and David Bressler, Progress in bio-based plastics and plasticizing modifications, DOI: 10.1039/C3TA12555F.

CHAPTER 5

Jörg Schnauß, Tina Händler **and Josef A. Käs**, Semiflexible Biopolymers in Bundled Arrangements, doi: 10.3390/polym8080274.

CHAPTER 6

Dietrich Braun, "Origins and Development of Initiation of Free Radical Polymerization Processes," International Journal of Polymer Science, vol. 2009, Article ID 893234, 10 pages, 2009. doi:10.1155/2009/893234.

CHAPTER 7

Eric Jean Fischer, Giuseppe Storti and Danilo Cuccato, Aqueous Free-Radical Polymerization of Non-Ionized and Fully Ionized Methacrylic Acid, doi: 10.3390/pr5020023.

CHAPTER 8

Hidetaka Tobita, Model-Based Reactor Design in Free-Radical Polymerization with Simultaneous Long-Chain Branching and Scission, doi: 10.3390/pr3040731.

CHAPTER 9

Antonio Veloso , Wendy García , Amaia Agirre , Nicholas Ballard , Fernando Ruipérez , José C. de la Cal and José M. Asua, Determining the effect of side reactions on product distributions in RAFT polymerization by MALDI-TOF MS†, DOI: 10.1039/C5PY00838G.

CHAPTER 10

Koiry BP and Singha NK (2014) Copper mediated controlled radical copolymerization of styrene and 2-ethylhexyl acrylate and determination of their reactivity ratios. *Front. Chem.* **2**:91. doi: 10.3389/fchem.2014.00091.

INDEX

G

H

I

K

L

M

N

P

R